Dante and the astronomers

Mary Acworth Orr

Alpha Editions

This edition published in 2021

ISBN : 9789354549908

Design and Setting By
Alpha Editions
www.alphaedis.com
Email - info@alphaedis.com

Contents

PREFACE

An observatory on a mountain top is an ideal place in which to write on astronomy and poetry, but it has one drawback: the difficulty of obtaining books on special subjects. My husband's criticisms and help have been invaluable, and of books on modern astronomy there is no lack; but many others which I have wished to consult I have been unable to procure, and doubtless there are many more which I ought to have read, but of whose existence I am ignorant.

This defect has been partly remedied, however, by Mr. Wicksteed's great kindness in lending me a number of books on Dante subjects.

I am much indebted to my sister, Miss Orr, who has helped me in preparing these sheets for the press. The index is mainly her work.

Thanks are also due to several members of the staff of this observatory: Mr. Sitarama Aiyar made some calculations, Mr. Nagarajan Aiyar projected the map facing page 295, Mr. Krishna Aiyar prepared photographs of several illustrations, and Mr. Krishnasawmy typed the greater part of the copy for press. Mr. Raymond Beazley, the Trustees of the British Museum, and Messrs Macmillan, have kindly allowed me to reproduce illustrations from their publications.

To all these I offer grateful thanks.

M. A. EVERSHED.

KODAIKANAL OBSERVATORY,
South India.
October 1913.

INTRODUCTION

In a beautiful passage of the *Convivio* Dante describes how he first began to devote himself diligently to science and philosophy. When the gentle soul of Beatrice had passed to heaven, a great darkness fell upon him: the streets of Florence were to him as a deserted city, and his life empty and purposeless. It was long before he could find any comfort, but at last he bethought himself of studying a book by Boëthius, who when exiled, imprisoned, and unjustly condemned to death, had strengthened his soul with the "Consolations of Philosophy." This led him on to Cicero's book "On Friendship," in which Lælius explains how he is consoled for the death of Scipio.

Books in those days could only be had in manuscript, full of abbreviations, and often also of errors, and at first the young student found the Latin hard to master; but as he struggled on, half deciphering and half divining the meaning, the mists cleared a little, and the weight was lifted from brain and heart. With elation he discovered that obscure passages were becoming luminous, and to the exhilarating sense of conquest was added the joy of finding, beautifully expressed, thoughts which had already floated in his own mind, but dimly, as in a dream. He compares himself to one who, seeking silver, should light (not without Divine guidance) on a treasure of gold; for he found not only relief from his tears, but a door into a new world of literature, philosophy, and science. Henceforth, he tells us, he eagerly frequented the schools of the religious orders and the discussions of the philosophers; and how extensive and thorough was his learning we can see in his writings. In them we find a reflection of thirteenth-century thought in every field of intellectual research.

Among all his studies was one which evidently had a great attraction for him, even in the early days of the *Vita Nuova*, before learning had become a passion. Astronomy appealed to many sides of his nature. The beauty of the skies stirred his imagination; their suggestive symbolism touched his religious sense; the harmony of the celestial movements and the accuracy with which they can be foretold delighted his instinct for order and precision. He must have read, and perhaps possessed, some of the best text-books then available, and he grasped with singular clearness the phenomena observed and the theories taught in his day. His works are full of allusions to astronomy. In the *Vita Nuova* he finds pleasure in connecting the story of his lady with the revolutions of the spheres; in the *Convivio* he teaches the elements of the science; in the Vision of the *Divine Comedy* he journeys through the universe as it was depicted by mediæval astronomers; and throughout his works are scattered similes drawn from celestial phenomena and descriptions of "le belle cose che porta il ciel."[1]

Therefore, for full enjoyment and understanding of Dante's works it is necessary to have a rudimentary knowledge of astronomy.

Many of his readers think that Dante's astronomy is very complicated and difficult to understand. What makes it seem difficult is that in this age we are generally unfamiliar with the skies. We do not eat our breakfast or go to our office by the sun, nor do we watch the stars to see when grouse-shooting begins or the summer holidays end. If it is important for us to know at what hour the sun sets and lamps must be lighted, or if we wish to see a view by moonlight, we consult an almanac. When we think at all of the movements of the heavenly bodies, our notions are usually taken from diagrams and tables, not from what is actually seen in the skies. We only think, for instance, of the seasons as caused by the earth's journey round the sun, and the tilt of her axis: therefore, when Dante speaks of Venus as a Morning Star veiling the Fishes with her rays, or the horn of the Celestial Goat touching the sun, it conveys little, although the seasons of spring and of winter are as clearly indicated as if he had spoken of the blossoming of primroses or the fall of snow. When Cacciaguida, in the heaven of Mars, tells the date of his birth by counting how many times the planet had since then returned to his Lion, those who only think of Mars as circling round the Sun, and have never traced his path among the stars, are at a loss, and think the method very far-fetched. A short description, and especially a little individual watching, of the apparent movements of the heavenly bodies, would put us in a position to realize the meaning of a large number of Dante's astronomical descriptions and allusions, without any knowledge of any theory.

Others complain that the subject is dull. Dante's astronomy, when interpreted only by means of notes on single passages, is undoubtedly dull—as dull as the history of his own times learned in the same way. But when either subject is studied as a whole these passages acquire a special interest; and they in their turn give new life to the subject they illustrate.

Other readers say that Dante's astronomy is so entirely false and obsolete that it is not worth study. This is hardly true. Where Dante speaks of appearances he is remarkably accurate, far more so than most modern artists and writers of fiction. Where he speaks of the heavens as he supposes them actually to exist, he is interpreting the appearances according to the astronomical theories of his day, with which he was very well acquainted. This interpretation was not correct, but it was an ingenious and beautiful system, and very successful in so far as it enabled astronomers to calculate the positions of sun, moon, stars, and planets for any date. Its main outlines can be explained in a few pages, with the help of a couple of diagrams, but when presented thus, especially to those unfamiliar with the skies, it seems very strange and artificial. To appreciate it at its true worth, we must know just what are the phenomena it was intended to explain, and trace its gradual development out of man's first clear perception that the movements of sun, moon, and stars follow unchanging laws.

The story of this development is of enthralling interest, and after the system had been completed by one of the greatest mathematicians the world has seen, its later history reads like a romance. Though of classical Greek origin, it was almost wholly lost to Europe for many centuries, it returned at last in Oriental dress, and its final form was given by a devout and learned Dominican friar.

It was at this time that Dante was born, and the scholar-poet immortalized the Ptolemaic system of astronomy in his verse, adding to its popularity in his own day, and making it known to thousands of readers since, who might otherwise scarcely have heard of it.

Dante's astronomy, therefore, is of wide and deep significance. To study its history is to learn a chapter in the development of the human intellect; to see the universe with his eyes is to know how it appeared, not only to his contemporaries but to men in many lands and many centuries. The system of Ptolemy was already a thousand years old when Dante studied it, and it continued to be taught long after Copernicus had introduced a truer one; nor has it ever been completely swept away, for much that it taught was accurate. The new astronomy has developed from the old, and bears traces to this day, in its phraseology, its written symbols, and its methods, of the many races and ages which have contributed to its progress.

This book, therefore, is divided into two parts. In the first, I put before my readers the elementary facts which form the foundations upon which all astronomy is based, the movements of sun, moon, stars, and planets, so far as they can be easily observed by the naked eye; then follows a sketch of the attempts which were made to interpret these observations from very early days until Dante's time. Unnecessary technicalities are avoided, but we shall try to enter into the thoughts of past generations concerning the stars, to see why they were interested, how they worked, what hindered and what helped them in their search for truth.

In the second part, we shall examine Dante's works, and see how familiar he was with the movements of the skies, and how well he understood the theories which in his time were held to explain them. We shall see how astronomy was generally regarded in his day, what books he read, and which authors influenced him most. We shall see how false is the assertion often made that in the Middle Ages men studied astronomy only for the sake of astrology, and how closely the science of the stars was connected with religion and the loftiest speculations of philosophy.

We shall also examine in particular some difficult passages connected with astronomy which occur in Dante's works, but my aim is not so much to explain all the astronomical references as to put the reader in a position to attempt an explanation himself.

My greatest ambition is to share with others the pleasure I have had in learning what Dante knew and thought about the stars, and who were the master builders who had erected through the ages the system so vividly pictured in his immortal poem.

FIRST PART

THE STORY OF ASTRONOMY FROM PRIMITIVE
TIMES UNTIL THE AGE OF DANTE.

I

APPARENT MOVEMENTS OF THE HEAVENLY BODIES AS SEEN FROM EARTH.

The stars appear to us like points of light, differing greatly in brightness, and scattered very irregularly over the dome above us. All are moving, some much more quickly than others, yet a little attention shows that they do not change their relative positions, and therefore that all must share in one connected movement. If, for instance, any one group be singled out, and looked for again some hours later, it will be evident that it has moved considerably as a whole, yet the stars composing it have kept the same places with regard to one another.

Careful and prolonged observations prove that to observers in the northern hemisphere one star has hardly any perceptible movement, that those nearest to it sweep round it in small circles, and those further away in larger and larger circles, parts of which are hidden below the horizon. All these circlings are performed in the same time, and therefore the stars near the stationary point move more slowly in their small circles than those further away.

All this is precisely what we should see if the sky were a great hollow sphere, turning about the earth on an axis which runs close to the almost stationary star—known therefore as the Pole Star. The direction is from east to west, and a complete revolution is made in a day and night.

We can plot the stars on a globe, and draw an equator on it, which will everywhere be at an equal distance from the poles, and we may add other circles, as on a terrestrial globe: then the position of each star can be referred to these circles as towns on earth are found by latitude and longitude, and the path of any moving body, such as a comet, may be traced.

The stars fade out when the sun rises, but he too sweeps across the sky as though carried round by the same sphere, and he sets like them, in the west. Has he a fixed place on the sphere, keeping always the same position relatively to the stars? No, for in the place where he has just set we do not always see the same stars. Night after night those which were clear in the western sky as soon as it was dark enough to see them, grow closer to him, till at last they are lost in his twilight beams. Thus the sun, though sharing in the daily east to west movement, has a slow movement of his own on the sky-sphere, slipping back from west to east, until in a year he has accomplished the whole round, and sets again among the same stars.

Moreover, this peculiar movement of the sun is not a mere lagging behind the stars, for his west to east motion is combined with a north and south motion. If we note the star-groups which are just behind him when he sets (or just before him when he rises), we shall find that they form a great circle round the globe, half of which lies north and half south of the celestial equator. The Greeks named this circle the Zodiac, or "Path of the Animals," because the star-groups forming it were mostly called by the names of animals (the Ram, Lion, Fishes, etc.). When the sun is in the most northerly part of the zodiac it is summer in the northern hemisphere; when he is in the most southerly, it is summer in the south. (See Map).

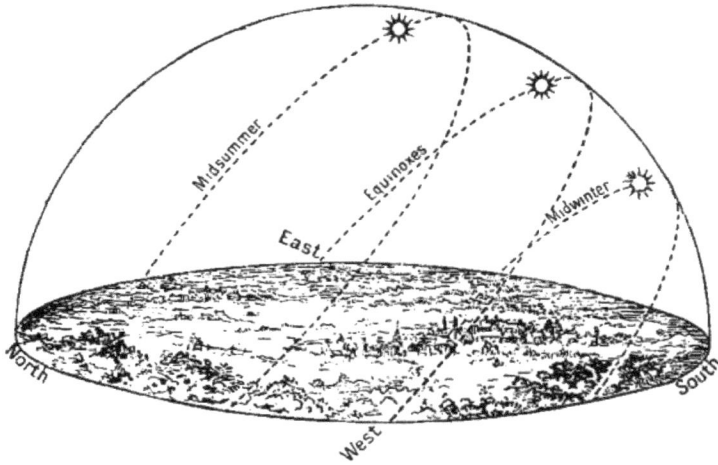

Fig. 1. The Sun's path in the sky at different seasons.

This north and south motion of the sun may be noted more directly in another way. Seen from any given place on the earth, each star rises and sets at the same points of the horizon always, and has the same course in the sky; but the rising and setting points of the sun, which on about the 21st of March are due east and west, travel daily further north, and the sun mounts daily higher in northern skies until about the 20th of June; then he returns towards the south, passing the east and west points again about September 23, and reaches his furthest point south about December 21. (The dates vary slightly owing to Leap Year). The dates on which the sun reaches his furthest north and furthest south points in this yearly journey are called the "solstices," because his motion seems to be checked, and he pauses or "stands" before reversing his direction; the dates on which he passes the midway point are the "equinoxes," because at those points he is on the equator, and makes day and night equal all over the earth.

The time taken by the sun to pass from one vernal (spring) equinox to another is 365 days, 5 hours, 48 minutes, 45 seconds. Since this slow motion along the zodiac is from west to east, contrary to the rapid east to west motion which he shares with all the stars, he takes a little longer to complete a daily revolution than they do; and if we reckon a solar day as consisting of 24 hours, a "sidereal" (or star) day is equal to 23 hours, 56 minutes, 4 seconds.

These are very elementary facts, but they are the fundamental facts of astronomy, and without recollecting and holding them clearly in mind we cannot understand Dante's allusions, nor see the fitness of any astronomical system, ancient or modern. To those who have only read about astronomy in books, and have not watched the skies, they may be puzzling, and I would beg these readers to make a few simple observations for themselves, as this will help them more than any written explanation can ever do to see the heavens with Dante's eyes. To appreciate the connected movement of the whole sky, some bright stars near the Pole should first be watched, such as the Great Bear and Cassiopeia, or for those in the southern hemisphere the Southern Cross, Canopus, Achernar. Their motions should be compared with those of bright stars near the equator, such as Orion, Virgo, or Aquila. The constellations of the zodiac should be studied, and notes made of the seasons at which each disappears in the rays of the sun.

The sun's north and south movements can be easily recognized by noting at what points of the horizon he rises or sets at different times of the year; and the different heights to which he rises in the sky are most simply observed by marking the length of the shadow of some tree or pole at midday. Or if some rough kind of gnomon[2] be made, even a flat piece of wood, laid on a sunny window-sill, with a long nail driven vertically into it, the movement and varying length of the shadow, from hour to hour, and from day to day, will make one realize vividly the diurnal and the seasonal movement of the sun. This device, in one form or another, was probably the first astronomical instrument invented, and by its means ancient astronomers in many lands solved important problems.

It is not necessary to explain that the daily apparent movements are caused in reality by the earth's rotation on her axis, and the yearly apparent movements by her revolution round the sun. These are the book-learned facts which for the most part obscure our perception of the very things on which they are based. I would ask the reader to do his best, for the moment, to forget them.

The movements of the moon among the stars are much more easily observed than those of the sun, since we can see the stars at the same time, and her revolution is much more rapid. She also is apparently carried round with the daily east to west movement, and she also has a west to east motion of her own, but so fast that it takes her round the star sphere in one month, instead of one year. This revolution also takes place in the zodiac. She is first visible as a fine crescent, just following the sun, in the west, after he has set; next night she is markedly further from the sun, on her eastward course, and is a larger crescent; she continues increasing her distance from the sun and the size of her disc, until, as full moon, she is rising in the east when the sun sets opposite her in the west, and setting when the sun rises. After this, she begins to wane, and, still travelling in the same direction, rises later and later at night, and sets in the day; she draws gradually nearer to the sun on the western side, till at last, as a fine crescent with the horns turned in the other direction (*i.e.* always away from the sun), she appears just before the rising sun in the east. Then for a short time she is lost in his rays, till she emerges as a new moon on the sunset side again.

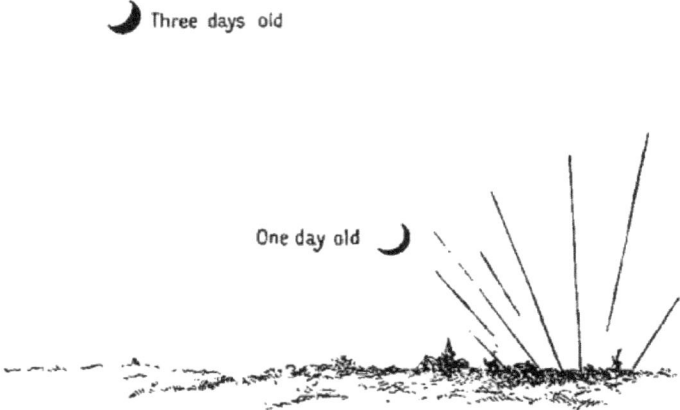

Three days old

One day old

Fig. 2. The Moon at Sunset.

The moon completes a revolution among the stars in 27 days, 8 hours; but it takes her a little longer to come up with the sun again, since he has meanwhile been moving in the same direction along his yearly path; and the 'synodic' month, or period from one new moon to the next, is 29 days, 13 hours.

As well as the moving sun and the moving moon, there are five other bodies, visible to the naked eye, which move among the stars. They look like stars, but their movements would lead us rather to class them with sun and moon. They also are in the zodiac, and they also, while carried round with the universal movement from east to west, revolve slowly, each in its own period, from west to east. But their motions are more complicated than those of sun and moon. Two, which we call Venus and Mercury, are never seen very far from the sun, and they oscillate from side to side, sometimes appearing before him near sunrise, and sometimes after him at sunset. Mercury keeps closest to the sun, and is not so bright, and therefore less easy to see; but Venus is a brilliant object when she gradually swings out further from the sun, remaining longer each evening after sunset in the western sky. Then she gradually draws back, closer to the sun, is lost in his rays, and a few days after begins to appear on his other side, as a Morning Star, visible in the east before sunrise. Here she swings out again, like a pendulum, to her furthest distance west, and then draws in again, just as she did on the sunset side of the sun.

In this way, swinging slowly from side to side of the sun, Mercury and Venus make with him the circuit of the zodiac, completing a revolution from west to east in about a year. The average period of Mercury's oscillation, counting, for instance, from one Greatest Western Elongation (*i.e.* furthest distance from the sun on the west) to the next, is 116 days; that of Venus is 584 days.

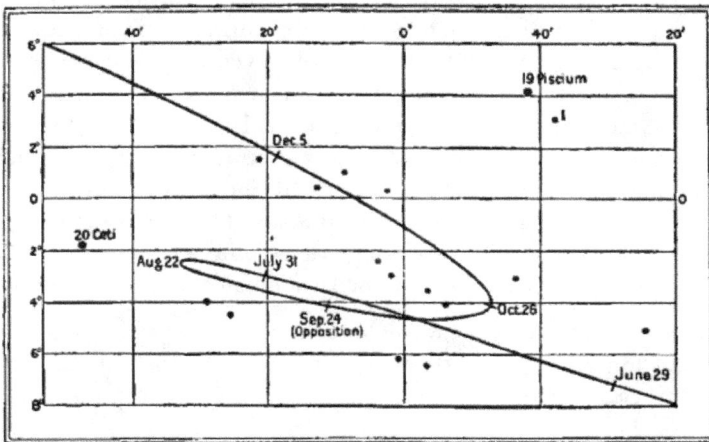

Fig. 3. The Path of Mars among the Stars, 1909.

The other three "wandering stars"—or "planets,"[3] as they were named by the ancient Greeks—Mars, Jupiter and Saturn, are also often seen as morning or evening stars near the sun, but they do not always accompany him, like Venus and Mercury. They may be seen at any distance from him, even exactly opposite, so that they rise as he sets. They keep as strictly to the zodiac, however, and travel in it from west to east, in periods of approximately two, twelve, and thirty years respectively; and their paths are also complicated by oscillations. Periodically they slacken speed, stop, and go back a little distance among the stars, then they slacken, stop, and advance again. These changes are technically called direct motion, stations or stationary points, and retrograde motion.

It must have originally taken many years of patient watching to discover and distinguish all these planets. In these days, by means of an almanac and some knowledge of the constellations, they may easily be found and traced. Mars and Venus move quickly during part of the time they are visible, and if sketches be made of their positions among the stars, and their paths marked for a few weeks, a very good idea may be gained of the motions of planets as seen in the skies.

Once again, it is not necessary to explain here that these movements of the planets are due partly to their revolution round the sun, and partly to the Earth's motion. Nor need we, for our present purpose, consider them in any detail: all that is important to realize is the general character of the movements, and their likeness to those of sun and moon.

The distances, and therefore the sizes, of all the heavenly bodies are completely beyond measurement, except with instruments and refined methods; their physical nature could only be guessed at before the discoveries of universal gravitation and spectrum analysis, in the 17th and 19th centuries of our era. All that can be observed by naked eye astronomy is difference of brightness and colour; as for instance the contrast between ruddy Mars and white Jupiter; the steadier light of all the planets as compared with stars; and the interesting fact that the moon shines by reflected sunlight, which is made evident by the connection between her phases and her position with regard to the sun. Her surface, too, is clearly seen to be diversified by dark markings of definite shape, but on no other body in all the sky can we make out the least detail without a telescope.

The movements of the heavenly bodies, therefore, which still form one of the most important parts of astronomy, were almost all that could be studied by ancient astronomers, and gave them the only key they had to the problems of the universe.

To sum up:—The chief apparent movements of the heavens, visible to the naked eye, are eight, viz:—

The daily revolution of the entire heavens, carrying with it every visible celestial body, in a little less than 24 hours; the revolutions of sun, moon, and five naked eye planets, in seven different periods.

The first of these is from east to west, and is by far the most rapid. The axis of revolution passes through two points which we call the celestial poles, and the motion is parallel to the celestial equator.

All the others are in the main from west to east, though the progress of the planets is complicated by periodical retrograde movements. All take place in the zodiac, which is a series of constellations forming a great band round the heavens. The path of the sun is a great circle through this, called the Ecliptic (because eclipses can only happen when the moon is also on it); and the paths of moon and planets are slightly and variously inclined to it.

Thus the daily path of a star is affected only by the simple uniform movement of the entire heaven (in reality the rotation of the Earth) but the daily path of a planet, or of the sun or moon, results from a combination of this general movement with its own peculiar movement, which is generally in the opposite direction.

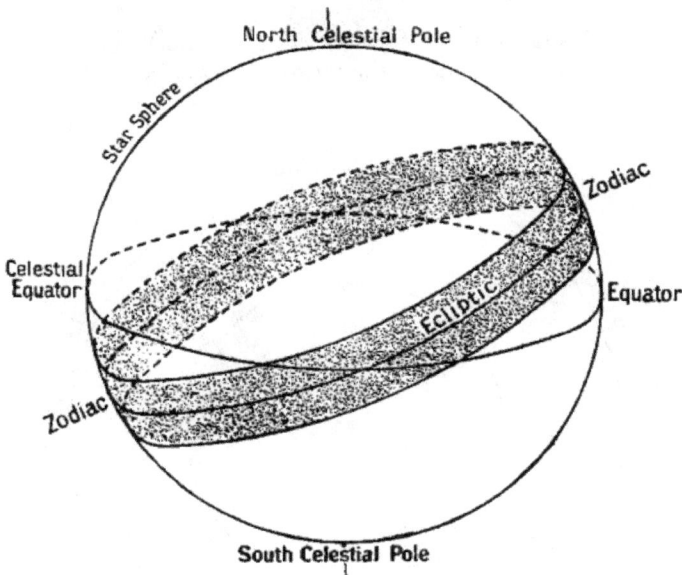

Fig. 4. The star sphere.

If it is difficult to conceive a body moving simultaneously in two different directions, an earthly analogy will make it easy. On a great moving platform, such as that which encircled the Paris Exhibition in 1900, there are fixed posts etc. which revolve exactly as the whole platform revolves and do not move about amongst themselves. These are like the fixed stars on the (apparently) revolving sphere. But human beings are free to add their own movements to that given them by the platform on which they stand. One man turns his back and walks steadily and very slowly in the opposite direction, and so he neutralizes part of the platform movement and is not carried onward quite so quickly as the stationary posts: he is the sun. A woman walks as he does, but much more quickly, so that she rapidly passes many posts, although all the time she is being carried backwards with them: she is the moon. Children run backwards and forwards: they are the planets. Finally, if all these people are also constantly crossing the platform slowly from right to left and back again, their movements will be oblique to the platform movement and will imitate the north and south movements of sun, moon, and planets.

It is in this fashion that the movements of the skies present themselves to careful observers on this seemingly stationary earth; and in the youth of the world these apparent movements were believed to be real. The ancients thought that the sky was actually revolving round a steadfast earth, while the sun and moon and certain other "wandering stars" had in addition various motions peculiar to themselves.

The table of periods which follows (see pp. 22-23) will be found useful for occasional reference. Some of the terms used will be explained later.

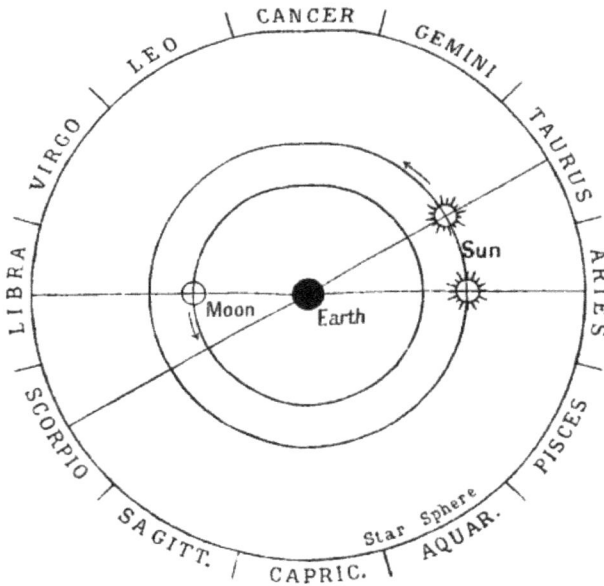

Fig. 5. Diagram illustrating Synodic and Sidereal Periods.

The arrows show the direction of the Moon's monthly and the Sun's yearly revolutions in the zodiac, as seen from Earth.

When the Moon is opposite the Sun, for instance in Libra while he is in Aries, she is full. In 27½ days she returns to the same place among the stars, and this is a SIDEREAL MONTH. But the Sun meanwhile has moved into Taurus, and not until the Moon has reached Scorpio, opposite to him, will she be full again, and complete her SYNODIC MONTH (29½ days).

REVOLUTIONS OF SUN, MOON, AND PLANETS AS SEEN FROM THE EARTH.

	Days.	Hrs.	Mins.	Secs.
Tropical Solar Year: period from one vernal equinox to another, or from one summer solstice to the next, &c. (Our civil year is based on this).	365	5	48	45·5
Sidereal Year: period between two successive returns of the sun to any star on his path	365	6	9	8·9
The difference between these two kinds of year is due to Precession of the Equinoxes, *vide infra*, p. 23.				
	Days.	Hrs.	Mins.	Secs.
Solar Day: period between two successive passages of the sun across the meridian (noon)	—	24	0	0
Sidereal Day: period between two successive passages of a star across the meridian	—	23	56	4
The difference between these two kinds of day is due to lag of the sun behind the stars, his daily motion westward being slightly retarded by his slow yearly motion eastward.				
	Days.	Hrs.	Mins.	Secs.
Mean Synodic Month: period between two full moons	29	12	44	2·8
Mean Sidereal Month: period between two successive returns of moon to any star on her path	27	7	43	11·5
The difference between these two kinds of months is due to the fact that while the moon is making her revolution among the stars, the sun is				

	Days.	Hrs.	Mins.	Secs.
also moving slowly on in the same direction.				
Mean Anomalistic Month: period between perigee and perigee (*vide infra*)	27	13	18	37·4

PLANETS:—Mean Synodic Revolution: period between two successive conjunctions with the sun, and mean zodiacal revolution: period of revolution round the zodiac.

	Mean Synodic Revolution.		Mean Zodiacal Revolution.	
Mercury	116	Days[4]	1·0	Years
Venus	584	”	1·0	”
Mars	780	”	1·88	”
Jupiter	399	”	11·86	”
Saturn	348	”	29·46	”

PRECESSION OF EQUINOXES: 50·25 seconds of arc in one year; that is, 360 degrees (a revolution among the stars of the zodiac) in 25,800 years nearly.

II

THE BEGINNINGS OF ASTRONOMY.

Note.

As these pages are passing through the press, a letter from Mr. Maunder appears in *The Observatory* for August 1913 on "The Origin of the Constellations," and this should be consulted by anyone interested in the subject. Mr. Maunder points out that Ptolemy gives us much more precise information than Aratus regarding the southern limits of the ancient constellations, and that the changes which he says he ventured to make in their traditional forms are extremely insignificant.

Mr. Maunder further observes that the celestial equator of Aratus cannot give any clue to the origin of the constellations (as R. Brown suggested), but only to the date of the work from which Aratus copied, when some astronomer had drawn the equator through the constellations. A slight alteration of the text, Mr. Maunder says, would give a correct equator for the date B.C. 1000.

See also Mr. and Mrs. Maunder's article in *Monthly Notices of the Royal Astronomical Society* for March 1904.

Proctor's "Origin of the Constellation Figures" is in his book *Myths and Marvels of Astronomy*.

The sky appears to us like an arch, embracing all our lives, Dante says.[5] From the dawn of intelligence man must have recognized his dependence upon the all-embracing heavens, especially the sun, without which life would be impossible. The consciousness expressed itself in many ways: in adoration of the sky, the sun, moon, and hosts of heaven; in superstitious fear which regarded events on earth as directly controlled by the heavenly bodies; in careful watching and recording of their movements for useful purposes. Thus, long before astronomy became an exact science, and was studied simply for its own sake, patient observers had laid the foundations, and were familiar with many of the movements we have been describing.

These are of great importance to primitive man. Sun, moon, and stars are invaluable as guides, especially at sea, and we know that the ancient Greek mariners used to steer their ships by observations of the Great Bear, while the Phoenicians preferred to use the Little Bear for this purpose. But the strongest and most universal incentive to careful and prolonged study of the skies is our complete dependence upon them for the measurement of time.

In the earliest period of their history, the Jews, the Greeks, and probably every other nation, divided the day simply into morning, noon, and evening, according as the sun was rising, or apparently stationary, or sinking, with regard to the horizon; and the passage of some bright stars indicated the time at night. But at a very early period the first of all astronomical instruments was invented, by which the sun's varying height can be measured: hence the time of noon, the dates of equinoxes and solstices, and the length of the solar year can be determined. The gnomon in its simplest form is a pole set up vertically on a smooth level surface, on which its shadow as cast by the sun can be observed. The moment of shortest shadow marks the middle of the day, the shortest midday shadow marks the summer solstice, the longest the winter solstice, the equinoxes falling between. The instrument also indicates the points of the compass, for the sun is always due south in northern latitudes at midday: hence the Latin word *meridies* (French *midi*) means south as well as midday, and the Meridian in astronomy is a line which passes through the north and south points and the zenith, and is crossed by the sun at midday.

The gnomon was said to have been introduced into Greece by Anaximander about 600 B.C., and the Babylonians claimed to be the inventors, but it was probably invented independently by several races. The Chinese certainly observed the length of the shadow more than two thousand years B.C., and the very interesting fact has recently come to light that a tribe in a hitherto unexplored part of Borneo use such an instrument, invented by themselves. They set up a post about 6 ft. high, and throw over the top a piece of string weighted at each end to show when it is vertical; the length of the shadow cast by the post is measured with a notched stick. By this means they tell the time of day; and they also observe the sun (presumably with the gnomon) to know the right season for planting their rice.[6]

However rough the first gnomon may have been, its importance can scarcely be overrated, for it introduced measurement and calculation into observation of the sun's movements, and it is the ancestor of our modern sextants, transit telescopes, and other instruments of precision.

It was also the beginning of the sundial. The course followed by the moving end of the shadow was traced on the ground, and divided into equal parts: hence arose the custom which the Greeks adopted from the Babylonians of counting twelve hours in every day, from sunrise to sunset, and twelve hours in every night, from sunset to sunrise, regardless of the varying lengths of day and night at different seasons. This is known as the system of "temporary hours." If we used it in England, the twelve hours of a midsummer day would take twice as long to pass as the twelve fleeting hours of a midsummer night; but in Greece the inequality is much less, and in the latitudes of Babylonia it is never striking. The skill and knowledge of the Greeks enabled them, later on, to construct dials of different kinds, which marked "equal hours," such as we use now; but the system of "temporary hours" did not altogether die out till after the invention of pendulum clocks in the 17th century of our era.

Clepsydras, or water-clocks, were also used in Egypt and Babylonia, and ancient Greece; and there is still a large one in Canton, where a reservoir is placed in a tower, and the water falling, drop by drop, into a receiver whose depth is marked in figures on the wall, indicates the passing of time just as sand does in running through an hour-glass. These clocks cannot have kept very good time, however, or they would have been more used by the Babylonian and Greek astronomers who took pains to ascertain the exact positions of the stars. Owing to the diurnal revolution of the skies, the time at which any celestial body rises or crosses the meridian after another is an index of their distance apart, east and west on the sphere, and this is how it is reckoned by modern astronomers. But the ancients seem to have been never able to trust their clepsydras sufficiently to use this method, and only referred to them for approximate time.

The gnomon, valuable as it is for marking the sun's daily course, and the north and south part of his yearly motion, is a limited instrument. It cannot show his westerly motion on the sphere, nor is it of any use for the planets. To trace these motions, and the monthly journey of the moon, the first step is to distinguish the stars, by grouping and naming them, especially those which lie in the path of sun, moon, and planets. The invention of some kind of zodiac is probably older even than the invention of the gnomon, and also originated independently among different races. The germ of the idea may be found to-day among races low in the scale of civilization. The Australian aborigines are familiar with that unique star-cluster which we call the Pleiades, and know that its appearances and disappearances are periodical and coincide with the seasons. A Queensland tribe, for instance, has a legend in which the stars figure as six sisters who have been transported to the skies, and it is said that they sometimes appear before the sun in order to throw down icicles, an evident allusion to the fact that the Pleiades begin to appear just before sunrise in May, the Australian winter. The natives of Tahiti divide their year into "Matarii i nia" and "Matarii i raro," which means Pleiades Above and Pleiades Below (*i.e.* the horizon at the beginning of night). Sir Norman Lockyer has shown that the mysterious alignments of stones found at Stonehenge, Carnac, and other places, may have been so arranged in order to show the direction, some of the sun at his rising on certain dates, notably the morning of the summer solstice, and some of the Pleiades or other striking stars, whose rising just before the sun would enable ancient astronomers to fix the dates of important festivals.

Such observations as these, of which many instances might be drawn from many parts of the world, are first steps towards studying the whole path of the sun through the stars, and of forming a calendar with a name or number for every day in the year. Until the stars are known, and a calendar fixed, the motions of sun and moon cannot be learned in detail, the planets can scarcely be distinguished from the stars and from one another, and there are no settled dates from which to calculate their periods.

The first zodiac of which we have written record is a lunar one of 28 constellations, which is referred to in the "Canon of the Emperor Yaou," a Chinese emperor who began to reign in B.C. 2356. "Yaou commanded He and Ho, in reverend accordance with their observation of the wide heavens, to calculate and delineate the movements and appearances of the sun, the moon, the stars, and the zodiacal spaces, and so to deliver respectfully the seasons to the people." That these astronomers studied "the movements and appearances" of the sun by means of the gnomon as well as observations of the stars is plain from what follows, where directions are given for determining the solstices. One astronomer was commanded by the Emperor to "reside at Nankeaou and arrange the transformations of the summer, and respectfully to observe the extreme limit of the shadow. The day, he said, is at its longest, and the star is *Ho*: you may thus exactly determine midsummer." *Ho* (= fire) is the fiery red Antares in Scorpio. The star of the winter solstice, when "the day is at its shortest" was Maou, which is the Pleiades. Directions are also given for observing the spring and autumn equinoxes, when day and night are of medium length, and certain other stars are to be observed.[7]

The Hindus had also a lunar Zodiac, but with only 27 constellations, and the Arabs had their 28 "Mansions of the Moon." These 27 or 28 asterisms were evidently suggested by the moon's sidereal period of 27¼ days. But her synodical period, *i.e.* her revolution with regard to the sun, in which she runs through her phases, is much more convenient for marking a period of time for general uses, and this month of about 30 days has been almost universally adopted by primitive peoples, the first day being counted when the crescent new moon begins to be seen after sunset. From this custom arose another, that of counting the beginning of the day from sunset, but in various times and places other starting-points have been chosen—sunrise, midday, or midnight.

Twelve of these synodical lunar months are nearly equal to one solar year, and this doubtless suggested the solar zodiac of twelve constellations, each constellation marking the portion of sky passed over by the sun in a month. The Chinese "Yellow Path of the Sun" contained twelve animals, the Mouse, Cow, Tiger, Rabbit, Dragon, Serpent, Horse, Ram, Ape, Hen, Dog, and Pig. These animals were widely adopted by other nations—the Koreans and the Japanese, the Mongols of Tibet, the Tartars, and the Turks.

Another zodiac, however, was destined to have an even wider popularity, spreading, in the course of centuries, from Greece to Arabia, Persia, India, and China, where it finally superseded the native constellations; it crossed the Mediterranean into Africa, conquered the whole of Europe, and is used to-day over the whole civilised world.

The names of these twelve zodiacal constellations are familiar to us all:—

English Names.	Latin Names.
Ram	Aries
Bull	Taurus
Twins	Gemini
Crab	Cancer
Lion	Leo
Virgin	Virgo
Scales	Libra
Scorpion	Scorpio
Archer	Sagittarius
Capricorn	Capricornus
Water-Bearer	Aquarius
Fishes	Pisces.

Strange to say, we cannot tell with any certainty where, when, or by whom, this ancient series of constellations was devised and named. The earliest full description which we possess is by a Greek poet of the 4th century B.C., Aratus. A line from the prologue to his "Phenomena" was quoted by St. Paul in his address to the Athenians on Mars' Hill.

"From Zeus we lead the strain, he whom mankind Ne'er leave unhymned; of Zeus all public ways, All haunts of men are full, and full the sea And harbours; and of Zeus all stand in need. We are his offspring;[8] and he, mild to man, Gives favouring signs and rouses us to toil, Calling to mind life's wants; when clods are best For plough and mattock, when the time is ripe For planting vines and sowing seeds he tells. Since he himself hath fixed in heaven these signs, The stars dividing; and throughout the year Stars he provides to indicate to men The seasons' course, that all things duly grow."[9]

But Aratus did not know who had invented the names of the star-groups which he describes. "Some man of yore," he supposes,

"A nomenclature thought of and devised, And forms sufficient found. For men could not Or tell or learn the separate names of all, Since everywhere are many, size and tint Of multitudes the same, but all are drawn around. So thought he good to make the stellar groups, That each by other lying orderly, They might display their forms. And thus the stars At once took names and rise familiar now."[10]

It is, to say the least, exceedingly doubtful, whether the naming of star-groups was so promptly carried out by one individual, especially as Aratus' poem includes, besides the twelve zodiacal constellations, thirty-six others, which contain all the bright stars of the sky except those too far south to be seen in the temperate regions of our northern hemisphere. The spaces thus left blank were afterwards filled up, chiefly in the 17th and 18th centuries of our era, and the regions round the South Pole are now crowded with a mixture of birds and scientific instruments; but the names and the figures of the traditional forty-eight constellations still find undisputed places on our globes and star-maps.

Some of these figures are very strange and suggestive. We have a maiden with wings, a centaur shooting arrows, a flying horse, a water-snake with a crow and a cup on its back, a charioteer with a goat on his shoulder, a man strangling a serpent, another pouring water into the mouth of a fish, and a strange beast like a goat with a fish's tail. All had their meaning, doubtless, to their originators, but to us they are cryptic characters, hard to decipher. Among the zodiacal constellations only one is obvious, Libra the Scales, the sign in which the sun is when days and nights are perfectly balanced in length; but this is comparatively recent, for Aratus and his contemporaries give in its place the Claws of the Scorpion, the latter being an enormous monster extending over the space of two asterisms. The figures may have been religious symbols, or an illustration of some myth concerning the sun's yearly course, or each of the twelve may have indicated the weather or the occupation suitable to the month it represented. The ear of corn in the hand of the Virgin, and the juxtaposition of three watery figures in Capricornus, Aquarius, and Pisces, suggest the latter explanation. Many different ideas probably played a part in the origin of these mysterious constellation-forms. The Greeks, and after them the Romans, when adopting the old constellations, sometimes adopted also the old myths which still clung about them; sometimes they ascribed legends to them from their own mythology. Thus, the kneeling figure with his foot upon a dragon, became and remains the hero Hercules, although Aratus only describes him as a man toiling at some unknown task, and says he is called simply the Kneeler. Successive generations of astronomers altered some of the figures, but probably only to a slight extent.[11]

The poem of Aratus enjoyed an immense popularity in classical times and throughout the Middle Ages, and no doubt helped to stereotype the forms whose origin was already forgotten when he wrote. He was not an astronomer, however, and the poem is only a popular paraphrase of a lost work by Eudoxus. This Greek astronomer had lived a hundred years earlier, and it is thought that he himself copied from an older source. The attempt to discover this source has been the object of many ingenious conjectures, and much research among ancient monuments and writings. Some of the old constellations are met with in Isaiah and Job, in Homer, on tablets found at Nineveh, and an immense antiquity is sometimes claimed for them. Dupuis, writing at the end of the eighteenth century, thought he had conclusively proved that the figures of the zodiac were designed in Egypt 15,000 years ago![12] Miss Plunkett, in her "Ancient Calendars and Constellations" assigns them to the seventh millenium before Christ.

THE OLD CONSTELLATION FIGURES.

NORTHERN HEMISPHERE.

SOUTHERN HEMISPHERE

THE OLD CONSTELLATION FIGURES ACCORDING TO ARATUS.

In Ptolemy's Catalogue Equuleus and Corona Australis are added to these.
(From Peck's "Constellations and How to Find Them.")

An ingenious theory, suggested independently by Schwartz and Proctor, and developed by Mr. E. W. Maunder,[13] is founded on an examination of the space round the South Pole which was left blank by the ancient constellation designers. From its extent, Proctor concluded that they cannot have seen further south than about 40° from the South Pole, and therefore that they must have lived in a latitude of about 40° north of the equator (say, central Asia or Asia Minor); from the position of its centre, which must have been the Pole, he concluded that the date was about B.C. 2200. For the centre of the circular patch seems to lie near the star *Delta Hydri*, which was the South Pole star at that time. (This movement of the Pole among the stars, due to "precession," will be explained later). This date does not differ much from that found by Robert Brown, from the position of the celestial equator among the stars, as described by Aratus; he says B.C. 2084.[14] Schwartz gave B.C. 1400; Mr. Maunder, from additional considerations of the positions of various constellation figures, says that they must all have been originally designed about B.C. 2800.

Unfortunately the descriptions of Aratus are neither very precise nor consistent with one another, and he is our oldest and our main authority for the forms and positions of the ancient constellations.[15] It may, however, be taken as practically certain that they had been designed many centuries before he wrote, and that the Greeks received them from the Babylonians. Orion and the Pleiades, the Great Bear and Arcturus, and perhaps many others, were familiarly known in the Levant as early as the tenth century before Christ; and we find traces of our zodiac, or its beginnings, in Babylonia at least as early as the eleventh. It is always to Egyptians and Babylonians that the Greeks referred as their predecessors and teachers in astronomy, but the native constellations of Egypt seem to have been different, and so far as we know at present the Babylonians began earlier and made greater progress in star-lore than any other nation before Greece. The latest results of expert investigation of astronomical tablets discovered in the ancient clay libraries of Babylonia and Assyria, tend to show that astronomy was of native growth there, and developed very slowly.[16] Star-worship and the need for a calendar led their inhabitants to observe the skies thousands of years ago; but their early work was naturally vague and rude.

This star-worship and star-study seems to have been learned by the Semitic Babylonians, and their descendants and rivals the Assyrians, from a race with whom they met and mingled in the grey dawn of history, but whose existence was unknown to us before the middle of last century.

[*To face p. 36.*

The Moon-God of Ur.

From a cylinder-seal in the British Museum, dated about B.C. 2400.

Reproduced by permission of the Trustees.

Fig. 6. The triple star-sign of the Babylonians.

This people, who belonged to a totally distinct family of nations, and are known to us now as Sumerians, had settled near the mouth of the Persian Gulf, when it ran further inland than it does now, and more than five thousand years ago used a kind of writing on soft stones (later, on bricks) which had obviously arisen from some form of picture writing, and ultimately developed into cuneiform. Their reverence for the heavenly bodies is shown by the fact that the familiar star sign, which appears on very early Sumerian inscriptions, denotes their word for god or lord, and on the monuments of Babylonia and Assyria we meet constantly the triple sign This, we learn from the inscriptions, stood for three great deities, the Moon-god, the Sun-god, and the goddess of the planet Venus. Our illustration shows an inscription in early Babylonian script, and a scene which represents the vassal of a king of Ur (Abraham's "Ur of the Chaldees") being led into the presence of the Moon-god.[17] It is believed to date from about B.C. 2400. The Babylonians were an intensely superstitious people, and a large part of their omens were drawn from observations of the skies. Every city from this period onward had its ziggurat or great tower formed of several superimposed cubes, usually seven in number, diminishing in size and probably crowned by the shrine of the local deity. It is not certain what purposes were served by these towers, but the successive platforms may well have been the observatories from which the Babylonian priests, gazing through the clear air and over the level plains, watched, year after year, and century after century, eclipses of sun and moon, risings and settings of stars and planets, and all the changing pageant of the skies, which to them were eloquent of peace and prosperity, or of war and misfortunes in their land.

Although this illusory art chiefly occupied the early Babylonian astronomers, they made some observations of real value, and gradually acquired true knowledge concerning the movements of the heavenly bodies.

Tablets a few centuries older than the Chinese *Canon of Yaou* containing lists of the Sumerian names of twelve months, show that this people had established a luni-solar year. Fortunately for the progress of astronomy the year does not contain an exact number of months, or even of days: at the end of twelve lunar months, a few more days and hours must elapse before the sun has returned to his original place among the stars, and before the round of the seasons is completed. Therefore the first rough approximation had to be constantly corrected if calendar festivals were to recur at the same seasons; and thus the priests, who in early times were usually the calendar makers and keepers, became gradually better and better acquainted with the movements of sun and moon, and the appearance of star-groups. It is interesting to compare the different ways in which various races have solved the problem of calendar-formation.

The Chinese had a year of twelve months, and added an intercalary month occasionally, in such a way that the average length of the year was brought up to 366 days. The written character for "intercalary" in both Chinese and Japanese is a compound of the characters for "gate" and "Emperor," because in ancient days the Emperor used to perform the ceremonies proper to each of the twelve months in the special room of his palace dedicated to that month, but in the intercalary month he performed them in the doorway of the palace.

The Egyptians and the Arabs seem to have given up the attempt to harmonize the two periods, though both of these nations reckoned twelve months in their years. The Egyptians counted thirty days to each month, and added five days more at the end of the twelfth, so that the months can have had no connection with the moon: the year had, in fact, been calculated from the position of the sun among the stars, beginning with the morning on which Sirius rose just before it. This "heliacal rising" of Sirius heralded the great event of their year, the overflow of the Nile. The Arab year, on the contrary, was purely lunar, for it consisted of twelve months which were alternately of twenty-nine and thirty days: they therefore corresponded pretty closely with the moon's phases, but had no connection with the sun or the seasons. The Mahomedans still use this lunar year.

The new moon festivals of the Hebrews prove that the moon was important to their calendar, but the three chief feasts of First-fruits, of Ingathering, and of the Passover, were so closely connected with the seasons that their year must have been luni-solar. It consisted of twelve months, one of which was sometimes doubled, but how they decided when this was necessary is nowhere described in the Old Testament. Some think, that as an offering of first-fruits was to be made on a certain day of a certain month, the month preceding it was doubled in every year in which it was evident that the crops would not be far enough advanced for the first-fruits to be gathered so soon: in this way no direct observations had to be made of the sun's movements, but the year was accommodated to them by observations of the seasons.[18]

The Babylonian calendar is the most interesting of all, for it was the most intimately connected with star-observation. At first an extra month seems to have been added to the usual twelve, in an irregular way, whenever found necessary, judging by a tablet of the great king Hammurabi, who united all the cities of southern Babylonia under one rule, and gave them the famous Code of Laws, communicated to him by the Sun-god. The tablet runs as follows:—

> "Thus saith Hammurabi: the year having gone wrong, let the coming month be registered by the name of Ululu the second. And instead of the payment of taxes being made on the 25th day of Tasritsu, let it be made on the 25th day of Ululu the second."

Hammurabi reigned about B.C. 2200. A thousand years or more after this, we find that royal decrees for correcting the calendar were never necessary, for the astronomers had invented more than one system for keeping the year right. One of these was to observe, like the Egyptians, the heliacal rising of certain stars. The little group of three stars in the head of the Ram, which we call Alpha, Beta, and Gamma Arietis, was found very convenient for this purpose. When it rose just before the sun in the month Nisan, the observers knew that all the twelve months would fall in their right seasons, but when it remained invisible (hidden in the morning twilight) until the following month, the calendar was evidently running ahead of the sun, and that year was lengthened by adding a thirteenth month. This is the meaning of the directions given on a tablet now in the British Museum:—

> "The asterism Dilgan[19] rises heliacally in the month of Nisan. Whenever this asterism remains invisible, let its month be forgotten,"

that is, let it be taken over again, as if it had not already been counted. Similar directions are given for some other asterisms and their corresponding months. But a second method, which was peculiar, so far as we know, to the Babylonians, was that of using the moon as a pointer to indicate the place of the sun. Whereas the sun's place among the stars can only be inferred, the moon's can be plainly seen, and her phase indicates her distance from the sun at any time. A tablet of unknown date, belonging to the last millenium before our era, or a little earlier, gives the following directions:—

> "When on the first day of the month of Nisan the asterism Mulmul (the Pleiades)[20] and the Moon are seen together, the year will be normal. When on the third day of Nisan the asterism Mulmul and the Moon are together, the year will be full" (that is, will contain 13 months).

Each Babylonian month began when the new moon was first visible after sunset; if at this moment she was seen with the Pleiades, it is clear that the sun, which had just set, was not far west of the cluster; if however, it was not till the moon was three days old that she was seen with the Pleiades, she would then be some distance above the horizon at sunset: consequently the sun was some distance west of the Pleiades. In this case he would also be west of Dilgan, the Ram's Head, so those stars would rise after him in the morning, and be hidden in his light: therefore, both the morning and the evening observation combined to show that his course was not completed, and that the year must be lengthened by the addition of an extra month.

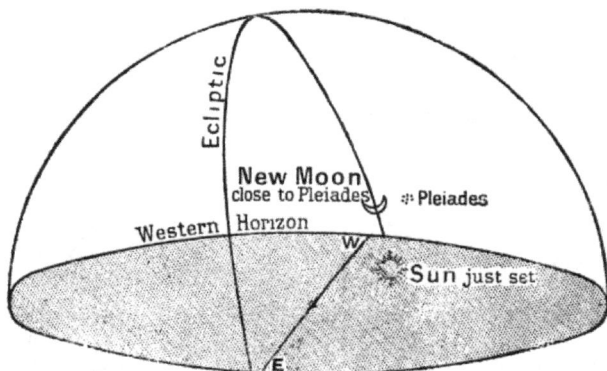

Fig. 7. First Year, normal. New Moon near the Pleiades after sunset on the 1st of Nisan.

The position of the young moon (which always closely follows the sun) showed that the sun was not far west of the Pleiades; and about 1000 B.C. this proved that it was near the vernal equinox. The sun's position is given for about half an hour after sunset, when the Pleiades would first be visible.

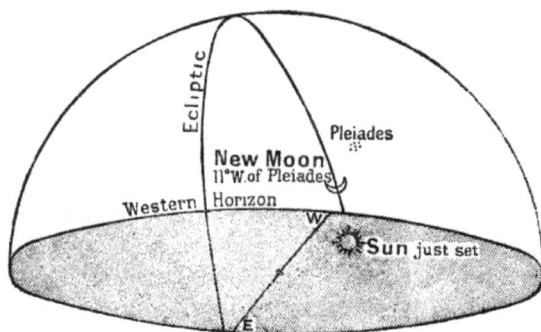

Fig. 8. Second Year, normal. New Moon not far from Pleiades on the 1st of Nisan.

It takes the sun 365 days to return to the same place among the stars, but the Babylonian year of 12 lunar months (each of 29 or 30 days) was 11 days short of this: therefore on the 1st of Nisan in this year the sun had still 11 days' march before him ere he returned to the position of Fig. 7. This is equal to about 11°, so the young moon was also about 11° west of her former position, near the Pleiades. But as she travels about 13° eastward every day, she would be near the Pleiades on the following evening, the 2nd of Nisan, so this year was also counted normal.

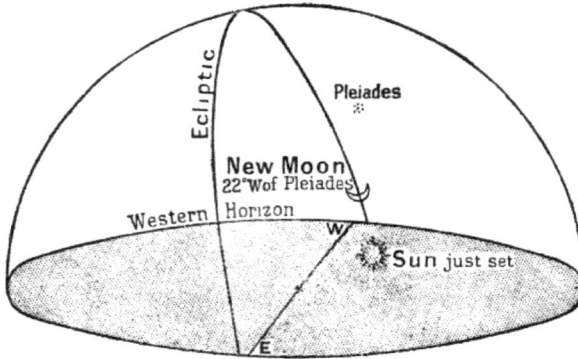

Fig. 9. Third Year, "full." New Moon distant from the Pleiades on the 1st of Nisan.

The sun is now 2 × 11 = 22 days' march, or about 22°, short of his first position, and the young moon consequently about 22° west of the Pleiades, so she will not come up with them until the 3rd Nisan, after travelling 2 × 13 = 26°. The year was therefore "full," that is an extra month of 29 days was added, which is more than the 22 days needed to enable the sun to reach his first position by the 1st of Nisan in the fourth year.

It appears, therefore, that the extra month must have been added once in three or four years.

Several lists of stars and star-groups indicating the months in this way have been found, the early lists containing only a few, the later twelve. If our zodiac originated with the Babylonians, there is little doubt the idea took its rise from these monthly stars, but it is not possible, with our present knowledge, to say when these old astronomers first linked the isolated stars into a continuous series of twelve star-groups and connected the idea of the month with the invisible group among which the sun was known to be shining, instead of with the stars seen east or west of him, or in conjunction with the crescent moon.[21]

Fig. 10. The Scorpion.

From a boundary stone
(now in the British Museum) set up
in the reign of Nebuchadnezzar I.,
king of Babylonia, about 1100 B.C.

Fig. 11. The Goat, with Fishes' Scales.
From a Babylonian boundary stone.

A Scorpion with immense claws, and a Goat with fishes' scales appear
several times on monuments at least as old as 1000 B.C. and it is very
probable, although this fact alone would not prove it, that they were then
used as constellation figures. It has been definitely proved from inscriptions
that before 600 B.C. the name of Scorpion was applied to some stars of our
present Scorpion, that there was a Lion corresponding with ours, and the
principal star in that asterism, which was called "The King" by Greeks and
Romans (Basiliskos and Regulus), bore a name with the same meaning in
Babylonia; the Celestial Bull seems to have been the group of the Hyades,

and the Great Twins were the two stars Castor and Pollux. The last two identifications seem to show how the single stars or small groups of the monthly lists were expanded into the large zodiacal constellations, for the Hyades cluster is in our present Bull, and Castor and Pollux are in our Twins.

Under the great Assyrian kings who in the 8th and 7th centuries B.C. made Nineveh the capital of their empire, Babylonian astronomy flourished exceedingly, and it made much progress through all the political changes which followed, until the beginning of our era. The motions, phases, and eclipses of the moon were carefully studied and could be accurately predicted, the positions of many stars were determined; the zodiac was divided into twelve equal spaces, which afterwards became 36 by sub-division (the constellations being too unequal in size for convenience); and finally the whole circle was marked out in 360 degrees. The movements of all the naked eye planets were well understood, their positions being constantly compared with those of a number of standard stars, mostly in the zodiac; and after watching and recording these for a number of years the astronomers were able to calculate where each planet would be found at future dates. Tables have been found on clay tablets of the 2nd century B.C. predicting the heliacal risings and settings, and the stations and retrogressions etc., with considerable accuracy.

When astronomy had reached this stage of accurate prediction, it was no longer in its infancy, but was fairly on its way to become a true science.[22]

Nevertheless, the astronomy of the Babylonians, advanced as it was, seems never to have progressed beyond the empirical stage. With them, there seems to have been no desire to group the facts they so patiently and skilfully collected into a system, and form a theory to explain them.

And this must be said of other ancient nations also. The Egyptians made careful observations, especially of the heliacal risings of different stars, by means of which they determined the length of the year, as we have already mentioned, and oriented their temples and pyramids. They worshipped the sun in all his aspects, and their astrology so much resembles the Babylonian that it is believed to have been derived from it. The Babylonians seem to have been more interested in the planets than any other nation of antiquity, but they were known also in other countries. The Chinese recorded comets, and all races were greatly interested in eclipses, which they were able to predict with some accuracy, having discovered that they occur in cycles. Yet we find no more rational attempt to explain these phenomena than the Hindu legend of a great dragon that attacks the sun, or the Egyptian story of a sow that swallows the moon; and their cosmogonies can only be regarded as poetical descriptions or survivals of early childlike notions of the universe.

The Hindu world resting on the back of an elephant, and that on a tortoise, is no doubt but an allegory. The Egyptians pictured the earth as a great parallelogram, long from north to south but narrow from east to west, like their own land, with the sky over it, upheld by huge pillars or lofty mountains. The stars were set in this domed lid of the world, but sun, moon, and planets were floating each in its own boat on a great celestial river which ran just below the summits of the mountains, and whose course was hidden towards the north. The bark of the sun came nearer to Egypt in the summer, because at that time the celestial river overflowed its usual bank, like the Nile. The red *Doshiri* was said to sail backwards, referring no doubt to the retrograde movement of Mars.

In Eridu, one of the oldest cities of southern Babylonia, on the Persian Gulf, the great abyss of the ocean was looked upon as the origin of all things, and it was believed that it encircled the earth like a great river. Later on, we find the world described as a great mountain, resting on the watery deep, and under the mountain is the abode of the dead. It is entered from the west, which surely was suggested by the setting of the heavenly bodies in the west. The vaulted sky above the earth has divisions: the rim of the lowest part rests upon the supporting watery deep; above it are the upper waters (the source of rain); and above this again is the dwelling-place of the celestials. The sun issues forth each morning from a door in the upper heaven, or from the mount of sunrise, and enters another heavenly door, or the sunset mountain, at night.

The similarity to these Babylonian ideas of the Hebrew "firmament," the "waters above the firmament," and the "of the great deep," in the book of Genesis,[23] and Ezekiel's "Sheol" in "the nether parts of the earth,"[24] has often been noted.

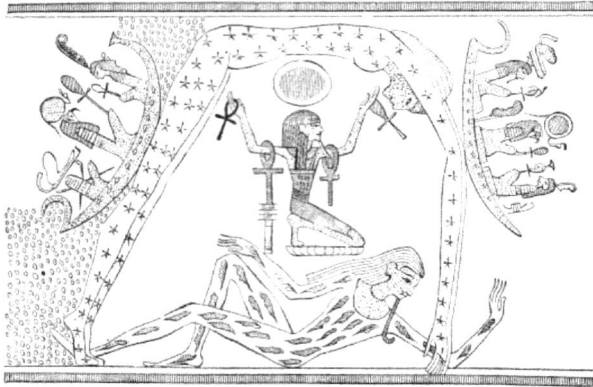

[To face p. 46.

The Boat of the Sun travelling over the sky.
From an ancient Egyptian papyrus.

The recumbent figure covered with leaves symbolizes the
earth; the figure leaning over Earth, covered with stars, is the
sky; the boat of the rising sun and of the setting sun floats over
it. The central figure represents Maon, the Divine Intelligence
which preserves the order of the universe.

*(Reproduced from Flammarion's 'Astronomical Myths,' by permission of Messrs.
Macmillan & Co.)*

To sum up:—

If we include as astronomy any observation of the heavenly bodies
which leads to a recognition of order and periodicity in their movements
and a power of forecasting their positions, then every race and age has had
its astronomers, rough though their methods may be at first. With growing
civilization more refined methods are used; the gnomon is invented for
studying the movements of the sun; the changing positions of moon and
planets are noted by means of certain stars; finally, all the visible stars are
grouped into constellations, and it is recognized that a great band of star-
groups crosses the sky, which forms the pathway alike of sun, moon, and
planets; the length of the month and of the year are determined more or
less accurately, and when an unvarying calendar has been formed, the
celestial cycles can be better recorded and studied. But in all this there is as
yet no scientific motive properly so called, no curiosity regarding the
phenomena for the simple pleasure of knowing and understanding them,

no attempt to group them into a system or to explain their underlying causes. The primitive idea that the heavenly bodies exist for the convenience of earth-dwellers is illustrated by the Egyptian hieroglyph for night, which consists of the sign for sky combined with a star suspended like a lamp; the other idea that they are mysterious divinities is shown by the Babylonian star-sign for a god or king, . The ancients found that the stars were of great use, especially for measuring periods of time; they recognised also in them a marvellous order and regularity, of which they dreamed that they found an echo on earth, and endeavoured to divine the future by watching the skies. Can we doubt that they were also attracted by the beauty that calls all men through all ages to lift their eyes and look upward?

HYMN TO THE SETTING SUN.

Sung by the Priests of Babylon.

Sun-god, in the midst of heaven, At thy setting May the latch of the glorious heavens Speak thee peace. May heaven's door to thee be gracious, May the Director, thy beloved messenger, direct thee.

Lord of E-bara, may the road of thy path be prosperous, Sun-god, cause thy highway to prosper, Going the everlasting road to thy rest. Sun-god, thou art he who is judge of the land, Causing her decisions to be prosperous.

From a lecture by T. G. PINCHES. (*Nature*, Dec. 31, 1891).

III

GREEK ASTRONOMY.

"Great men! elevated above the common standard of human nature by discovering the laws which celestial occurrences obey, and by freeing the wretched mind of man from the fears which eclipses inspired! Hail to you and to your genius, interpreters of heaven, worthy recipients of the laws of the universe, authors of the principles which connect gods and men!"

PLINY
(Apostrophe to Thales and Hipparchus.)

1. HOMERIC GREECE.

To turn from the astronomy of Egypt and Assyria to the astronomy of the Greeks is like coming to a sudden bend in a river which has flowed through level country for many miles in a slow majestic course, and finding beyond the bend a series of rapids and waterfalls. Instead of patient age-long accumulation of observations, instead of a mystical adoration of stars, supposed to be beyond man's power to understand, we find that the Greek's first instinct is to inquire into the meaning and the origin of what he saw, even before he had taken time to investigate. Behind the varied splendours of earth and skies which fascinated his bodily eyes, his intellect divined laws and forces which held the whole together in a wonderful harmony. Then, as fresh facts, or a fresh point of view, thrust itself upon him, a new explanation must be attempted, and thus many complete systems of the universe were evolved. Not the name only, the idea of Cosmos was Greek.

Homer *c.* 900 B.C.

Hesiod *c.* 800 B.C.

The first ideas of astronomy among the Greeks were as primitive as those of any other race in its early stages. They evidently had no conception of the sky as a sphere, or of the revolution of the stars as a whole, round fixed poles, though they watched the motions of certain bright star-groups, and called them by the names that we use now (however these names may have reached them), as we see in Homer and Hesiod. Ulysses, guiding his raft cunningly by night, keeps on his left the Bear, also called the Wain, which turns round in her place and keeps watch on Orion, and never bathes in ocean; he watches also the Pleiades and the "slow-setting Ploughman" (Boötes)—an apt description, as anyone may see who watches

Arcturus, the brightest star of Boötes, when low on the western horizon. Being a northern star, its motion seems very slow, and makes so small an angle with the horizon that for a long time Arcturus glides above it before finally dropping below; whereas the Pleiades, or any other stars near the equator, move very quickly and almost at right angles to the horizon, and so drop below it quite suddenly. It is Ulysses also who warns his companion, when they are setting out to spy upon the Trojan camp, that two watches of the night are already past, "for the stars have gone forward." The stars also announced the seasons, for Hesiod says that the time of harvest is indicated by the heliacal rising of the Pleiades, and when Orion with Sirius stands in mid-heaven, and Arcturus rises in morning twilight, it is time for the vintage.

Homer and Hesiod both mention Venus, as a morning star "the brightest of all the stars, which comes to herald the light of dawn," and also as an evening star, apparently without recognizing that it was the same star; as they do not mention any other planet we do not know if the others were known to the ancient Greeks.

The first appearance of the new moon's slender crescent was watched for from hill-tops, and celebrated by sacrifices, and this—as with other ancient nations—fixed the first day of their month.

Mimnermus c. 580 B.C.

Day seems to have been divided into three parts, morning, midday, and evening, according as the sun was rising, or nearly stationary, or sinking. The sun was thought to rest upon and slide over the solid dome of the sky, otherwise perhaps it would have fallen to the ground; and at night it was supposed to go behind Mount Atlas, and then to travel behind high northern mountains to its rising place in the east. This primitive explanation of its movements is so poetically described by an early poet, Mimnermus, that I cannot resist a quotation, though the lines can hardly be regarded as an astronomical fragment. They may be freely rendered thus:—

Endlessly toiling Helios speeds. No rest for him or for his steeds When Dawn has climbed the height. Soon as he lays his weary head Upon the golden wingèd bed Made by Hephaestos' might, It bears him sleeping o'er the seas, Far from the fair Hesperides, Through realms of darkest night; Till in the Ethiopian land He sees his horses ready stand; And when the child of light, The rosy-fingered, early-born, Has ushered in another morn, He mounts his chariot bright.

The earth, as pictured on the shield of Achilles, was flat and round, just as it appears from a height, and of course Greece was the centre, just as

Egypt was the centre of the Egyptian, and Babylon the centre of the Babylonian cosmogonies. It was a small earth: a few countries lay round the Middle Sea, and further to the south was the land of the Ethiopians where the Sun passes overhead and burns the inhabitants black; there was another sea to the north, over which the Argonauts sailed, and in the extreme east was the Lake of the Sun, out of which he rose every morning. This was a great gulf of the River Oceanus which encircled the whole earth. Its sources were in the furthest west, just beyond the Pillars of Hercules, and thence it flowed north, east, and south, finally returning into itself. A branch from near the source, called the Styx, flowed down into the underground world of Hades, the abode of the dead, and beneath this again was Tartarus, where were imprisoned the Titans who had fought against Jove.

Above the flat earth the blue dome of heaven was spread like a tent, and across it travelled

'The never-wearied Sun, the Moon exactly round, And all those stars with which the ample brows of heaven are crowned."

What a compact little universe, and how important a part of it was man! But as thought developed, the universe expanded.

2. THALES AND ANAXIMANDER.

B.C. 585.

Thales *c.* 600 B.C.

In the sixth year of the war of the Lydians against Cyaxares, king of the Medes, just when a battle was about to begin, day was suddenly changed into night by an eclipse of the sun, and Herodotus adds: Thales had told the Ionians of this before, and in what year it would happen. This does not necessarily imply any accurate understanding of eclipses on the part of the Ionian philosopher. He had visited Egypt, and may have learned from the priest-astronomers there that eclipses recur in cycles and so can be predicted. But Thales was not content with cycles. He wanted to know, not only that eclipses would happen at such and such times, but *how* they happened. Perhaps from reports of Babylonian observations, perhaps from questions put to Egyptian astronomers, he learned that solar eclipses only happen when the moon is new and in the ecliptic, that is, in the same part of the sky as the sun, and that the black body then seen on the sun has always a rounded edge. These no doubt were the arguments on which he founded his assertion that solar eclipses are caused by the moon passing in front of the sun; and he further added that this shows the moon to be of an "earthy" nature, that is, not made of fire or any substance either luminous or transparent, but of opaque matter, probably having weight and substance, and not altogether unlike what we know on earth. He is said to have stated also that the moon receives her light from the sun, a conclusion which would follow from a little attention and thought bestowed on her phases.

Besides his speculations regarding the moon, Thales took pains to note the sun's movements as accurately as possible, by means of gnomons, with a view to discover the exact length of the solar year, and it was he who advised the Greeks to adopt the Phoenician method of directing their course at sea by the Little Bear instead of the Great Bear, which appears to have been the constellation used in Homeric times.

Thales imagined that Ocean did not merely encircle Earth, but that the whole Earth, which was a thin flat disc, floated upon the Ocean.

This zealous observer must have had something of the absent-mindedness of his great successor, Newton; for it is told of him that while star-gazing he fell into a well!

B.C. 611-545. Anaximander

It is evident that the moon's passing in front of the sun implies a lesser distance from us, and it must have been this which suggested her place in the scheme of Anaximander. This scheme is rather difficult to understand, from the allusions and quotations of later writers, for we have no original writing by Anaximander; but we can gather enough to show that already in the sixth century B.C. the Greek philosophers were asking themselves what was the explanation of the movements and appearances of the heavenly bodies, how they were supported in the sky, what force moved them, how large and how distant they were, and what they were made of. Anaximander asserted boldly that sun and moon were larger than the whole earth: he thought the sun might be 27 and the moon 19 times as large. How he reached this conclusion it is impossible to say. The Egyptians had already tried to measure the apparent size of the sun as compared with the circumference of the sky, by noting how long it took to set from the moment the lower rim touched the horizon to the moment when the whole disc disappeared: this, divided by the 24 hours taken by the sun to traverse the 360 degrees of the sky, gave the sun's size in degrees (it is about half a degree), and had they been able to find the actual distance of the sun from the earth, they could have deduced its actual size, but this they had no means of determining.

There are two possible ways of seeing that the moon is smaller than the sun, although they usually appear to us the same size. Anaximander may have seen or heard of an "annular" eclipse, in which the moon is rather more distant from us than at a total eclipse, and therefore her dark body just fails to cover the whole sun, and a bright ring of light surrounds her. More probably he realized what was implied in Thales' explanation of solar eclipses, and concluded that the moon must be smaller than the sun, because she looks no larger although she is nearer to us.

His scheme is the first of which we have any knowledge in which the movements of the heavenly bodies are explained by supposing them not all in one sky together, but placed in a series of heavens, one above the other: hence it is of peculiar interest to the Dante student, for in it we trace the first attempt towards the theory of the Revolving Spheres. It is true that the Babylonians and Hebrews divided their heaven into three parts, one above the other, but this was only to divide the place of atmospheric phenomena from the dwelling-place of the gods, and sun, moon, planets and stars all moved in the same heaven. Here, in the universe of Anaximander, we find one heaven, the lowest, for air, rain, etc., another for all the stars, a higher heaven for the moon, a yet higher for the sun, and above all the region of fire, the brightest, lightest element, whose nature it was to ascend and which therefore is outside all, as it was the nature of earth to descend and therefore to be at the bottom. Probably either in or above the heaven of celestial fire was the heaven of the gods, for, as Aristotle remarks, all our ancestors, indeed all who believe in gods at all, whether Greek or of any other race, place the dwelling-place of the gods above in high heaven, as the unchanging, unmoving region of eternity.[25]

Unfortunately we do not know what would be of great interest, whether Anaximander also provided separate heavens for the planets, or found a home for them in the heaven of the stars. Perhaps he hardly knew of their existence, or said with Aratus who wrote nearly 800 years later:—

"Of these I dare not speak with certainty, As of the fixed stars' orbits."

The successive heavens were in layers, as it were, one above the other, "like the bark enclosing a tree," but they were transparent and invisible. The heavens themselves were not in motion, carrying the stars, sun, and moon; Anaximander had an ingenious mechanical scheme of wheels or rings to carry them inside their respective heavens, which doubtless was clear to himself, though unfortunately it is not at all clear to us. Some writers have maintained that these heavens were spheres, but for several reasons it is difficult to believe this, and probably the sky was still to Anaximander, as to the Homeric Greeks, a slightly flattened hemisphere,[26] only divided into these layers, and instead of ending at the horizon it continued a little below, to allow of the passage of the heavenly bodies between setting and rising. Perhaps it was for this reason that he gave the earth a greater thickness than the disc of Thales, comparing it to a short thick pillar, three times as broad as high, the top of which only was inhabited. His Cosmos, then, would be something like the diagram, with regard to the disposition (though not the relative sizes) of Earth and the heavenly bodies.

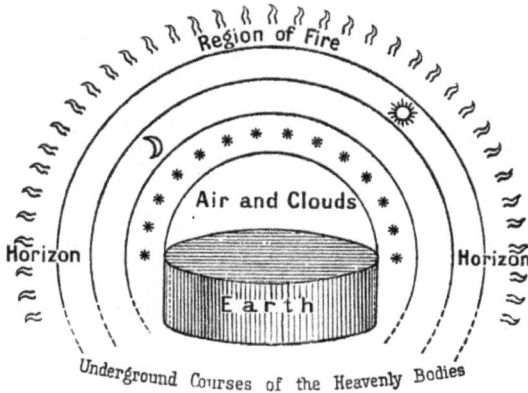

Fig. 12. The Universe according to Anaximander.

Somewhat timidly the barriers have been thrust back. The earth goes a little deeper into the dark unknown, the sky is wider and higher, the heavenly bodies are much larger and more distant and go under Earth's surface; but they are cautiously upheld by solid domes, and worked by wheels. Earth is still the floor of the World, and Heaven—now a series of heavens—the vaulted roof above.

3. LATER FLAT EARTH SYSTEMS.

It is interesting to see how long this timidity persisted among the Greek philosophers, especially of the Ionian school, in spite of the fact that other schools had advanced much bolder ideas, as we shall presently see. Quite a number of universes were constructed somewhat after the pattern of Anaximander's, with Earth as floor of the world; but some placed the stars beyond moon and sun, some definitely included the planets, though they do not seem to have explained their motions; and there were various ways of supporting the flat earth, and of supporting and moving the heavenly bodies.

Anaximenes *c.* 550 B.C.

Empedocles *c.* 450 B.C.

Anaxagoras died 428 B.C.

Anaximenes, a follower of Anaximander, having doubtless pondered the fact that very heavy bodies can float in water if only they are the right shape, and that Earth itself was supposed by Thales to be floating on the Ocean, suggested that the moon is so broad a disc that she floats in the ether, "like a leaf," and of course the same would apply to the sun. Two later philosophers, Empedocles and Anaxagoras, held that a great whirlwind swept continually round the Earth, which both kept the heavenly bodies from falling down upon it and drove them across the sky.

Equally diverse were the opinions as to the nature and composition of the heavenly bodies. Most philosophers of this age believed that they were of pure fire, or else that they were vessels containing fire, which was extinguished, or in one way or another became invisible to us, during eclipses and when they set. Others held, as we have seen with Thales, that they were of an earthy nature; and Anaxagoras, seeing a meteorite which had fallen from the sky during the daytime, thought he actually held a piece of the sun in his hands, and concluded that the sun was an enormous mass of iron, "much greater than Peloponnesus," and shone because it was red-hot. But the popular idea still was that the sun was a god, or the chariot driven by a god across the sky, and Anaxagoras was banished from Athens for his impious words.[27] The markings on the face of the moon were thought to prove that she was of mixed composition: she was made of air mingled with only a little fire, or earth mingled with fire; but according to Democritus the markings were shadows of mountains on her surface, and Anaxagoras is reported to have said that the moon was inhabited, and the markings were "plains and valleys."

Anaxagoras suggested that the stars were fragments torn off the circumference of the earth by the encircling whirlwind, and that they glowed with the heat caused by friction, though they were too distant for us to feel this heat, being far beyond the sun. The Milky Way was a source of speculation: some said it was the former path of the sun, and still burning from his heat, but Democritus explained it as caused by the shining of innumerable stars, too faint and close together to be distinguished separately.

The doctrines of the different philosophers as to origin and first stages of the universe do not concern us here, but we must mention that of Empedocles, as his views are directly referred to by Dante.

This philosopher was the first to assert that everything consists of the four elements, earth, air, water, and fire, pure or in combination; and the combinations he supposed to be brought about by two forces, one attractive, the other repulsive, which he named Love and Discord. Of these, one alternately predominates at different ages of the world, and thus its history is divided into periods of different character.

A great step forward was taken when it was realized that the sky is not a hemisphere, ending at the horizon, or even extending a little way below, but that it surrounds Earth in every direction, like a sphere. This idea probably originated with the Pythagoreans, or it may have occurred independently to several thinkers, when the diurnal motion of the heavens was better observed, and geometrical conceptions understood and applied. Now it became no longer necessary to extinguish and rekindle the stars, nor to send the sun round swimming on River Ocean behind northern mountains, or creeping through strange underground regions, through the night. It was clearly recognized that the visible course of each heavenly body was part of a circle, the whole of which we could see if we could only travel fast enough and go to the underside of the earth.

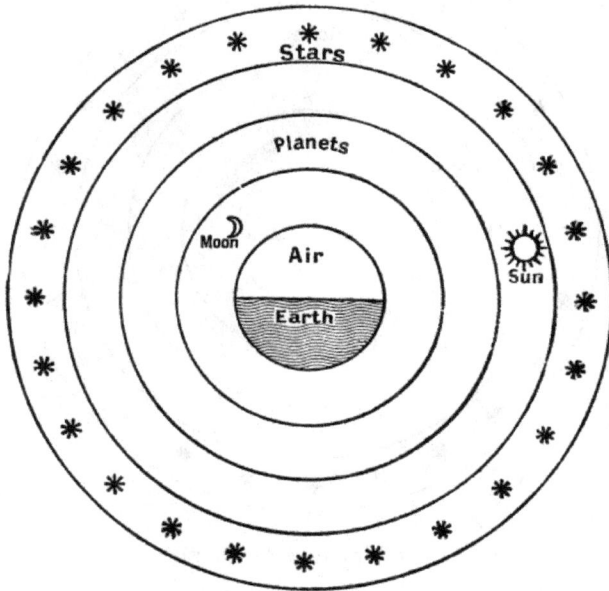

Fig. 13. The Universe of Leucippus.

Leucippus c. 450 B.C.

Democritus c. 430 B.C.

But this was just what never could be done, for outside the schools of the Greeks in Italy (Pythagorean and Eleatic), Earth still had an uninhabitable underside. Distinguished men like Leucippus and

Democritus sought to combine the belief in an all-surrounding spherical heaven with a flat supported earth which might still give them a solid floor beneath their feet. Leucippus made the earth a hemisphere, with a hemisphere of air above, the whole surrounded by the supporting crystal sphere which held the moon. Above this came the planets, then the sun, and probably the stars were outside this. His disciple, Democritus, on the other hand, retained the disc-like earth, raised a little at the rim, to secure its contents, and made it divide the sphere of air into two parts, so that it rested upon air, and air was also in the sky above. The underside of the disc was not inhabited, no doubt because no one could stand upside down. His order of the successive heavens is not quite the same as that of Leucippus, as he puts the moon and the Morning Star together, and the rest of the planets beyond the sun.

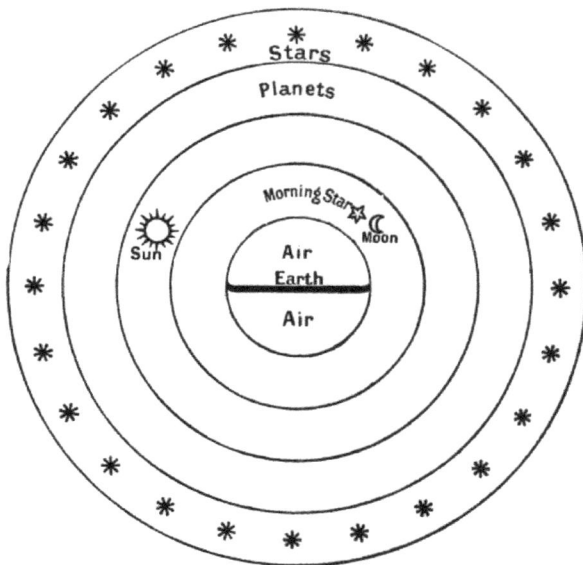

Fig. 14. The Universe of Democritus.

This scheme gave the universe a beautifully symmetrical form, which must have pleased the Greeks, but now they were puzzled to know why the heavenly bodies did not circle symmetrically with regard to the central earth. Why was not the pole in their zenith and the equator on the horizon? They could only guess that it must have been so at first, and that the disc had slipped out of position, either through some irregularity in its weight, or in the density of the underlying air. Compare Milton's—

"Some say He bid His angels turn askance The poles of Earth twice ten degrees and more, From the Sun's axle; they with labour pushed Oblique the centric globe." *Paradise Lost*, Bk. X., 563-566.

All these theories and guesses may seem to us very crude and fanciful, and we may compare them to the eager questionings of intelligent children, too impatient to consider whether the answers given are satisfactory explanations or no. But we must remember that all we know of the early cosmogonies is from allusions and descriptions by later writers, who often—like Aristotle, for instance—only quote to condemn. "If each could defend his own opinion, may be we should see that there is truth in all." (Conv. IV. xxi. 25-7).

At least we find a keen and disinterested desire to penetrate the causes of things, and a fertile imagination, without which science can make no advance: moreover there was a progress in true knowledge. It was discovered that the (apparent) diurnal paths of sun, moon, and every star were circles, although only a part of the paths could be seen; and that, although all were seen projected on a sphere, their actual distances from earth were very varied.

It is disappointing to find no record of observations of the planets, and from the almost random way in which they were placed in the heavens it seems that but little attention had been paid to them as yet. In fact, Seneca tells us that Democritus knew neither their number nor their names. They were often classed with comets, and thought to be entirely erratic, and the Greek mind was more attracted towards those phenomena which were seen to be orderly.

4. PYTHAGORAS AND HIS FOLLOWERS.

Socrates. As the eyes are appointed to look up at the stars, so are the ears to hear harmonious motions, and these are sister sciences. That is what the Pythagoreans say, and we, Glaucon, assent to them?

Yes, he replied.

Pythagoras *c.* 540 B.C.

About the same time that Anaximander was inventing solid hemispheres and rings to hold and move the heavenly bodies round about a flat earth, Pythagoras was founding a school in southern Italy which gave to the world a very different scheme. One of the characteristics of his school was secrecy, its methods were oral, and his later followers were fond of attributing to their master everything which had gradually grown out of his teaching: it is difficult therefore to say with certainty what he himself taught. It has often been stated by modern writers that he anticipated

Copernicus, and discovered that the earth revolves round the sun. Though this is a mistake, we may venture to believe that Pythagoras taught that the earth is a sphere, hanging freely in space.

We are so familiar with this idea from childhood, that it is difficult to imagine what a tremendous innovation it was. Pythagorean noviciates, doubtless after solemn initiation and preparation, were told: This earth, which seems to you the floor of the world, with heaven stretched over it like a tent, is a round globe, with men like you living on the other side of it, and yet they do not fall, and earth does not fall, for it is poised in the centre of the world, and has no tendency to fall in one direction rather than in another. Earth, itself a perfect sphere, is in the centre of an infinitely greater sphere, the star-set heaven; and within this seven heavenly bodies move in perfect circles, each at its proper distance and pace, all needing no support and no force to drive them, for harmony is the motive power of the Cosmos. Their distances are proportional to the intervals between musical notes, and as they circle they make heavenly music, which we should hear did we not always hear it, like one who lives beside a waterfall[28]. There is no below, and no above, for above is below and below above to our antipodes: there is but the centre, where we live, and Heaven is all around.

How did Pythagoras reach this great and startling truth of the round unsupported earth?

His school relied more on experiment and observation than the Ionian, and the colonizing Greeks of Italy had travelled. They might have noticed the curvature of the sea, and the varying height of the Pole Star according to latitude. We know that in early days the Greeks were struck by the remarkable fact that the brilliant star Canopus (second only to Sirius in brightness), which was invisible in Greece, could just be seen close to the southern horizon in Rhodes, and was well seen in Egypt. Then the moon may have helped once more. When it was understood that lunar eclipses only happen at full moon, when we are between her and the sun, and that they may therefore be explained by the earth's shadow falling on the moon, then, since the edge of that shadow is always a circle, it is demonstrable that the body throwing that shadow can have no form but that of a ball.

Sun and moon are obviously round: it was guessed that they also are globes rather than discs, and the spherical shape of all heavenly bodies was a doctrine of the later if not the earliest Pythagoreans.

Whatever may have been the steps which led to these two great discoveries that Earth is a sphere, and that the apparent path of every celestial body is a circle, the sphere and the circle were soon accepted as the only forms suitable for celestial bodies and their orbits. The founder of the school was a great mathematician, and it is not strange that these forms should have commended themselves to his disciples. The sphere, which has its surface everywhere similar, and its contents greater than those of any other figure with equal surface, was the "most perfect" of solids; and the circle, which has no beginning and no end, is alike in every part, and presents ideas of haunting suggestiveness to the geometer, was the "most perfect" of lines.

In the system of Pythagoras we first find the five planets distinctly enumerated, and playing as important parts as sun and moon. Number was the principle of this universe, and the planets with sun and moon made the sacred number of seven. Among the Greeks, and through the middle ages, all these bodies are spoken of as planets or "wanderers," in distinction from the "fixed" stars which do not appear to move amongst themselves. These seven "planets" represented the seven notes of a musical scale, and the star sphere made up the octave. Pythagoras is said to have been the first to teach that Phosphor and Hesperus, the morning and the evening star, were the same. When, however, we ask what was the order of the planets in his scheme, we meet with many conflicting reports, and a serious difficulty suggests itself. If the planets had really been observed with care, it must have been seen that their motions could not be accounted for by simple circular movement. The large oscillations of Mercury and Venus on either side of the sun would strike an observer before he thought of tracing their movements among the stars, and noting that they made a circuit of the zodiac. Similarly, the other planets are most conspicuous, rising after sunset and remaining long visible through the night, at the very time of their retrograde movements, so these must have been noticed if a long enough series of observations had been made to distinguish them from one another. The only solution of the difficulty seems to be that Pythagoras, on the journeys into Egypt and Babylon which he is said to have made, learned that there exist planets to the number of five, which move in regular periods, and he may also have learned the length of their zodiacal periods at the same time, or perhaps these were only known to his school much later. If the order assigned to them was that which was finally and generally accepted by the ancient world, the periods must have been known, for this is the only possible clue to the order Moon, Mercury, Venus, Sun, Mars, Jupiter, Saturn. The planet with shortest period (the moon, with a month) was naturally placed by the Greeks nearest to earth, with the smallest circle to traverse, and so on outwards.

It is very possible, however, that the early Pythagoreans, at least, did not venture to assert more than that there were five planets, without assigning to them any order, for Aristotle tells us that their universe was divided thus:—

From Earth to Moon was the Ouranos, or sky, within which exists all that is changing and corruptible.

Cosmos, the place of ordered movement, was the region of Sun, Moon, and Planets.

Olympos, the place of pure elements, held the stars; the region of Celestial Fire came beyond this, and the Apeiron, the Infinite Space, or Infinite Air, from which the world draws its breath, was outside all.

Philolaus towards end of 5th century B.C.

The diagram shows, then, the earliest form of the Pythagorean universe. But they did not remain content with this. Out of it grew a most interesting scheme (referred to by Dante), which is usually attributed to one Philolaus, of whom hardly anything is known, not even his date.

It seems to have struck Philolaus as a difficulty that the seven planets, which were circling round Earth in the same direction but at very different distances and speeds, and also the immense sphere of stars beyond, were all sweeping together at the same time in an opposite direction, and at the almost incredible pace of one revolution in a day. The brilliant idea occurred to him: Leave the stars at rest, let the seven planets revolve in their seven orbits, the nearer to the centre the faster, and let earth herself revolve fastest of all, viz. in twenty-four hours, in the same direction. If she keeps one face always turned towards the centre, like the moon, this will account quite as well for the apparent diurnal revolution of all the heavenly bodies, and the change of day and night on the earth.

Philolaus did not make Earth remain stationary and simply turn on her axis, which would have had just the same effect on the apparent motions of the heavens; for it seemed more natural that she should revolve as did the rest. The five naked eye planets are all mentioned by name in his scheme.

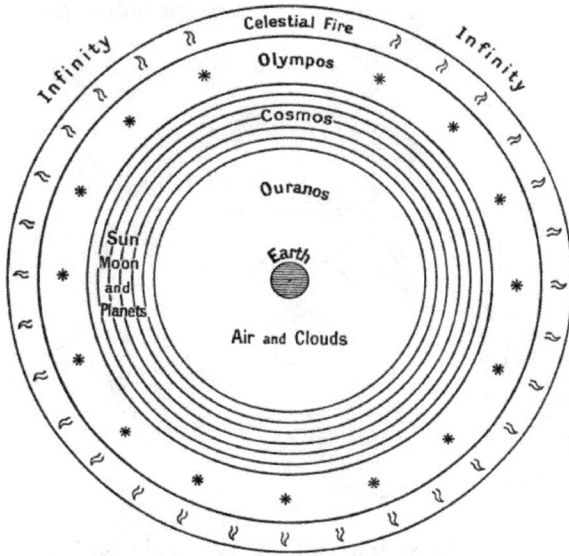

Fig. 15. The Universe of Pythagoras.

Was the centre, deprived of Earth, to be left empty? No, the centre was the Watch Tower of Zeus, the Hearth of the Universe, and here they placed the purest element, fire. It was invisible to us, because we live on the side of the earth-sphere turned away from the centre; and also invisible to us was another planet, Antichthon, or Counter-Earth, for this revolved within our orbit, and also in twenty-four hours. It was added to the system, because the addition of Earth as a heavenly body spoiled the sacred number of seven, but by adding Antichthon, and counting the star sphere as another, the total was brought up to ten, another sacred number.

The objection was made that, if Earth is moving in space, this must bring about a change in the apparent sizes of sun and moon, as Earth is nearer or farther from them, but the Pythagoreans were quite ready to believe that all the heavenly bodies are so distant that this journey of Earth makes no difference to their apparent size or brightness. The planets were thought to be worlds like ours, and inhabited; and it was even guessed that plants and animals on the moon must be fifteen times as strong as ours, apparently because there the average day consists of nearly fifteen of our days (of twenty-four hours), and the nights are equally long.

It was the braver of the Pythagoreans to shake the steady earth from her centre, and set her whirling in the depths of space, that they realized, as no one had done before, how large she must be; for Greece and the surrounding lands, the Middle and the other seas, instead of making the whole of the earth, were now understood to be only a portion of a great globe.

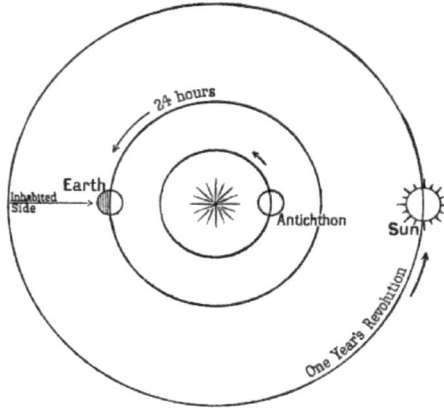

Fig. 16. The System of Philolaus: night on earth.

Only the side turned away from the centre is inhabited: consequently the Central Fire and Antichthon are invisible.

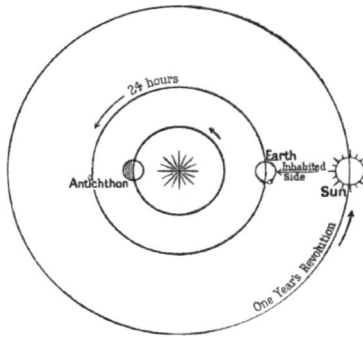

Fig. 17. Twelve hours later: day on earth.

Earth has made half a revolution and her outer side is now lighted by the sun, which has only moved about half a degree forward in its yearly orbit. Antichthon has also made half a revolution, therefore remains invisible.

Here, then, is a conception of the Universe widely different from Homer's. The little flat disc has become a great round ball, a planet among planets, swiftly moving through space; the crystal dome that tenderly covered it like a bell-glass over some fragile flower, has lifted, and the vast sphere is seen, infinitely distant, and studded with enormous stars. Man himself is now a tiny creature on a great earth, and his world but one among many, but if he is humiliated by his insignificance, is he not elevated by the vastness of his outlook?

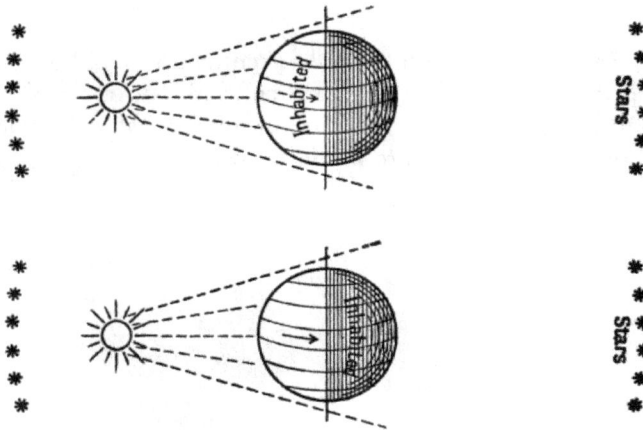

Fig. 18. Earth and sun according to Heracleides.

In the upper figure it is day, in the lower, night, on the inhabited side of Earth. The sun is on the equator, as at the time of equinox.

Heracleides c. 370 B.C.

But not even here did the Greeks stop. It was taking a less startling step than they had already taken, to reach the truth that Earth was merely rotating on her axis once in a day, and so causing the apparent diurnal revolution of the heavens. This step was taken (it is said) by a Pythagorean called Hicetas of Syracuse, who is quoted as saying that the earth, "while it turns and twists itself with the greatest velocity round its axis, produces all the same phenomena as if the heavens were moved and the earth were standing still." We are told also that "Heracleides of Pontus and Ecphantus the Pythagorean let the earth move, not progressively, but in a turning manner like a wheel fitted with an axis, from west to east round its own centre."

A brilliant guess,[29] which seems fully justified by facts, has recently explained the personalities of these two mysterious Pythagoreans of unknown date, Hicetas and Ecphantus, whose names have been coupled for centuries with that of Heracleides, as teaching the rotation of Earth on her axis. It seems that they resemble the Shadow in Hans Andersen's tale, which became a man and lived apart from the man to whom it originally owed its existence, for it is now thought that they were speakers introduced by Heracleides into one of his dramatic dialogues to discuss astronomy.

Heracleides, therefore, was the sole author of this remarkable discovery.

In this way Earth was restored to her central position, but as a rotating sphere, and the later Pythagoreans apparently tried to reconcile their new scheme with the old by calling Antichthon the uninhabited hemisphere of Earth, and placing the central fire within the earth.

IV

GREEK ASTRONOMY.

SECOND PERIOD. B.C. 400 TO A.D. 150.

"Chiamavi il cielo, e intorno vi si gira, Mostrandovi le sue bellezze eterne."

1. PLATO.

The Ionian school of philosophy died about the middle of the fifth century B.C., and the Pythagorean towards the end of the fourth. But meanwhile a new school of astronomers was growing up. The philosophers still laid down general principles, founded on abstract reasoning, which they believed must regulate the nature and movements of the heavenly bodies, but astronomy began to be regarded as a branch of mathematics, not of philosophy, and the mathematicians, leaving problems of ultimate causes to the philosophers, devoted themselves to observation and calculation. They carefully studied the peculiar motions of each planet, and their chief aim was to represent these geometrically by some scheme which should include them all, and make it possible to predict the places of the heavenly bodies in the sky for any given date.

One cause of this great progress in methods was no doubt the natural intellectual growth of the Greek race, as they discovered that their eager curiosity concerning nature could only be satisfied by patient investigation. The value of observation was taught, in the latter half of the fourth century, by the philosophy of Aristotle, and a great impetus must have been given by the campaigns of Alexander, in which the Greeks saw distant countries, new climates, strange peoples and customs.

Callisthenes *c.* 330 B.C.

But a potent cause of the advance in astronomy seems to have been the closer connection between Greek astronomers and those of Egypt and Babylon. The astronomer Callisthenes went with Alexander to the East, and received a letter from Aristotle praying him to send to Athens the Babylonian eclipse records which were centuries old; and Aristotle mentions, when speaking of the motions of the planets, that the Babylonians and Egyptians had furnished trustworthy information about each one of them. Even before this, we find that the Greek descriptive names of the planets were changed for names of Greek deities which are believed to correspond with the Babylonian gods and goddesses who presided over the planets. Thus Plato speaks of "the star sacred to Hermes" as well as Stilbon the Glitterer, and he is the last to use commonly the name of Phosphor for the planet which henceforth was known as Aphrodite among the Greeks, and Venus among the Romans, corresponding with the Babylonian Ishtar; and so on with the rest. Instead of vague records of journeys in Egypt or Babylonia, we have a definite statement that Eudoxus, who was the founder of the new school, went to Egypt about 378 B.C., with letters from the king of Sparta to the king of Egypt, and we are told that he studied the planetary motions under a priest of Heliopolis. It seems highly probable, to say the least, that Eudoxus was the first Greek to appreciate the value of those methods of observation and continuous recording of phenomena which he found among the Egyptians, and to understand the wonderful regularity which was hidden behind the seeming irregularities of the "wandering stars." He was also, apparently, the first Greek to write a detailed description of the forty-eight ancient constellations.

But if Egypt and Babylonia gave to Greece records of celestial phenomena, and set the example of accurate and long-continued observations, Greece made the new knowledge her own, and transformed it. The legend that Eudoxus applied his mathematical skill to the ancient monuments of the Egyptians, and showed them how to calculate the height of the Great Pyramid by measurements of its shadow, is typical of the history of Greek treatment of Oriental astronomy. One geometrical theory after another was invented to represent the planetary motions, was compared with the skies, and rejected or improved, and meanwhile observation became much more close and accurate; new instruments were introduced, new methods of calculation invented, new motions discovered which had to be accounted for; finally, five hundred years after Eudoxus' visit to Egypt, the result of all this labour was summarized in a truly epoch-making work, which remained the standard treatise on astronomy until the time of Copernicus.

Plato c. 427-347 B.C.

Eudoxus was born at Cnidos, in Asia Minor, but at the age of twenty-three he went to Athens, and studied under Plato. It is said to have been Plato who inspired the young man with the idea of devoting his brilliant mathematical powers to solving the problem of celestial motions, and with this view he went to Egypt. The story is easy to believe when we recall the many passages in the *Dialogues* in which Plato uses the splendid imagery of the skies to illustrate his philosophic doctrines, dwelling especially on the perfect though little understood symmetry of the celestial motions, and it will be remembered that astronomy was one of the subjects to be learned by the rulers of his ideal state.

It is true that Glaucon is gently but decidedly snubbed by Socrates in the *Republic*, for suggesting that the study of astronomy is valuable because of its use in navigation, husbandry, and the arts of war. This is "vulgar praise," but has there ever been nobler praise of astronomy than that which Socrates himself then proceeds to give? Although he believes that true knowledge, knowledge of realities, is only to be obtained by the exercise of pure reason without the aid of sense, he considers that the study of celestial motions is one of the best means of training the mind to reach those heights, and he does not hesitate to say that sight was given to us in order that we might look at the skies. For the embroidery of heaven, says Socrates, though wrought upon a visible ground, is the fairest and most perfect of visible things; and it is displayed to our mortal eyes as a pattern of the eternal realities which are granted to the vision of the soul.

In the *Timaeus* this idea is elaborately developed, and it undoubtedly had an effect on Plato's contemporaries, although his direct influence on astronomy cannot be compared with that of Aristotle. The *Timaeus* was widely read also in the Middle Ages, during the long period when Plato's other writings were unknown, and it is quoted by Dante. We are often reminded of him when reading the astronomical and quasi-astronomical parts.

Timaeus, who is introduced to Socrates by Critias as "the most of an astronomer among us, and one who has made a special study of the nature of the Universe,"[30] describes the Creation as he conceives it most probable that it took place. He assumes a chaos to begin with, where there is no order, and no matter which can be distinguished by name, but all is confused and seething with random restless motions.

Of this, in order to produce something which should express his own goodness,[31] the Creator formed the four elements,—earth, water, air, and fire,—and of them he made a world, which became a fair and intelligent being, animated by a living soul. He made it in the most perfect form, that of a sphere, polished and smooth on the outside, "as if from a lathe." The soul was placed in the centre, and hence diffused throughout the whole bodily frame. It is the cause of the harmonious motions of the stars, and of these there are two kinds: the motion of the Same (the diurnal revolution of the whole heavens) is in the noblest direction, simple and uniform; the motion of the Diverse is in the opposite direction and diagonal to the first, and it is divided into seven parts (the seven orbits of the planets), which bear certain definite ratios to one another.

Timaeus does not name the planets, but in the *Republic* Socrates names some, and indicates the rest by their colour or other characteristic,[32] so we know that the order which he assigns to them, counting outwards from the central earth, is: Moon, Sun, Mercury, Venus, Mars, Jupiter, Saturn.

It was in order to make the world like its eternal pattern that the Creator made a "moving image of eternity," which we call Time, in the revolutions of the heavenly bodies; and to make it visible he "lighted a fire which we now call the sun, in the second of these orbits, that it might give light to the whole of heaven [note that the stars shine by reflected sunlight, as well as moon and planets], and that the animals who were by nature fitted might participate in number: this was the lesson they were to learn from the revolutions of the Same and the Like. Thus, then, and by these means, the night and the day were created, being the period of the one most intelligent revolution. And the month was created when the moon had completed her orbit and overtaken the sun; and the year, when the sun had completed his own orbit. The periods of the other stars [the planets] have not been understood by men in general, but only by a few, and they have no name for them, and do not estimate their comparative length by the aid of a number, and hence they are hardly aware that their wanderings, which are infinite in number and admirable for their variety, make up time. And yet there is no difficulty in seeing that the perfect number of time completes the perfect year when all the eight revolutions, having their relative degrees of swiftness, are accomplished together, and again meet at their original point of departure, measured by the circle of the Same moving equally."

The heavenly bodies, according to Timaeus, are all divine intelligent beings. In form they are perfect spheres, like the world of which they form part, and they are composed of fire. The stars have two motions, for each rotates on its own axis while it is carried round the centre on the rotating star sphere.

Earth is also a sphere, immoveable at the centre of the World. Of her Timaeus says: "The earth, which is our nurse, encircling the pole which is extended through the universe, he made to be the guardian and artificer of night and day."

This passage has given rise to the idea that Plato believed the apparent diurnal revolution of the heavens to be caused by earth's rotation on her axis; but the word here translated "encircling"[33] may mean—as that does—either motion or situation round about something, and the whole context ascribes the diurnal movement so clearly and emphatically to the heavens, that it seems evident Plato could only have meant that earth was guardian and artificer of day and night by virtue of her position. The only strong argument in favour of the other meaning is that Aristotle, when speaking of Earth as supposed by some to be central in the Universe but moving, quotes Plato and the *Timaeus*. It might easily happen, however, that Aristotle knew from other sources, perhaps from conversation with Plato, that at some time the latter had inclined towards belief in Earth's motion, and remembering the ambiguous expression in the *Timaeus* he quoted it from memory as a statement of Plato's belief.

There is some evidence that late in life Plato accepted the doctrine of Philolaus that Earth was not only in motion, but in motion round a Central Fire. There is a legend that he bought the books of Philolaus at a great price, and Theophrastus, a disciple of Aristotle, is reported by Plutarch to have said that "Plato when old assigned to Earth another place, the central and nobler place being reserved for something else more worthy of it." However this may be, he does not teach either theory in his writings. His views seem to be quite the same as those of Pythagoreans of the old school, whom he sometimes quotes.

After describing the creation of the Universe, Timaeus relates that the Creator deputed the gods whom he had made (including the stars) to create living beings on the earth, he himself creating directly only their immortal part, which he made of the same essence as the World-Soul, but diluted. Then follows the passage which came to the mind of Dante when he met the first spirits of Paradise in the moon.

> "And when he had framed the Universe, he distributed souls
> in equal numbers to the stars, and assigned each soul to a star;
> and having placed them as in a chariot, he showed them the
> nature of the Universe, and the decrees of destiny appointed for
> them, and told them that no one should suffer at his hands, and
> that they must be sown in the vessels of the times severally
> adopted to them.... He said that he who lived well during his
> appointed time [on earth] would return to the habitation of his
> star, and there have a blessed and suitable existence." If he lived
> ill, he would be a woman at his second birth, if a bad woman,
> then a beast, and as long as he continued to do ill he would "not
> cease from his toils and transformations until he followed the
> original principle of sameness and likeness within him.... When
> he had given all these laws to his creatures ... he sowed some of
> them in the earth, and some in the moon, and some in the other
> stars which are the measures of Time."

The creation of man's body and all the remainder of the *Timaeus* does not concern us here, except that when speaking of the highest use of man's faculty of sight, we realize how near Dante and Plato are in their feeling for the revolving heavens:

"God invented and gave us sight to this end, that we might behold the courses of intelligence in the heaven, and apply them to the courses of our own intelligence which are akin to them, the unperturbed to the perturbed; and that we, learning them and being partakers of the true computations of nature, might imitate the absolutely unerring courses of God and regulate our own vagaries."

2. EUDOXUS.

With words like these ringing in his ears, Eudoxus went from the Greek philosopher to the Egyptian priest, and studied "the courses of intelligence in the heaven." Legend says that the sacred Egyptian Bull licked his garment, and the priests no doubt were encouraged by this omen to divulge their secrets to a person so highly favoured by the gods. They prophesied that he would have a short but very illustrious life.

After a year, or perhaps more, spent in Egypt, Eudoxus returned to his own city, set up an observatory of his own, received pupils, and worked out an exceedingly ingenious and original planetary scheme. He did not accept (if he knew of them) the risky theories of the Pythagoreans as to Earth's motion, but assumed a central stable earth, round which circled the stars and the seven planets, according to the teaching of Plato and the general belief among educated Greeks of his day. But Egyptian observation and Greek geometry enabled him to describe for the first time the complicated movements of the planets, and to represent them by an imaginary mechanism.

This was a series of spheres, or hollow balls, fitting inside one another, and gradually diminishing in size like the ivory boxes of a Chinese puzzle, or the coats of an onion. Their size was stupendous, for the outer one, which contained all the rest, was nothing less than the sky we see, and was encrusted all over with stars. Of the inner smaller spheres, one bore, fixed in it like a jewel set in a ring, the sun; and six others bore, in the same way, the moon and the planets, one in each. All these hollow spheres were symmetrically placed so that all centred in a single point, and at this point was a solid sphere, exceedingly small in comparison, which was the earth. The star sphere, without moving from its place, rotated round this central Earth, and this caused the diurnal motion that we see in the stars. Each planet-bearing sphere rotated also, but the special characteristic of Eudoxus' system is that each of these was surrounded by its own complete set of spheres, bearing no planet, but all attached together, the poles of one sphere resting on the surface of the next, and moving with different speeds, in different directions, and with differently inclined axes: these motions being all communicated to the innermost sphere on which the planet was fixed, the net result was the movement of the planet as we see it in the sky. Each planetary set was quite separate from the rest, and did not interfere with their movements, although each set was enclosed within the next larger. Since all the planets have a diurnal motion like the stars, as well as their own proper motions, each set had to be provided with a sphere which moved exactly like the great all-enclosing star sphere.

Thus, the sun had one sphere turning like the star sphere, and within this was a second, on which the sun was fixed, which turned round in a year, in a west to east direction. The sun, carried along by the combined motion, travelled through the sky with the daily and yearly motions, as we see them.

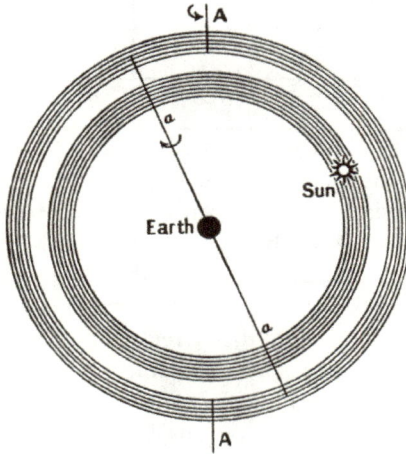

Fig. 19. The spheres of the sun in the system of Eudoxus.

The outer sphere turns on its axis A A in a day and night; the inner on its axis a a in a year, in the opposite direction.

The planetary spheres were much more difficult to arrange. Eudoxus used four spheres for each, and these had in every case to be carefully adjusted to the very different periods and amplitudes of the planetary oscillations. It must be confessed that the scheme failed with the difficult case of Mars, and was not quite satisfactory with Venus, but it represented remarkably well the movement—so far as then known—of Sun and Moon, Saturn, Jupiter, and Mercury. It was certainly a feat for those days, whether we consider it merely as the solution of a mathematical problem, or as an embodiment of astronomical knowledge. The periods of the planets as known to Eudoxus, stated in round numbers only, are given in the following table. They are taken from Simplicius, who describes the system of Eudoxus, but as in the so-called Papyrus of Eudoxus the synodical revolution of Mercury is given as 116 days, the same as the modern value, Eudoxus may have had much more exact data. It will be seen that his synodic period for Mars is the only one which is totally wrong, and the large error is difficult to explain.

Planet.	Synodic Period.	Modern Value.	Zodiacal Peroid. [34]	Modern Value.
Mercury	110 days	116 days	1 year	1.0 year
Venus	19 months	584 ”	”	1.0 ”
Mars	8 months, 20 days	780 ”	2 years	1.88 ”
Jupiter	13 months	399 ”	” 12	11.86 ”
Saturn	13 months	378 ”	” 30	29.46 ”

It is disappointing, after the splendid hypotheses of the Pythagoreans, to be back again on a central stationary Earth among mechanical contrivances for moving the heavenly bodies, which remind us of Anaximander's series of hemispherical heavens and heavenly wheels, but at least the earth is spherical, owing to the Pythagoreans, and the sky extends like a sphere all round, and we shall never have a flat Earth or a hemispherical sky again among the Greeks. We do not know whether Eudoxus regarded his spheres as convenient mathematical abstractions only, or whether he reasoned that the stars were evidently set in an invisible uniformly rotating sphere, and Plato considered this kind of movement the most suitable for all heavenly bodies; that therefore he would try whether a series of similar spheres interacting on one another would account for the complicated motions of a planet, and finding that they would, taught that they must truly exist. In any case the basis of his system was a detailed knowledge of planetary motions hitherto unapproached by the Greeks, and its chief merit was that it challenged comparison with the skies.

3. CALIPPUS.

Calippus c. 330 B.C.

The challenge was soon taken up, for twenty or thirty years later one of the pupils of Eudoxus, Calippus of Cyzicus, undertook to improve the system. The defects in the theories of Mars and Venus had evidently been discovered, for Calippus added another sphere to each of these, as well as one to Mercury, which would be quite enough to bring the theories into better agreement with the facts.

With regard to the sun and moon, Calippus had paid special attention to their movements, for he made an improvement in the old luni-solar cycle of Meton, to which we shall return later. Eudoxus had ignored a very important fact discovered by Meton and Euctemon about B.C. 430, viz.: that the seasons are of unequal length, showing that the sun takes unequal times to pass over the four arcs of his orbit lying between the four points of the vernal and autumnal equinoxes and the summer and winter solstices.

"Why," exclaims a later writer[35] "are there unequal numbers of days in the four seasons, seeing that the course of the heavenly bodies must be regular, not being swayed by human passions or affairs?"

Calippus considered this question very seriously, and made a careful determination of the length of the four seasons. It seems at first sight impossible to reconcile their inequality with uniform circular motion of the sun round the earth, but he found that he could do it by adding two more spheres to the sun's set, rotating uniformly but so arranged that their motion, added to the others, would result in an actual velocity in the sun itself varying just in the way required by the facts. The same had to be done for the moon, for the same reason, so the number of spheres, which Eudoxus had made twenty-seven, was brought up to thirty-four (including the star sphere). The varying velocity of the five planets had not yet been perceived.[36]

4. ARISTOTLE.

Aristotle B.C. 384-322.

Calippus went up to Athens about 330 B.C. to lay his scheme before the great master, Aristotle, and it had his cordial approval. But Aristotle definitely accepted the spheres as things having a concrete existence, for he says (*De Cælo* II, 12) that we must regard them as heavenly bodies like the stars and planets, and that they are composed of the same celestial stuff.

He made one change, when incorporating the system in his scheme of the Universe. He was not satisfied that each set of spheres should work quite independently of the rest, and thought that the outermost sphere of stars ought to communicate its motion to those below (*i.e.* nearer Earth); and no doubt it did seem rather clumsy to have a separate sphere in every set rotating in exactly the same manner as the star sphere. But how could the impulse be communicated without disturbing the other movements? Aristotle introduced below each set another set of "unrolling" spheres, as he called them, which successively neutralized the rotations of all spheres in that set except the one with diurnal rotation, hence this movement alone was communicated to the set next below. This seems, however, more clumsy than the defect it was intended to remedy. Aristotle was perhaps led to it by his wish to give greatest importance to the star sphere; and if so, he acted on the principle which he blamed in the Pythagoreans, of making deductions not from things as they are seen, but as, according to his own ideas, they ought to be.

For indeed Aristotle, in spite of his own doctrines, and the great impulse which he gave to truly scientific methods of observation and experiment, could not rise altogether above the prejudices of his age, and consequently his Cosmos is a curious mixture of sound reasoning, based on observation, and of metaphysics, the latter predominating. For instance, it is only at the end of his second book *On the Heavens*, after he has "proved," from purely metaphysical reasons, that Earth must necessarily be spherical and at the centre of the World, that he adds in support of his assertions the fact that the curved line of Earth's shadow seen on the moon during eclipses is always round, that stars vary in visibility as we change our horizon, and that astronomers say that the celestial phenomena occur as they would if Earth were at the centre of the World.

Nevertheless, Aristotle's teaching had so overwhelming an influence, not only throughout this epoch, but in the age of Dante, and the latter was so greatly influenced by him, both directly and indirectly, that it is exceedingly interesting to know his ideas about the Cosmos. We find them in the two books *On the Heavens*, in the *Meteorology*, the *Metaphysics*, and some other works. A special treatise on Astronomy, to which he refers[37], is unfortunately not extant.

The form of the Universe, Aristotle says, must be a sphere, because a sphere is the most perfect of solids and a solid is more perfect than a surface or a line, because it is in three dimensions, and three means completion, perfection.[38]

The Universe had no beginning, and will have no end; and this conclusion, drawn from reasoning, is supported by the belief which all have who believe in gods, "whether Greeks or not Greeks," that the gods, who are immortal, live in the highest heaven, which is therefore also immortal; and by the fact that no one, throughout the ages, so far as we know, has ever seen any change in it.

But it is of finite dimensions, for no infinitely great body could rotate in a finite time; and it is the only universe which exists or ever can exist: outside is neither space, nor void, nor time. For space is that which is or may be occupied by matter, and time is the measure of motion occurring in matter, and no matter exists or can exist there. Therefore that which exists there is not in space nor is altered by time, but lives for ever the best and the self-sufficing life (*i.e.* the purely spiritual).

As matter has three dimensions, so motion is of three kinds: viz. (1) in a straight line down, that is, towards the centre of the World; (2) in a straight line up, that is, towards the circumference; (3) in a circle round the centre. Thus simple heavy bodies such as all kinds of earth, have a simple motion downwards; simple light bodies such as fire, move upwards; and when they reach their respective goals they remain where they are, unless disturbed by external force—earthy things on the earth, fiery vapours in the upper atmosphere. Composite bodies have composite motions, the motion proper to the predominant substance predominating. But for the heavenly bodies the only possible motion is in a circle, where there is neither beginning nor end, no goal and no limit, hence this motion is eternal.

Thus Aristotle solved for himself the problem of the early philosophers: how the stars in the sky remain there, for ever circling round us, and never falling to the ground. There is no need, he says, to assume an Atlas to support the sky on his shoulders, as in the old myths, nor a whirlwind such as Empedocles suggested, nor a Soul of the World, as Plato said; for the heavenly bodies are not heavy things like Earth to need support, and they are not moved by force, but are eternally in motion from the nature of their being.

In the same way he disposes of the difficulty of supporting Earth, having first "proved" that because there is an ever-circling spherical Heaven, there must also be an ever-resting spherical central Earth; there must be its opposite, the ever upward-striving Fire; and there must be the intermediate pairs of opposites, Air and Water. It is indeed, he says, a strange thing, and one to set any thoughtful man thinking, that the smallest clod of earth, when thrown up into the air, immediately falls down, and presumably would never stop falling if the earth were suddenly removed from beneath it; yet here is Earth herself, so large and heavy, not falling, but remaining steady in one place. But the explanations given by philosophers are more difficult than the fact they seek to explain. Xenophanes of Colophon said that the earth roots in the infinite, which simply saved him the trouble of considering further; others that the earth rests on water, which is our oldest tradition, said to be derived from Thales; but on what then does water rest and how can water, which is lighter than earth, support it? Do we not see that even small pieces of earth sink in water, and larger ones still more quickly? Anaximenes, Anaxagoras, and Democritus said that Earth rests on air, through her flat shape, as a leaf can float on the wind, and they added that the air cannot escape because the flat earth fits close down upon it, like a lid, which is also the reason that it can support the earth, because it is compressed, and they brought forward many proofs to show that air, when compressed and still, can support great weights. Others, like Anaximander, said that Earth rests because she is in equilibrium, for there is no reason why she should move in one direction rather than another. But all earthy things (says Aristotle) do not merely remain at the centre when there, they move thither whenever displaced. They are not suspended like a hair which is powerfully but uniformly stretched, and so never breaks, nor like a man who is equally hungry and thirsty, and has meat and drink at equal distances from him and therefore starves.[39] No, the truth is that Earth, and every particle of Earth, tends naturally towards the centre of the Universe, and rests when at its goal. We must remember that on every part of the sphere of Earth, heavy bodies fall vertically to its surface, showing that it is not to the surface in general that they fall, but exactly to the centre, which is also the centre of the Universe.

Of course it was no solution of the mystery, but only moving it a step back, to say that the stars circle because it is their nature to do so, and heavy bodies fall to Earth because that is their nature. But the interesting point is that the Greeks did reason about these two motions, and compare them; that they clearly grasped the fact that "weight" simply means a tendency to move, and that the motion of falling bodies at Earth's surface is invariably towards Earth's centre, accelerating as it appproaches the surface (*De Cælo* I. 8). It is only the fresh mind of a little child, or of a really intelligent man, which is forcibly struck by the mystery of everyday sights, such as stones falling, and stars *not* falling but eternally moving in the sky. The force which makes bodies move towards one another we still call "heaviness," *i.e.* gravity, and the mystery of its ultimate nature and mode of action is still unsolved. It is so weak and so often complicated by other forces that, except in very delicate experiments or with the mind's eye, we can only see it in action when bodies fall to the ground; and thus Nature guarded for centuries the secret that every tiny particle on the whole earth attracts every other, and also the earth itself, as surely as the earth attracts them. The further grand secret concerning this force Eudoxus had unwittingly set out to discover, with his planetary periods learned from the Egyptians, and his three motions of the moon. For when this study was far enough advanced, the necessary data were at hand for Newton, as he pondered the mystery which had baffled Greece; and he was able, from the moon's motions, to verify his guess, that even the heavenly bodies are in truth always falling, falling, towards one another, exactly as Aristotle's "smallest clod of earth" fell to the ground.

From the theory of the three simple motions, it obviously follows that Earth must be at the centre of the world, that her particles must be arranged in a spherical form round the central point, also the sphere must be at rest.

But all are not agreed about this, says Aristotle. All who consider the Universe finite say Earth is at the centre; but the philosophers in Italy, the so-called Pythagoreans, on the contrary, say that in the centre is Fire, and that Earth, which is one of the stars, is in motion round the centre, and so causes day and night. They also assume a Counter-Earth, merely from pre-conceived ideas, not from observation of facts. And some agree about the Central Fire, from pre-conceived ideas, because they think that the noblest should have the noblest place: fire is nobler than earth, and boundary nobler than what is bounded, and circumference and centre are both boundaries; therefore (they say) Fire and not Earth is at the centre. Moreover, the Pythagoreans say that the most important part of the Universe is the best guarded, and that the centre is such a part, and they call it the Watch Tower of Zeus.

To these metaphysical reasons Aristotle replies that the centre is not a true boundary, it is rather an end than a source, it is the material, the limited, while it is the circumference which limits, encloses, and gives the form. Besides, the centre of a thing is not necessarily the centre of its being; as with animals the centre of their life (meaning the heart) is not the centre of their body. So the philosophers need not disturb themselves to put Earth out of her local centre, but they would be wiser to examine that other centre of the Universe (meaning the sun), and find out what is its nature and its place, for it also is a point of origin, and noble.

He continues "Some also assert that though Earth is at the centre, it is wound and moving round the axis which is extended through the Universe, as is written in the *Timaeus*." Plato's actual words in the *Timaeus* will not bear this interpretation, as we have already seen (p. 85). It is a little surprising that Aristotle does not mention the names of Ecphantus the Pythagorean and Heracleides of Pontus in this connection, since the latter was his contemporary, and perhaps the other also, for they are mentioned together as teaching the doctrine of Earth's rotation on her axis.[40] Also Aristotle seems hardly fair to either this or the Central Fire theory, in that he only answers the metaphysical reasons of the Pythagoreans, and omits to mention that either would unify the diurnal celestial motions in a much simpler way than all his "unrolling" spheres. If he had not especially mentioned that Earth's motion was supposed by the Pythagoreans to cause day and night, we should be inclined to think that he did not understand that the period was twenty-four hours, and that its effect would be to produce the apparent diurnal rotation of all the heavens.

The passage, however, has been a cause of endless controversy from the earliest commentators of Aristotle to the present day, and such a thorny question would have been avoided altogether in this book were it not that it is actually quoted by Dante in the *Convivio*.

As to the size of the earth, Aristotle held that it was not a large sphere, and small when compared with the stars. For, he says, if we take quite a short journey to north or south, our horizon changes markedly, so that the stars above us look quite different, and we do not see the same stars; for some which are well seen in Egypt and near Cyprus are not visible at all in northern parts, and those which in the north are always in the sky, set when we go south. And therefore, he adds, those who say that the regions near the Pillars of Hercules are connected with India, so that the ocean is one, are not saying anything altogether incredible; and their proof is that there are elephants both in the extreme east and the extreme west. He does not mean, evidently, that there was no sea at all between, but only that one could quickly travel from one to the other, always going west: there is no immense stretch of land or sea between west Africa and east India, nor are they the extremities of a flat disc-like earth. Aristotle tells us, moreover, that the mathematicians, who have tried to measure the circumference of the earth, find that it is about 400,000 stadia. This is the first time we hear of an attempt to measure the earth, but unfortunately we do not know what stadium was used, nor what was the method employed.

Aristotle's Cosmos is arranged as follows:—

Upon the central spherical Earth rests water, and above this is air, but these intermingle more or less, and are not sharply divided; in the same way, though fire rises highest of the four elements, there is not a distinct sphere of fire, but the higher part of the atmosphere is chiefly composed of it. It is in this upper fiery atmosphere that shooting stars are produced: hot and dry exhalations rising from Earth take fire there, but are quickly consumed. Comets have their origin in the same place, when large masses of vapour rise and are directly below the sun (had the Greeks noticed that comets' tails are always streaming away from the sun?). Aristotle also explains that the Milky Way is formed from these constantly-rising vapours, but under the influence of the stars, for it always has the same position amongst them, and that is where the most numerous and brightest stars congregate.

Thus within and below the fiery atmosphere constant changes are taking place, and all things are perishable, but as soon as we reach the lowest of the heavenly spheres, the moon's, we enter another world. All is changeless, eternal, divine. Motion is in circles, space is filled with ether, the heavenly bodies as well as their spheres are of an ethereal substance.

The Pythagorean idea of music made by the spheres, Aristotle dismisses as very pretty but unfortunately not true. For if in truth these immense spheres made a sound as they moved, even if we could not hear it (as they said) we should feel it, for even earthly thunder bursts rocks asunder! And there is no reason why they should make any sound, for nothing moves out of one place: the spheres are simply rotating, which is the natural movement for a sphere, unless it rolls along, which they are not doing. If nature had wished the spheres or the stars and planets to move forward, she would not have treated them worse than terrestrial animals, in giving them no limbs by which they could progress! The stars and planets have no motion themselves of any kind, but are simply carried along by their rotating spheres, as we can plainly see by the moon turning always the same face towards us: hence they make no more noise than a ship's mast set in a ship, or the whole ship as it glides down a river.

Stress is laid both by Plato and Aristotle on this absence of any motion of translation in the heavenly bodies and their spheres; both insist that a movement of rotation, in which the moving body continually occupies the same place, is the only movement existing in the heavens. One wonders whether the spheres of Eudoxus suggested or resulted from this idea.

Aristotle does not enter into detail about the separate planetary motions, in any extant work, but explains as the general principle that the outermost, the prime movement of the whole universe, is simple, and the most rapid, while the inner are complex, slower, and in the contrary direction; so that the planet nearest to the prime movement (Saturn) is longest in making his own revolution, because most affected by it, and the others less so in proportion to their distance. He refers his readers to the mathematicians, and quotes the Egyptians and Babylonians as having furnished satisfactory proofs of the relative positions of the planets, by such observations as occultations of other planets by the moon, which show that she is below them (*i.e.* nearer to us). The passage is quoted by Dante[41], in which Aristotle describes how he himself once saw an occultation of Mars. "When the moon was a half sphere, she passed beneath Mars and he disappeared under her dark side, but came forth again on her bright illumined side. And the same kind of thing," he adds, "is reported to happen with the other planets also, as those tell us who for a vast number of years have made observations, viz. the Egyptians and Babylonians."

There is one point which is elaborately discussed in the *De Cælo*, which seems very curious to us, but the main point must be noted here, since it is of some interest to the Dante student. Aristotle tells us that he considers the sphere of the Universe to have a top and a bottom, and that the Pole which is not seen by us (the south) is at the top. One cannot help thinking that Dante had this in mind when he chose the southern hemisphere for the mount of Purgatory, whither, after all their mistakes and wrong-doing on this underside of the earth, souls go to purify themselves on the upper side, under the stars of the southern pole.

In his book on metaphysics, Aristotle gives a very brief sketch of the spheres of Eudoxus and his own "unrolling" spheres; and says that all these planetary movements prove the existence of Essences, eternal and immoveable themselves, who cause these movements. And it has been handed down to us in a mythical way, from the most ancient teachers, that these eternal Essences are gods. Above all these must be a First Mover, the Primum Movens Immobile, who is one, eternal, and enjoys for ever the kind of existence which we only experience in our best moments. Upon this First Mover depend the whole heaven and all nature.

5. ARISTARCHUS.

With a system of revolving spheres accepted by the mathematical astronomers, and sanctioned by the great philosopher Aristotle, it may be thought that we are within sight of our goal, the system of Greek astronomy which was to dominate the scientific world for many centuries, including the age of Dante. But so exacting had the careful observers become that the system of Eudoxus must be completely transformed, by aid of two quite new hypotheses, before it would satisfy their demands. Also, about half a century after the consultation of Calippus and Aristotle in Athens, a strange new theory was propounded, the boldest and strangest of all.

Aristarchus *c.* B.C. 281.

Nearly two thousand years before Galileo was summoned before the Inquisition, and forced to recant upon his knees his "most damnable heresy" that the earth goes round the sun, Aristarchus of Samos was accused of impiety by his countrymen for the same crime. But he met an even sadder fate than Galileo—neglect. His daring scheme was almost ignored by his contemporaries, and but for a casual mention by Archimedes and by Plutarch, we should know nothing about it.[42]

He was, however, very famous as a mathematician, and also as an observer. Ptolemy quotes his determination of the summer solstice of the year b.c. 281, and this tells us the date at which he flourished. He was renowned for a very ingenious method by which he tried to discover how much further from us the sun is than the moon. When the moon is half full the angle sun-moon-earth is a right angle, and if the angle sun-earth-moon be measured, by pointing the astrolabe first to sun and then to moon, the third angle, at the sun, may be computed, and then the ratio sun-earth to moon-earth will be known. The method is perfect theoretically, and if the sun were comparatively near, say about ten times the moon's distance, it would be practicable; but the distance is really so much greater that the angle at the sun almost vanishes, and a very small error in estimating it causes an error equal to many millions of miles in the result. It is also impossible to determine from looking at the moon the exact time when the division between light and dark is a straight line. Aristarchus made the angle at Earth 87° instead of 89° 50', and this gave the sun a distance of about 19 times as far as the moon, instead of 400 times, which is the true value.

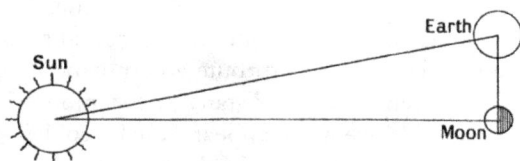

Fig. 20. Method of Aristarchus for finding the distance of the sun.

The foolish fad (as they thought it) about Earth's motion, held by this otherwise great man, is described quite clearly by the two writers above-mentioned. He suggested that the stars might be immoveable, and Earth be turning on her axis at the same time that she moves in a circle round the sun. Moreover realizing all that this implies with regard to the immense distance of the stars, he said that the circle in which Earth revolves round the sun, compared with the sphere of the stars, is as the centre of a sphere compared with its circumference. That is, not only Earth, but Earth's whole orbit, shrinks to a point when compared with the infinite distance of the stars.

We have unfortunately absolutely no information as to the way in which Aristarchus was led to these remarkable truths, and can only make conjectures from what we know of his times. Evidently the Central Fire theory was a suggestive step, and so was another theory which was afloat about this time, and has been called the "Egyptian system" on the authority of Macrobius in his commentary on Cicero's *Dream of Scipio*. No one really knows where it arose, but it is ascribed with much probability to Heracleides of Pontus. According to this, the two planets Mercury and Venus circled round the sun, and all three together circled round Earth, which still remained the centre of the Universe and of the other celestial motions. It was an idea which might have occurred to any unprejudiced observer, since the oscillations of Mercury and Venus from side to side of the sun are more striking than their movements through the stars. They never go far from him in the sky, like the other planets, and seem to belong to him.

Further, the clear understanding of the periodic motions of Mars, Jupiter, and Saturn, and the accurate observation which had been introduced by Eudoxus, must have revealed the fact that the loops in the orbits of these planets are connected with the apparent movement of the sun; and the varying brightness (especially noticeable with Mars) is inconsistent with the assumption of unvarying distance from Earth.

The great importance of the sun, above all the other planets, had of course always been recognised, and it is possible that Aristarchus was struck by an unconscious suggestion in Aristotle's advice to the Pythagoreans to examine the nature and place of the sun, "that other centre of the Universe," for that was also a point of origin, and a noble one.[43] His own work, too, would lead him to attribute a commanding position to the sun, for as it was nineteen times as distant as the moon, it must also be nineteen times as large (since they appear equal), so he must have felt certain that it was a very great size, probably much the largest of all bodies in the Universe.

It was the grandest and truest of all the Greek astronomical theories. But it was not accepted, it was hardly even discussed, and we can scarcely be surprised at this. Such an improbable theory needed many more convincing proofs than Aristarchus could bring forward. Observation, calculation, comparison of theory with facts, this was what was needed before safe ground could be won for belief, and in this Eudoxus was a true pioneer. But freedom of thought and courage in imagination is needed also in science, and Aristarchus seems to have been almost the last to possess this.

Seleucus B.C. 160.

Aryabhata born A.D. 467.

A certain Babylonian, named Seleucus, who lived about B.C. 160, perhaps on the Tigris, and an Indian astronomer, Aryabhata, of the fifth century A.D., both taught that Earth turns on her axis, but Aristarchus stood alone in suggesting that she also has a movement of revolution round the sun. The possibility of Earth's motion was alluded to but seldom by classical and mediæval writers, and then almost always as a foolish fancy hardly worth discussion;[44] until at last Copernicus, hardly daring to publish his bold idea in sixteenth century Europe, for fear of persecution, sought and found among the followers of Pythagoras, and in Aristarchus of Samos, kindred spirits with his own.

6. THE SCHOOL OF ALEXANDRIA.

The spheres of Eudoxus would not work. The system was already overladen, and more variations in velocity were becoming known. Observation shows, too, that the planets, especially Mars and Venus, vary greatly in brightness, in regular periods which correspond with their movements, and this could not be accounted for by any possible number or arrangement of spheres if all were to remain concentric to Earth, so that every planet remained always at a uniform distance. Yet Aristotle had said that Earth must be at the centre of the Universe. The new philosophers of the Stoic school agreed with him, and among mathematicians only Aristarchus dared to disagree.

But they were not long at a loss. The homocentric spheres[45] were thrown aside, and during the third century two new hypotheses were suggested (beside that of Aristarchus), and although we do not certainly know by whom, there is little doubt as to the place in which they originated. By a strange fate, Egypt, the home of astronomy many centuries before, became the seat of the latest, most brilliant, and most successful astronomical school of ancient times, and the knowledge won there in the five centuries between B.C. 300 and A.D. 200 spread, following in the wake of Alexander's conquests, over the whole of the civilized world.

It was a purely Greek school, however. Greece lost her independence under Alexander, and was finally crushed by Rome in 146 B.C., but never was Greek learning and culture so much honoured and sought after as in this age. In Egyptian Alexandria, Greek men of science found a welcome, and opportunities of research which did not exist in any other place in the world. In the Museum, founded and liberally endowed by the royal Ptolemies, was a great library whose custodians were bidden to obtain every book that had ever been written, and it is said that when any stranger arrived with a new book it was taken from him and copied for the Museum, and the copy returned to the owner. Within the great marble colonnaded building were lecture-halls, and reading rooms, and laboratories; there were gardens for botanists and zoologists, and observatories for astronomers. These astronomers were all Greeks, and though now living in Egypt they do not seem to have learned anything more from the Egyptians. Perhaps the priests resented the intrusion, and kept their secrets jealously to themselves; perhaps Eudoxus and his immediate followers had learned all they had to teach. This seems the more probable because, although the Greeks of this age did use Babylonian records of eclipses to form a lunar theory, they complained of the insufficient accuracy of all available planetary records. Moreover, they were not hindered from learning astrology: the sacred books of the Egyptians which taught its principles were translated into Greek about 300 B.C.; but it was always treated by them as quite a separate branch of study. Their geometrical methods were entirely their own. They introduced a system of notation which greatly simplified calculation[46]; their discovery of the principles of spherical trigonometry inaugurated a new era in astronomy; and they invented a new class of astronomical instrument.

We know what these instruments were like, and it is even possible to give an illustration; for, from descriptions in Ptolemy's *Almagest*, we find that the "well-made copper circles," the gnomons, and the celestial globes, which were set up in the Square Portico of the Museum, were of the same pattern as instruments which existed in Pekin, in the ancient observatory on the ramparts, until they were looted by the Germans during the late

Chinese war. The Alexandrian instruments were not supported on their stands by beautiful bronze dragons, but on the other hand the circles were more accurately divided, which after all was of more importance from the astronomers' point of view. Fig. 21 gives a general view of the Pekin Observatory, and Fig. 22 one of their astrolabes dating from the 13th century A.D.

At first glance there seems to be here absolutely nothing like our modern observatories. Ancient and mediæval astronomers had indeed no telescopes, being ignorant of the properties of lenses: therefore they were unable to study the features of any heavenly body except the moon, and they had no way of finding out anything about their physical constitution; but they had many ways of measuring their distances and motions, and even the angular sizes of sun and moon, and their instruments were the forerunners of our sextants, micrometers, and transit instruments, our chronometers and sidereal clocks.

[To face p. 114.

PEKIN OBSERVATORY.

From a photograph taken in 1888, and published in the "Bulletin de la Société belge d'Astronomie".

The gnomon has been already described[47], and it was one of the most valuable instruments used by the Greeks. The Pekin gnomon at the right of figure 21 was more than 40 feet in height, and on the top had a little plate of copper which was pierced by a hole as fine as the eye of a needle: the observations made with this were much more exact than observations of the end of the shadow, which must always be vague, and the Chinese records of the sun's movements made with this instrument between 1270 and 1280 A.D. are of great value in modern research. Ptolemy explains that his gnomon was made accurately vertical by the use of a plumb-line, and that one way of testing the level of the surface on which the shadow fell was to flood it with water.

Clepsydras, or water-clocks, were used by the Greeks, and many kinds of sundials for telling the time by day. Tables were also made of the risings of bright stars which served for clocks by night.

The instrument in the middle of the platform is a quadrant, and beyond this on the left is a large celestial globe, which, however, only dates from the Jesuit missionaries of the seventeenth century. Ptolemy says that his globe was made the colour of the night sky; Sirius being marked in his proper place, all other stars were placed relatively to him, and in their own colours as nearly as might be; the Galaxy was drawn, and the figures of the constellations outlined. The globe was arranged to turn on either the poles of the ecliptic or of the equator; circles of wood represented the horizon and meridian, and the pole could be arranged at any altitude according to the latitude of the place.[48]

But perhaps the most interesting are the astrolabes, and these owe their origin to the Greeks. The essential part of any instrument for determining angular distances is a divided circle and a pointer: the pointer is directed first to one object then to another, and the angle between them is then read off on the circle. In the astrolabe, the pointers themselves were also circles, provided with little perforated rods for "sights." (These are not visible on the instrument in Fig. 22, and have probably been broken off.) There were two fixed circles, set in the plane of the ecliptic and perpendicular to it. Three other circles could be rotated round the poles of the ecliptic. One of these was directed (by means of the sights), to some body whose position was already known, another to the body whose position was to be ascertained, and the angle between them was read off on the ecliptic circle; on the third the angular distance north or south of the ecliptic circle could be read. This last and the ecliptic circle were both divided into 360 degrees, and as many fractions of a degree as space and skill would allow.

The equinoctial astrolabe was similar, but the fixed circle was in the plane of the equator, instead of the ecliptic. One of each of these is seen in the view of the Pekin Observatory.

But how did the old astronomers know how to find the ecliptic and the equator in the sky, and set their circles in those planes? This they did by means of the sun's motion. The gnomon told them the day of the equinox (see p. 25), and on that day the sun was in the equator: therefore, if a circle was set up so that the shadow of the upper part fell symmetrically upon the lower, with a little line of light each side, it must be exactly in the plane of the equator. In the Square Porch such a circle was erected, a large one of copper, and when once correctly adjusted it was a standard plane, and also showed the date of the equinoxes, as accurately as the gnomon itself. Since the ecliptic is the path of the sun as seen in the sky, it is obvious that it could be determined from a number of different observations of his position at different times of the year.

[To face p. 116.

A PEKIN ASTROLABE OF THE 13TH CENTURY, A.D.

From a photograph taken in 1888, and published in the "Bulletin de la Société belge d'Astronomie".

Finally, accurate solar tables were drawn up, showing the sun's position in the sky in degrees for different dates, and then from these it was possible to find the places of planets and stars. They could not of course be compared directly, but the position of sun and moon were compared during the day, when both were in the sky, and then after dark the planets and stars were compared with the moon, allowing for her motion among the stars in the meantime. Or secondly, when the moon was eclipsed, and therefore known to be in the ecliptic and exactly opposite the sun, the places of stars could be found directly.

This very brief description will give some idea of the chief instruments and methods used, and when we see how very rough and elementary they were, and remember that the Greeks had to work out their observations without algebra, or decimal notation, we are amazed at their results, and their far-reaching ambitions.

Eratosthenes B.C. 276-194.

Already in the very early days of the Museum, Eratosthenes, a celebrated geographer, made a bold attempt to utilize observations of the

- 74 -

sun measuring the size of the earth. It was known that in Syene (the modern Assuan) on the day of the summer solstice at noon no shadows were thrown, and the bottoms of wells could be seen: evidently therefore the sun was in the zenith. Eratosthenes found that the sun's distance from the zenith in Alexandria at noon on the same day was 7° 12', or one-fiftieth of the circumference of the heavenly sphere, consequently the two towns must be distant from one another (assuming them to be nearly in the same meridian) one fiftieth of the circumference of the earth. The distance from Alexandria to Meroe was known, and from Meroe to Syene had been paced by the king's professional pacers; the whole was 5000 stadia. 50 times 5000 = 250,000. The figure always quoted by the ancients is however 252,000. If the stadium used by Eratosthenes was the measure generally used for long distances which have been paced, this estimate is equal to 24,662 miles, only about 200 miles less than the modern value. It was partly by luck that Eratosthenes got such a good result, for he was evidently only working with round numbers, and the extra 2000 stadia seem to have been added in order to make one degree equal to exactly 700 stadia. But in any case it was a highly creditable performance.

Euclid *c*. B.C. 300.

Apollonius *c*. B.C. 270.

There were celebrated mathematicians and geometers at Alexandria, whose work was most useful to astronomy, such as Euclid, and Apollonius of Perge. The latter is specially mentioned by Ptolemy in connection with the new theory of "moveable eccentrics," which was invented to account for the varying brightness of the planets, as well as their peculiar movements.

Fig. 23 explains this theory. Let P A be a great revolving circle upon which Mars is fixed. (In the hands of the Alexandrian mathematicians the spheres almost disappear, and they deal practically only with circles.) If the earth were at its centre, as Eudoxus demanded, Mars must always be at the same distance, but if we make the circle eccentric to Earth, by putting its centre at C while Earth is at E, then the distance and consequently the brightness will constantly vary, and Mars will be brightest when at perigee P (point nearest Earth), and faintest when in apogee A (point furthest from Earth).[49]

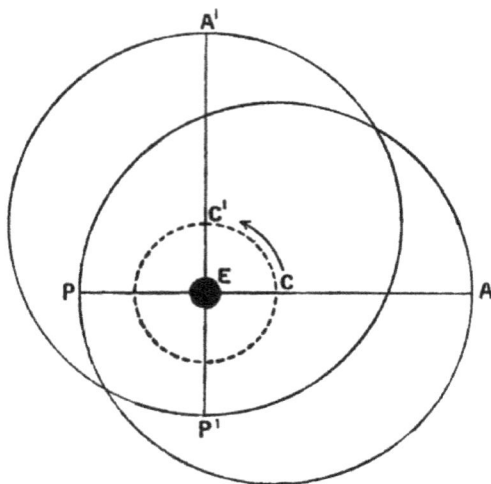

Fig. 23. The Moveable Eccentric.

But, as the Greeks had discovered, Mars attains his greatest brilliance at different points of the zodiac, so P must be made moveable, and it always happens when he is opposite the sun, therefore P must keep pace with the sun's apparent motion in the zodiac and P E always point towards him. This was accomplished by making P C A turn round upon the fixed point E, so that for instance when the sun had moved through a quarter of his circle (in three months) P A had moved to P' A', and the whole eccentric had moved into the new position shown in the diagram, its centre C being now at C'. In other words, the centre of the eccentric moves round Earth in the same time and in the same direction as the sun, that is in one year, and "with the signs" (from west to east).

At the same time, Mars is moving in an opposite direction on the eccentric, and without entering into all the details of the problem, we may add that the Greek geometers found that by determining the proper relative sizes of the large and the small circle they could make the two motions neutralize one another when the planet reached its stationary points, and the retrograde motion prevail over the direct when it retrograded. A similar arrangement was made for Jupiter and Saturn.

In this very ingenious way the varying brightness as well as the varying motions of these three planets were accounted for, without violating the principle of uniform circular motion, and without removing Earth from the centre of the Universe. She was also still the centre of planetary motion, in a certain sense; but to place the true centres of these planets' spheres *outside Earth and in the direction of the sun* was a very suggestive step, and may well have helped Aristarchus to his bold hypothesis. For he had only to put the sun, not at some indefinite point along the line E A, but exactly at the point C, and it became the centre of motion for Mars, Jupiter, and Saturn, just as in the "Egyptian theory" it was the centre of motion for Venus and Mercury. In this way he would arrive at the conception of the sun circling round Earth and carrying all the planets with him, (a theory which was held by the great astronomer Tycho Brahé in the sixteenth century A.D.). Then a flash of insight may have revealed to him the fact that this motion of the sun is apparent only, being but the reflection of Earth's own motion; for she is circling round the sun like all the other planets.

It is, however, only a guess that the Moveable Eccentrics played this part in the theory of Aristarchus. They did not long hold the field, because they were not applicable to Venus and Mercury, which are never seen in opposition to the sun. So they were thrown aside for another system, the Epicycles, which illustrates much more simply the stations and retrogressions of the planets, and can be used for them all.

Later on, when more irregularities of motion were discovered, it was found necessary to combine eccentrics and epicycles, and by means of this joint system it became possible at last to represent completely, and as accurately as they could be observed, all the apparent movements of the heavens. First, however, an immense amount of work had to be done, and new methods devised, both in observation and mathematics. The man who contributed most, in both ways, to make it possible, was Hipparchus.

7. HIPPARCHUS.

Hipparchus c. 140 B.C.

Of this great man we know scarcely anything but what can be gathered from the work he did, and this corroborates Ptolemy's description of him: "Hipparchus, lover of toil and truth φιλοπονον και φιλαληθεα." He lived about B.C. 140, since this is the date of the only book of his still extant, and his work was not done in Alexandria, though he may have studied there in his youth, and he used the Museum records. We count him among the Alexandrians, as he belongs to this era, but he seems to have been a private astronomer, who set up an observatory of his own in Rhodes, his native place. Here we seem to see him, surrounded by his primitive instruments and his papyrus books, patient, eager, modest, seeking no fame and no reward but the joy of his work. By day he would keep watch over the sliding shadow of his gnomon, would write up his observations, make long calculations, and devise new methods in mathematics, improve and modify his astrolabes and his clepsydras; at night he would spend long hours with moon, planets, and stars, making up for the defects and shortcomings of his instruments by the skill and care with which he applied them to measure positions in the sky. Nothing but the most loving and conscientious care could have raised his work to such a pitch of accuracy, and made such rude means suffice for such splendid achievements.

The book we possess, apparently an early one, is chiefly concerned with the positions, the risings and the settings, of stars, and at the end is a list of sixteen which came to the meridian at intervals of an hour: from this list and the knowledge of spherical trigonometry which he possessed, it would be possible to calculate the time at night to within about a minute.

Hipparchus was able to construct a satisfactory theory of the sun and to some extent of the moon, but he found more irregularities in the planetary motions than Eudoxus had suspected. The records of his predecessors were not accurate enough for him to construct a theory for the planets, and he soon realized that one life-time would not be long enough to collect all the data necessary, so, as Ptolemy tells us, "Hipparchus, who loved truth above all things," quietly set to work to make as good and as many observations as possible, leaving it to his successors to complete and explain them.

In the same spirit he undertook the laborious task, of which Pliny speaks with awe as a presumptuous scheme, even for a god, "*rem etiam Deo improbam*," of numbering the stars. Pliny says he was led to do this by the appearance of a New Star, which blazed out suddenly in the constellation of Scorpio in b.c. 136, just as Nova Persei did in Perseus in February 1901. He saw that even in the upper regions of the eternal heavens, which Aristotle had supposed absolutely changeless, changes may occur, and in order that even the least of these should not pass unnoticed, he set to work to note the number, brightness, and position of all he could see. This great catalogue of 1080 stars, copied by Ptolemy in his *Almagest*, was the basis for all succeeding catalogues, from Spain to Turkestan, until quite modern times. In it, for the first time, the places of the stars were not merely described according to their position in the constellation figures, but were noted in degrees on the sphere, as is done to-day.

Timocharis *c.* B.C. 280.

One day, when comparing his notes with those of Timocharis, who had worked at Alexandria about a century and a half earlier, he found that the brilliant star Spica, the Ear-of-Corn which the Virgin carries in her hand, had apparently moved nearer to the autumnal equinox by about 2°. (Two degrees is about four times the angular diameter of the sun). Of course he or Timocharis might have made a mistake, or Spica might really be moving among the stars, or she might be carried along with the rest by a slow movement of the whole star sphere. Apparently Hipparchus satisfied himself that he could rely upon Timocharis' observation, and took pains to verify his own; the second hypothesis could be disproved by the fact that Spica does not change her place perceptibly among her neighbours; and finally it became clear that her motion is part of a slow apparent movement of the whole heavens.

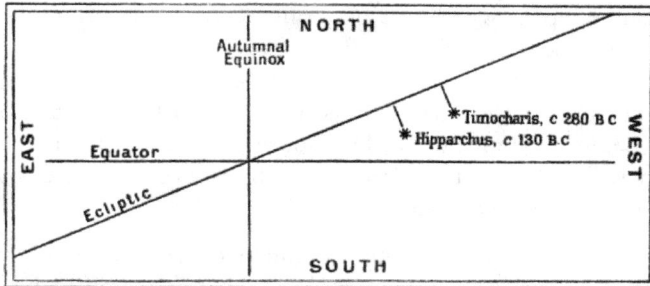

Fig. 24. The movement of Spica.

Here was a discovery of first importance, an unexpected reward of patient accuracy, of which the white Spica, flashing down from summer skies, may always remind us. Hipparchus had discovered the grand cycle which we call the "Precession of the Equinoxes," and before Spica returns to the same position in which he saw her then, when she led him to his great discovery, she will have been watched by generations of astronomers for another twenty-four thousand years. No notice of the cycle has been found as yet among the records of any other nation, although it seems as if the astronomers of Babylon and Egypt, and other countries where observations had been carried on for many centuries, must have been aware of it. We can only imagine that at long intervals of time they found that the stars had somehow changed, and made corrections accordingly, but without understanding the nature of the change. What Hipparchus thought about its cause we cannot tell: probably he left all speculations to future astronomers, and confined himself to noting the fact.

The displacement of Spica which he observed is shown in the diagram.

Both Timocharis and Hipparchus evidently measured her position indirectly by comparing it with that of the moon, which was eclipsed at the time,[50] and therefore known to be in the ecliptic and opposite the sun. To find the sun's distance from the equinox was an easy matter, since his yearly course had long been carefully studied, and the days on which he passed the equinoxes were regularly observed with the gnomon. Spica, then, had moved eastward along an arc parallel to the ecliptic, and since celestial latitude and longitude are referred to the ecliptic, we may define her apparent movement in astronomical language by saying that while her latitude had remained constant, her longitude had increased by about two degrees; and further, as the celestial equator is oblique to the ecliptic, this implied that her declination (position north or south of the equator) had also varied. The diagram shows that she had a less northerly declination than before.

At this rate, Spica, which was now only 6° from the autumnal equinox, would reach it in less than five hundred years, and thereafter would lie east instead of west of it; and that she has in fact done so, may be seen by consulting a modern star atlas. She is now 22½° of longitude east of the autumnal equinox, and nearly 11° south of the equator. Her south declination will continue to increase for about five thousand years, after which she will come north again.

Ptolemy says that Hipparchus examined other stars, and found that they also were increasing their longitude at what appeared to be the same rate as Spica. The yearly amount of the movement, derived from the Spica observations, is within a few seconds of arc of the true value, which is 50¼ seconds; but Hipparchus would not fix any value until it had been tested by further observation, and merely stated that it could not be *less* than one degree in a century *i.e.* 36 seconds per annum.

This is a very uncomfortable phenomenon for astronomers, since every star is for ever changing its measured position on the celestial sphere. Take three stars, one at the north pole, another on the equator, and a third in the southern hemisphere. After some years, the first will no longer be a pole star, the second no longer an equatorial star, and the third may have so far increased its south declination that it will be invisible at latitudes in our northern hemisphere where formerly it used to rise above the horizon. One compensation for this inconvenience is that if we know what star was near the pole, or which stars lay along the equator, on any given occasion, we can calculate the date. Thus it is believed that the Great Pyramid was built when Alpha Draconis was the Pole Star, that is, nearly 3000 years B.C.; and by a similar method Mr Maunder determines the epoch at which the ancient southern constellations were invented, as we have already seen.

The greatest inconvenience, and also the greatest historical interest, attaches to stars like Spica which belong to constellations of the zodiac, for if they are not stationary with regard to the equinoxes and solstices they are not such simple guides to the length of the solar year as the ancients supposed them to be. The scheme of the Babylonians for beginning their month Nisan when the stars of Dilgan rose just before the sun was an excellent one for a time, but if they had continued it for many centuries they would have found that their year was too long, and the months were all falling in the wrong seasons. This has actually happened with Hindus and Parsis, who now keep their New Year in the middle of our April, although when their calendar was fixed, about thirteen hundred years ago, the years began at the spring equinox. For the sun is like a runner in a circular race-course who thinks he has completed a lap when he returns opposite a group of spectators originally standing at the starting-point, but after several laps he finds that the spectators and the goal no longer coincide; either they, and also all the others surrounding the course are walking away from it, or an unseen hand has been moving the flag towards him, and so shortening the lap.

It is the flag which must count, in any case, not the spectators, and with the sun it is the equinox which must count, and not the stars, for this is the point at which he crosses the equator, making day and night equal, and from this we count the beginning and ending of our seasons. So our year is counted from equinox to equinox, and is twenty minutes shorter than the "sidereal," or star year, of the ancient Babylonians. Hipparchus, from observations of equinoxes and solstices, made the year 365 days 5 hours and nearly 55 minutes, which is only 6 minutes longer than the correct value.

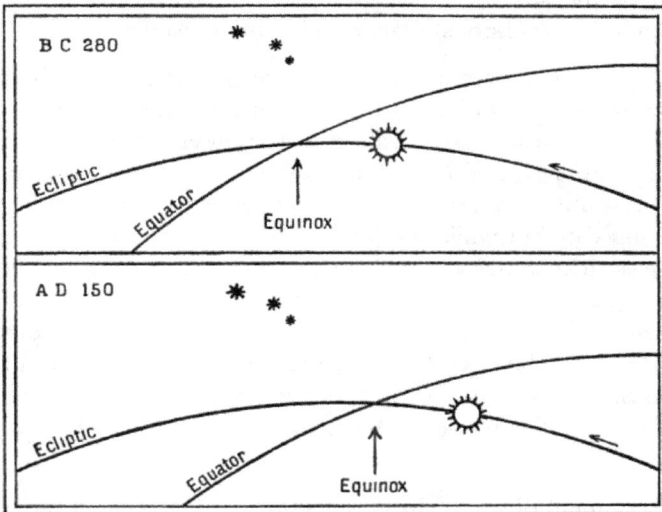

Fig. 25. The Sun and the Equinox.

After Hipparchus had made his discovery, astronomers agreed upon a somewhat clumsy and very confusing device, by which the zodiac was divided into twelve equal "signs" of 30 degrees, which bear the same names as the zodiacal constellations, but whose beginning is always reckoned from the vernal equinox. These twelve "signs of the zodiac," therefore, do not now agree with the twelve constellations of the zodiac, and our present "first point of Aries," which marks the vernal equinox, is in the *constellation* of Pisces.

What is the true cause of this strange phenomenon? Are the stars really all in motion, or is it the equinox which moves?

The successors of Hipparchus, who believed that the stars were fixed on a sphere, found no great difficulty in conceiving that this sphere had a very slow easterly motion, round the poles of the ecliptic, completing a revolution once in 36,000 years (*i.e.* one degree in century). To us, however, it is impossible to believe that the stars, which we have found to be at enormous and varied distances, are all revolving at one rate, parallel to the ecliptic. The ecliptic, to us, is simply the plane of Earth's own orbit, and as she moves in it she has a very slow "wobbling" motion on her axis, as well as the rapid spinning of the diurnal motion, like the "wobble" of a spinning top. The top has this motion because gravity is trying to pull it down from its upright position; the earth because the sun is trying to drag her slightly protuberant equator into the plane of her orbit.

The resulting motion is not a revolution of the earth, nor an apparent revolution of the star sphere round Earth: what really happens may be illustrated with the traditional orange and knitting needle.

Ignoring all motions but the one we are speaking of, let the points of the knitting needle (Earth's axis of rotation) trace out small circles in space, and the equator of the orange will be seen to alter the direction of its tilt, but without turning round (Fig. 26). Stick a pin in the equator, and others in north or south latitudes, between equator and pole; these will always remain facing you, but while the pole makes its small circle, the equatorial pin will be seen to move up and down, while the tropical and temperate pins trace out ellipses. These are the movements which we see reflected in the stars; and if Earth's diurnal rotation were suddenly to cease, while her revolution in her orbit and the movement of "precession" continued, we should see Spica, for instance, sink slowly lower in the southern sky and after ages rise again northwards, but there would be very little preceptible movement east or west.

The movement observed by Hipparchus, then, was not a movement of Spica and other stars, but a movement of the equinox. For the celestial equator is simply a reflection of Earth's equator in the skies, and as it keeps changing the direction of its tilt in the way described, it changes the point at which it cuts the ecliptic. This may best be seen by taking two rings or hoops (two large curtain rings, for instance), one of which just fits inside the other. Tilt the inner ring, so that half of it is above and half below the other ring, and they touch at two points, 1 and 2 (Fig. 27). The outer ring is the Ecliptic, the inner the Equator, and where they touch each other are the Equinoxes. Now move the inner ring, not sliding it round, nor making any difference in the angle between the two, but simply so that they touch at fresh points, 1' and 2'. In this way you may make the points of contact revolve entirely round. This is what the real equinoxes are doing: while the equator opposite the group of stars in figure 25 rises and falls, the equinox travels on, and finally returns to the same place.

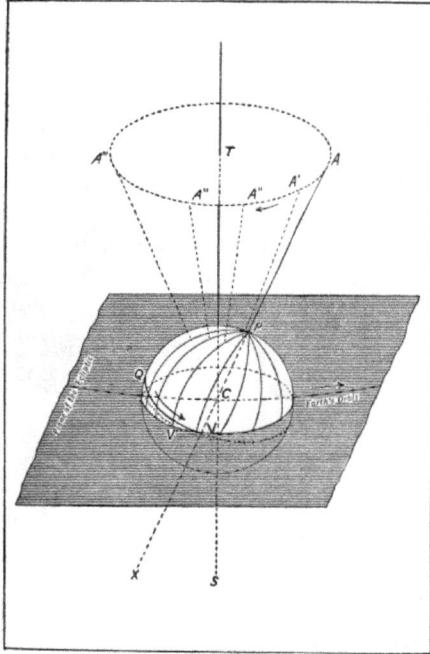

Fig. 26. The movement of Earth's axis,
which is the true cause of Precession.

V is the vernal equinox, at the intersection of the equator and plane of the ecliptic: APX the earth's axis, which always preserves the same inclination (23½%) to the plane of the ecliptic. As APX slowly revolves round T in the direction of the arrow, the vernal equinox is gradually shifted to V', and so on.

(From Young's "Manual of Astronomy," 1902.)

The phenomenon is called "precession of the equinoxes," because they thus move on to meet the sun in his yearly course.

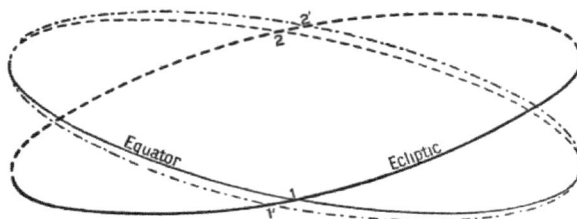

Fig. 27. Precession of the equinoxes.

The discovery of precession is what has chiefly made Hipparchus famous, but the invention of the astrolabe and of spherical trigonometry, both believed to be due to him, his star catalogue, and his many observations, more accurate than any made before, were so valuable as pioneer work that Ptolemy justly called him the Father of Astronomy. If Hipparchus could visit one of our observatories to-day, and see the clock-driven equatorials, the transit instruments, the beautifully divided circles read with microscopes, and the sidereal clocks, one wonders whether he would be more astonished at the advance on his astrolabes and clepsydras or at the homage paid to him as one in whose footsteps all astronomers are proud to tread.

8. PTOLEMY.

Claudius Ptolemæus. I know that I am mortal and ephemeral, but when I scan the multitudinous circling spirals of the stars, no longer do I touch Earth with my feet, but sit with Zeus himself, and take my fill of the ambrosial food of the gods.

For more than two centuries after Hipparchus very little original work was done in astronomy, and no one seems to have had the courage to take up his unfinished task and study seriously the difficult problem of planetary motions.

Posidonius *c.* 135 B.C. to *c.* 50 B.C.

Posidonius the Stoic, who lived for some years in Rhodes, made a fresh determination of the earth's circumference, basing it not on observations of the sun, like Eratosthenes, but of the star Canopus, which in his time was just visible at Rhodes while in Alexandria it rose "a quarter of a sign" (*i.e.* 7½ degrees) above the horizon. By his method, the earth was a little smaller, (240,000 stadia instead of 250,000), but it must have been difficult to measure the distance between Rhodes and Alexandria over the sea, and it is impossible to say when a star is exactly on the horizon. Posidonius also observed the tides in the Mediterranean, and showed that "Ocean follows

- 84 -

the movements of the heavens," and especially of the moon, having daily and monthly periods.

Geminus *c.* 70 B.C.

A little later Geminus wrote an Introduction to Astronomy, which was an excellent little book as far as it went, but although he was apparently a native of Rhodes, and speaks of Hipparchus, he seems to know nothing of his work, for he does not quote his careful determination of the length of the year, nor his discovery of precession.

Cleomedes *c.* 20 B.C.

Theon of Smyrna *c.* 100 A.D.

Nor were other writers, such as Cleomedes and Theon of Smyrna, better informed, and they added nothing new to the advance of astronomical science.

Ptolemy *c.* 140 A.D.

But at last a worthy successor arose at Alexandria, the immortal Ptolemy, whom Dante met in that Limbo of antique spirits which was almost Elysium, although on the brink of the Inferno.

We do not know when Ptolemy was born nor when he died, nor where was his native town: we only know that his first recorded observation was made in the eleventh year of the Emperor Hadrian, that is A.D. 127, and his latest in A.D. 150, and that he lived and studied in Alexandria. He had splendid opportunities for carrying on the work of Hipparchus, for besides the use of the instruments in the Museum Observatory, he had at hand all the Museum records, which included the writings of Hipparchus. Ptolemy was not so painstaking and accurate an observer as Hipparchus, but he was a very able mathematician to whom it was evidently a joy to handle figures and work out problems. By examining a number of observations spread over several centuries, and combining them with his own, he was able to accomplish the task in which so many others had failed, and to frame a system which embraced all the celestial motions then observed. The monumental summary in which he set forth this system contains a great deal of interesting information about his methods and instruments, and about the work of Hipparchus, for whom he always expresses the most generous admiration. The original name of his book was the "Mathematical System of Astronomy," but his admirers having called it the "Great System," *Megiste Syntaxis*, the Arabs affixed their article *al* and gave it the name it preserves to this day, of *Almagest*. It remained the standard treatise on astronomy until the *De Revolutionibus Orbium Celestium* of Copernicus appeared in 1543.

The name of his book indicates the scope of Ptolemy's work. It was to represent all the observed motions of the heavenly bodies by means of a mathematical system, so that they became amenable to calculation; but to explain the causes of these motions was thought to lie quite outside an astronomer's province. It was not for a mere observer and calculator to determine which motions were real and which apparent, else Ptolemy must have decided in favour of Earth's rotation, for he says that it would be much easier to account for celestial motions on this assumption. Nor was it his business to investigate the substance of the stars. Little did he dream that astronomers would one day solve such problems, and uphold their conclusions in the face of all the world: for him, as for his contemporaries, the decisions of the philosophers, and especially of Aristotle, were final, and his task was to describe what he saw in the light of their teaching.

He brings forward, in the introduction to his book, a few arguments against the absurd notion, taught by some, that Earth is in motion, turning on her axis or moving through space; but all that he proves is the immense difficulty, even to a trained mind, of accepting these theories, and the great authority of the philosophers who denied them on abstract principles. Educated Greeks might still discuss the nature of the heavenly bodies, as in Plutarch's delightful dialogue *On the Face of the Moon*, (written about half a century before the days of Ptolemy), and might ridicule the law of gravity, laughing at the absurdity of supposing that if the middle of a man's body were at the middle of the earth, his feet as well as his head would be "up," and that falling weights if they reached this point would stop short, or oscillate to and fro. Yet even they all agreed that the fixed stars do most probably move "in a circle of eternal and never-ending revolution," and that they are of a pure and eternal substance unlike Earth; and for the professional astronomers of Alexandria these axioms were assumed as the basis of all their work. Earth must be immoveable at the centre of the Universe because the heavy stuff of which she is made sinks necessarily to the centre and there remains in globular form without need of support; the heavenly bodies must be in motion, because, being of ethereal substance, it is their nature to revolve eternally in circles.

This being granted, we can feel nothing but admiration for the extent of Ptolemy's knowledge, the comprehensiveness of his scheme, and the skill and patience with which he overcame its difficulties.

Earth, according to Ptolemy, is but a point compared with the immense surrounding sphere in which the stars are set, and this turns always round us, communicating its motion (he does not inquire how) to sun, moon and planets, so that day follows night, and the heavenly bodies daily rise and set. For the slow movement of precession, which also affects all heavenly bodies, Ptolemy accepted the least value of Hipparchus, one degree in a century, only testing it in rather a perfunctory way, which was a great pity, for he might have determined it much more closely after an interval of some 250 years. But one man cannot do everything, and he doubtless thought it best to spend more time on the planets, whose intricacies had baffled Hipparchus and gave him also a great deal of trouble.

He retained the great spheres which were supposed to carry them round Earth, inside the star sphere, but the chief feature of his system is the use of small spheres, which were fixed on the larger, and therefore called "epicycles," while the large were known as "deferents" or carriers. The general principle of epicycles is very simple, as may be seen by comparing the two diagrams.

Fig. 28 shows the path of Mars as we saw it among the stars of Pisces in the year 1909. Throughout July the planet was travelling in its usual direction, "with the signs," but on August 22nd it came to a stop, then turned and travelled backwards "against the signs" until October 26th, when it stopped again, reversed its direction once more, and during the rest of the year moved rapidly forward.

Fig. 29 shows the principle on which Ptolemy would have explained this curious track. Each planet was supposed to be fixed on a small circle, the epicycle, and this was fixed upon a large circle, or deferent, upon which it travels in the direction shown by the arrow, at a uniform speed, returning to the same place in the sidereal period of the planet. Thus Mars, as seen from Earth, which is near C the centre of the deferent, makes a great circle through all the zodiac in two years, Jupiter in twelve, and so on. But meanwhile the epicycle is rotating round its own centre, C1, and when the planet reaches the point marked S, the two motions neutralize one another, so that it appears stationary, as Mars did on August 22, 1909. After this, the motion of the epicycle more than counter-balances the motion of the deferent, and the planet seems to reverse its direction until it reaches the point on the epicycle marked S'. After this the two motions are once more in the same direction, so the planet is seen to move rapidly forward, as Mars did after October 26.

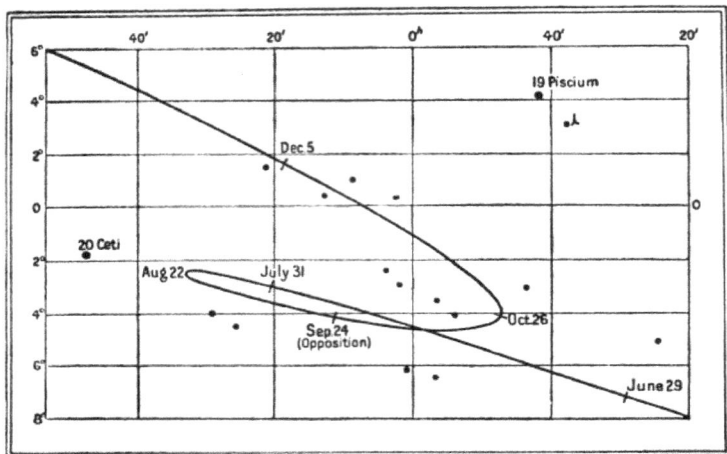

Fig. 28. The Path of Mars among the Stars, 1909.

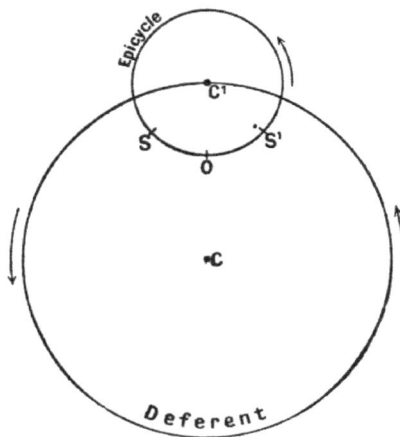

Fig. 29. The Epicycle.

Ptolemy's method of accounting for movements such as those shown in
Fig. 28.

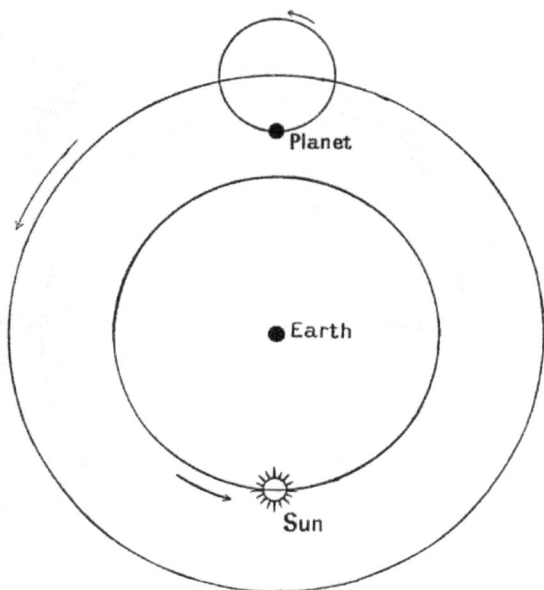

Fig. 30. Planet retrograding, and in opposition (*i.e.* in opposition to the sun.)

In this extremely ingenious way the strange planetary oscillations were accounted for, without violating the law of uniform circular motion, and in a more convenient and satisfactory way than by the concentric spheres of Eudoxus or the moveable eccentrics of Apollonius. Each of the five planets was provided with an epicycle and a deferent, and these were made of the proper relative size and given the right speed, so that the motions should correspond with what we see in the sky.

Ptolemy calls the oscillation a planet's "anomaly with regard to the sun," because (as we have seen when discussing the moveable eccentrics) it was known to be connected in every case with a planet's angular distance from the Sun. On September 24, when Mars was in the middle of his retrograde arc in Pisces, the sun was exactly opposite, in the constellation of Virgo. This is found to be always the case, not only with Mars, but with Saturn and Jupiter too. Whenever one of these planets has the position O on its epicycle, and therefore is retrograding, the sun will be found to be exactly opposite in the sky. Mars comes into this position, and is opposite the sun, once in 780 days; this, therefore, Ptolemy called the period of his epicycle, while a little less than two years was the period of the deferent. The two periods of Saturn are 378 days and 29½ years nearly; of Jupiter 399 days and nearly 12 years.

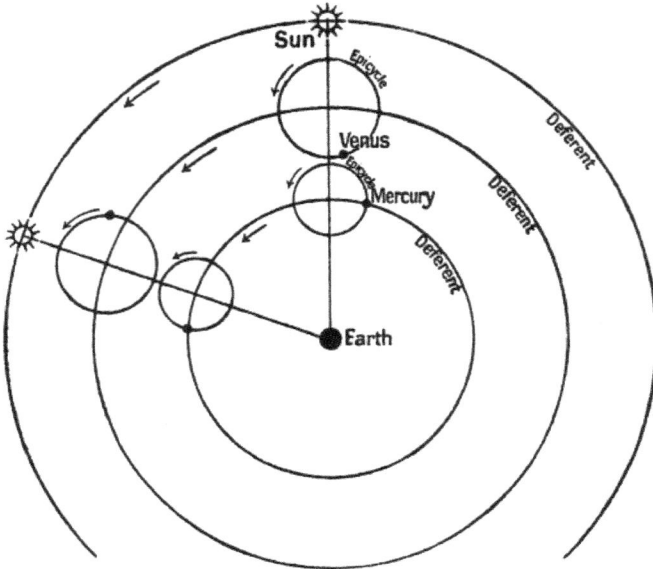

Fig. 31. Venus, Mercury, and the sun.

Venus and Mercury betray their dependence upon the sun in more striking fashion, for since these planets simply oscillated from one side to the other of the sun, their epicycles must be supposed to be keeping pace with him all the way round the zodiac. Fig. 31 shows their relation to one another. On September 25, 1911, Mercury was seen from Earth as a morning star as far west of the sun as it is possible for him to travel, while Venus, after shining as an evening star all the summer, had come into line with the sun and become invisible.[51] On December 7 following, the rotation of the epicycles (Ptolemy would say) had brought both planets to new positions, Mercury now being an evening star at his "greatest elongation east," and Venus a morning star. But the centres of the two epicycles always remain in a line with one another and the sun, and so their periods on the deferents are the same as his, viz. one year. The epicyclic periods, or intervals between two "greatest elongations" west or east, are 116 days for Mercury, 584 days for Venus.

We still explain the complicated course of the planets by resolving it into two approximately circular motions, but we know now that only one belongs to the planet itself, the other is Earth's own motion. The reason why the sun's position affects the position of every planet is simply that the epicycles of Mars, Jupiter, and Saturn, and the deferents of Venus and Mercury are reflections of Earth's yearly journey round the sun.

Ptolemy had by no means finished with the planets when he had provided each one with an epicycle to represent the "anomaly with regard to the sun." Hipparchus had noticed that there was another lesser irregularity, which seemed to be periodical likewise, although no former system had taken it into account; and this he called the "anomaly with regard to the zodiac," because the speed of each planet and the amplitude of its loop varies slightly according to the part of the zodiac it happens to be in. He suggested that this might be dealt with by combining the two theories of epicycles and eccentrics, and this suggestion Ptolemy adopted with success. He placed each deferent with its centre not exactly at the earth, but at a certain small distance which was different for each planet. (This is not shown on our small-scale diagrams, therefore Earth appears at the exact centre.) The true explanation of this irregularity is that each planet's path is not strictly circular, but elliptical.

Besides this, Ptolemy had to represent the planet's movements north and south (note how this varies in fig. 28.). This was partly managed by the aid of small wheels, rotating in such a way that they lifted and lowered the epicycle as required.

Although Ptolemy quotes Babylonian observations of lunar eclipses dating back as far as the eighth century B.C., the oldest planetary observations that he uses were made only four hundred years before his time, and they were probably Greek. Even these were generally very rough. For instance:—

In the 496th year of Nabonassar, on the 17th day of Choeac, in the morning, Mercury was three moon-breadths north of the tail of Capricorn.

In the same year, Phamenoth the 30th, Mercury was three moon-breadths south of the horn of Taurus which is also the foot of the Charioteer.

The first year of Nabonassar (a Babylonian epoch) corresponds with B.C. 747, so the 496th year is B.C. 251. The months used by Ptolemy are usually Egyptian. The later observations, made with an astrolabe, were much more precise; Ptolemy quotes one from Theon of Smyrna, which states that Mercury was 3° 50' in advance of the Heart of the Lion (Regulus), and for his own observations he also usually gives the sign, degree, and minute. Sometimes the planets had been observed so near stars that their positions could be very accurately determined by the aid of Hipparchus' star catalogue. Timocharis, on a certain morning in the 13th year of Ptolemy Philadelphus (B.C. 273), saw Venus beside the last star in the wing of the Virgin (Beta Virginis); Ptolemy himself saw her so close behind a certain star in Aquarius that she seemed to touch it with her rays; and in the 83rd year after the death of Alexander, Jupiter had been observed to eclipse the Southern Ass, that is the southernmost of the pair of stars on either side of the little cluster in Cancer which the ancients called the Manger.

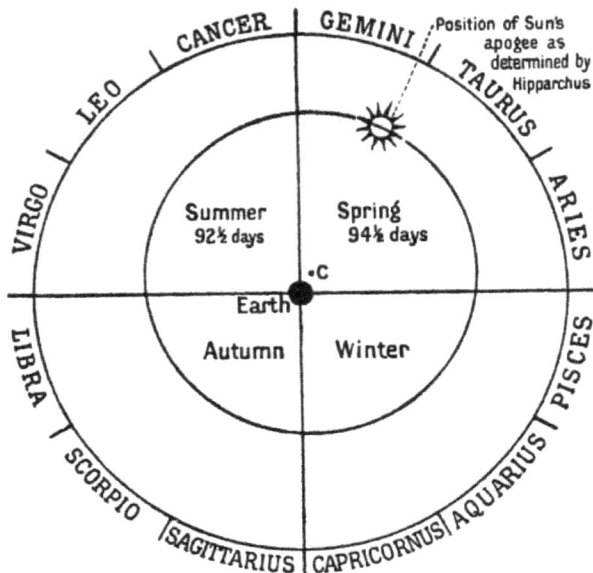

Fig. 32. The sun's Deferent. (Its eccentricity is exaggerated).

The sun was much easier to manage than the planets. Moving always in the same direction, he needed no epicycle, and remaining always in the ecliptic, no wheels. His one irregularity, the varying speed in different parts of the zodiac, which Meton had discovered and Calippus confirmed, Hipparchus accounted for by placing him on eccentric deferent; Ptolemy adopted this without change, and made the year the same length as Hipparchus had done. As the slowest motion was observed when the sun was in the sign of Gemini (at the time of year corresponding with the end of our month of May), the eccentric was placed as in the diagram; and the sun, supposed to be revolving uniformly round the centre C, had a slower motion as seen from Earth when he went north, because he was then more distant.

We know now that it is really Earth which revolves round the sun, but the elliptical shape of her orbit is not unlike the eccentric deferent which the Greeks gave to the sun, and they guessed quite rightly from his motion that he is further from us in one part of the year than in the other. This is easily proved to-day by the fact that he grows apparently smaller, as shown by photographs or measurements taken at different times of year, but the change was not perceptible to the rougher methods of the ancients.

It was a pity that Ptolemy did not take more pains to verify the work of Hipparchus on the sun, for the sun's "apogee," or point of greatest distance from Earth, has an extremely slow motion among the stars which ought to have been quite perceptible in nearly three centuries;[52] but here again he seems to have thought that where Hipparchus had given so much attention he might pass on to something else, and so missed an interesting discovery.

1911, July 1.1912, Jan. 3.

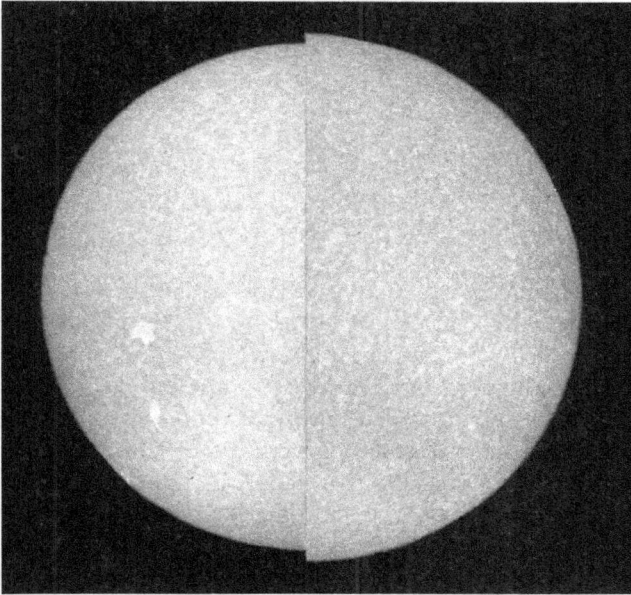

[To face p 146

Fig. 33. Apparent Variation in the Size of the Sun.

Two half photographs of the Sun, taken at Kodaikanal Observatory. The smaller was taken two days before apogee, the larger on the day of perigee.

Instead, he worked very hard at the moon, and added another periodical irregularity to those already known. Perhaps his feelings were somewhat mixed when this happened, his pride and pleasure in his important discovery being counter-balanced by the consciousness that it would still further complicate his lunar theory. Our satellite is acted upon by ourselves as well as by the sun, so that she suffers many perturbations, for Earth, though so small compared with the sun, is comparatively near. Ptolemy's discovery was a difference in her speed at full and new, as compared with her intermediate phases, and this periodic difference is called by modern astronomers the "evection." It was already known that her nodes, or the points at which she crosses the ecliptic, are in constant retrogressive motion, just like the equinoctial points, where the sun crosses the equator; but the moon's crossing points, instead of taking thousands of years to circle the zodiac, run round in about eighteen years. This was discovered early, because observations were chiefly made during eclipses, and at these times the moon is always at a node, that is to say, she is crossing the ecliptic, the sun's path; otherwise the eclipse could not happen. It was also known that she has a varying speed in the zodiac, and that her apogee, where the motion is slowest, instead of being apparently fixed, like that of the sun, also runs round the zodiac, but with a direct motion, and in a period of about nine years.

We need not enter into all the details of Ptolemy's arrangements for the moon, which are exceedingly complicated, but it is interesting to note that he does not explain her varying velocity by an eccentric, as with the sun and the planets. She has an eccentric, but Ptolemy needed it for representing his own discovery, the evection, so he gave her an epicycle, using it in quite a different way from the epicycles of the planets. This epicycle also revolved while moving on the eccentric, but in the opposite direction, and there was so little difference in speed between the two motions that it never brought the moon to a stop, nor reversed her direction, but simply increased and retarded her motion alternately during her monthly revolution. Thus, when the moon was at M, in what Ptolemy called the upper apsis (or arc) of her epicycle, or as we should say in her apogee, the motion on the epicycle was contrary to her motion on the eccentric, and made it seem slower. When the epicycle had travelled halfway round the eccentric, it had also made nearly half a revolution on its own axis: consequently the moon was at M^1, near the lower apsis, or perigee, and the motion on the eccentric seemed to be accelerated.

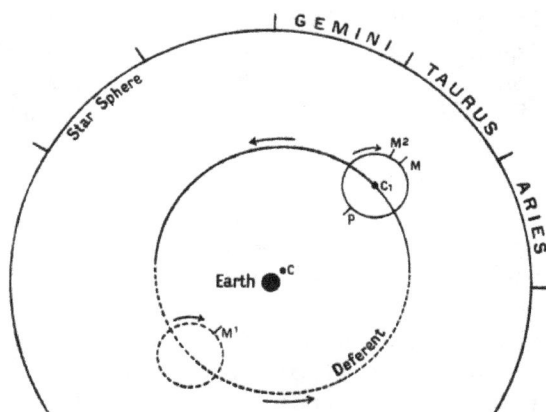

Fig. 34. The moon's epicycle and deferent.

The slight difference in speed between the two motions accounted for the continuous displacement of the apogee in the zodiac, as may be seen from the diagram. For suppose that when in apogee at M the moon is seen from Earth among the stars in the middle of Taurus. At the return of the epicycle to this place next month, she has not yet quite completed a revolution on her epicycle but is at M^2, and will not be in apogee until the centre of her epicycle is advanced 3° further in Taurus. After some time (five months), apogee does not occur until the moon is in Gemini, and it will be nine years before it occurs again in Taurus.

These are the leading features of the system by which Ptolemy represented the motions of the planets, the sun, and the moon. He is uncomfortably conscious that it may strike us as very complicated, and in his last book he makes a kind of apology. We must remember, he says, that we are not dealing with earthly machines which jar and wear, but with celestial bodies which have no weight, cause no friction, and are eternally the same. Delambre somewhat cruelly retorts that he need not have wasted time in writing such rubbish, but have been content to put his results into tables.

But what true astronomer could be content with tables and nothing more? He must try to understand their significance. Nearing the end of his great work, which had cost so much labour, and was so brilliant a success within its limits, Ptolemy allows us to see in this little paragraph that he felt he had but touched the hem of Nature's veil, and longed in vain to lift it. The circles which he manipulated so skilfully were only mathematical abstractions to him:[53] what was the reality behind? What were the stars? whence came their unwearied strength, their eternal calm? The power of

Egypt, of Assyria, and of Persia had declined, Greece was now laid low, and it was the day of Rome, but still Venus pursued her ancient path among the little stars of Aquarius, and when all faded in the solemn dawn, the sun arose in his ancient majesty. Then Ptolemy looked at his circles and triangles, and felt how inadequate they were; yet it was the nearest approach he could make to truth.

They did indeed represent very beautifully the celestial motions, and also in a general way the variations in brightness or diameter, but their unreality is betrayed by the fact that the latter was often grossly exaggerated, as for instance with the moon, whose epicycle had to be so large, in order to represent her motions, that at perigee she ought to appear twice as large in diameter as at apogee! Ptolemy can hardly have failed to notice this, though he does not mention it. With regard to the planets, he is careful to tell us that he knows no way of finding their real distances: the ratios of their epicycles to their deferents he had carefully computed in each case, but to estimate the diameters in stadia was utterly beyond his powers. He believed, however, that all were very much nearer than the stars, and more distant than the moon, and he found universal agreement among astronomers in placing Saturn, Jupiter, and Mars, beyond the sun, in order of the periods of their deferents, which were all longer than a year. Venus and Mercury, however, had periods the same as the sun: on which side of him, then, must they be placed? Some modern authors, Ptolemy says, thought they were beyond the sun, but he agrees with the most ancient, and places them between sun and moon. The general disposition of the heavenly bodies according to Ptolemy is shown in Fig. 35, with the periods of epicycles and deferents, and the directions of the several movements, but epicycles and spaces between deferents are all made equal. It will be noted that the sun has no epicycle and that the moon turns on hers in a reverse direction, that the centres of those of Mercury and Venus are in a line with the sun, while the lines joining Mars, Jupiter, and Saturn to the centres of theirs are in each case parallel with the line joining Earth and Sun. Beyond all the wandering stars is the sphere of the fixed stars, moving in its vast period of 36,000 years; and the whole system is carried round Earth in one great revolution of a day and a night.

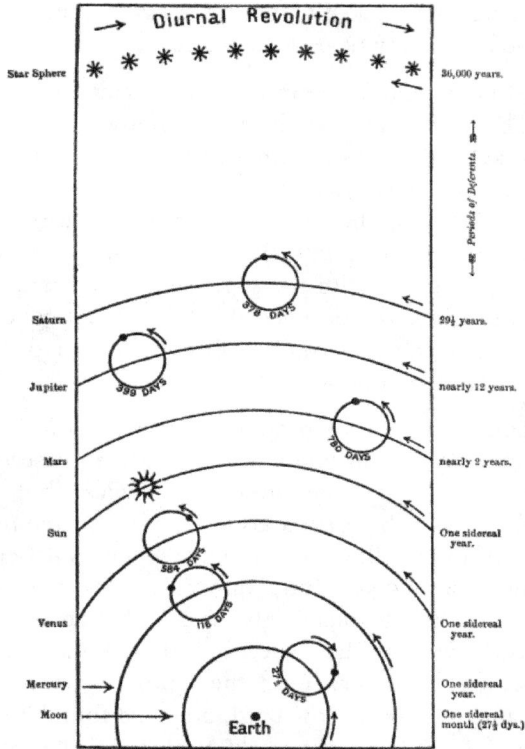

Fig. 35. The Ptolemaic system.

The arcs of circles are portions of the deferents which carry round the smaller circles (the epicycles) in the periods named at the side.

And here we have to note one astonishing fact. Although the planets were all beyond reach of measurement, it was not so with the moon, which the Greeks rightly recognized as our nearest neighbour. They had actually achieved the long-desired feat of measuring her real size and distance.

The fundamental principle on which they worked is easily explained. If we look out of a window at a tree in the garden, it appears against a background of some distant scenery; if we alter our position by walking to another window, the tree changes its place with regard to the relatively motionless background, and its apparent change of place bears a definite proportion to its distance from us. A tree near the house changes its position greatly, a tree at the far end of the garden much less. From careful measurements of the angle through which the tree appears to have moved and the distance we have walked between the two windows, the distance of the tree from the house may be easily deduced.

The tree in the garden is the moon, the distant landscape is the star-spangled sky. The space between the two windows must be increased to thousands of miles, but the astronomer need not walk it. If he makes his first observation when the moon has just risen, carefully measuring her position among the stars, the revolving Earth will carry him to a new position in a few hours, when with the moon high in the sky he can once more compare her position with the same stars, and from the change he then finds he can deduce her distance. The Greeks thought it was the sky, and not the earth which moved, but this makes no difference, as the question is one of relative motion only.

The problem, however, when applied to the moon, is a complicated one, implying not only skill in trigonometry and the possession of suitable instruments for measuring the necessary angles, but also accurate knowledge of the size of the earth and the motions of the moon, since her progress eastward among the stars during the interval between the two observations must be allowed for. Hipparchus and the astronomers of Alexandria were the first to qualify themselves for attacking this difficult problem, and the proof of their success is that Ptolemy's value for the distance of the moon is very near the truth as obtained by modern methods. The same method is now used, only it is found better to do what was not possible for the Greeks, namely to compare observations made at places several thousands of miles distant, for instance Greenwich and the Cape, instead of allowing the same place to be moved by the earth's rotation.

When the distance of the moon had been thus discovered, it was a very simple matter to find her real size from her apparent size.

The distances of all other heavenly bodies are too great to be determined by naked eye methods, for the displacement (technically called "parallax") is so small that it is quite invisible without a telescope. Hipparchus had tried repeatedly to measure the distance of the sun, but recognized that neither the method of Aristarchus nor any other was really conclusive. It was, however, the best attempt that had been made, and Ptolemy assumed that he knew the distance to be about twenty times that of the moon, so he gave the sizes and distances of these two bodies as follows, taking Earth's semi-diameter as unit:—

	According to Ptolemy.		Modern Values.	
	Semi-diameter.	*Distance.*	*Semi-diameter.*	*Distance.*
Moon	$^5/_{17}$ (=0·290)	59	0·273	60·3
Sun	5½	1210	109·4	23,439

In Books VII and VIII of the *Almagest* Ptolemy describes the 48 ancient constellations and the course of the Milky Way among them. The position of each star is noted as it appears in its constellation-figure, but the celestial latitude and longitude is also given, and this great catalogue is evidently taken from Hipparchus.

PTOLEMY'S FORTY-EIGHT CONSTELLATIONS.[54]

Ursa Minor	Pegasus
Ursa Major	Andromeda
Draco	Triangulum
Cepheus	Cetus
Auriga	Orion
Corona Borealis	The River (Eridanus)
The Kneeler (Hercules)	Lepus
Lyra	The Dog (Canis Major)
The Bird (Cygnus)	Canis Minor (Procyon)
Cassiopeia	Argo
Perseus	Hydra
Böotes	Crater
Ophiuchus	Corvus
Serpens	Centaur
Sagitta	The Wild Beast (Lupus)
Aquila	Ara
Delphinus	Corona Australis
Equuleus	Piscis Australis,

Aries, Taurus, Gemini, Cancer, Leo, Virgo, Libra,
Scorpio, Sagittarius, Capricornus, Aquarius, Pisces.

Ptolemy sometimes remarks on the colour of the brighter stars, and always mentions the brightness, or "magnitude" as we call it now, for the classification of Ptolemy (or Hipparchus?) was found convenient and accurate enough to be retained by modern astronomers, and the same system is now continued with the faint telescopic stars.

The stars ranked as "first-magnitude," or brightest of all, are fifteen in number, and as these are evidently the "quindici stelle"[55] alluded to by Dante in *Par.* xiii. 4, it will be interesting to give a list of them here.

PTOLEMY'S FIRST-MAGNITUDE STARS.

	Name and Description	*Modern Name and Meaning. in the Catalogue.*
1.	Arctouros, fire-coloured	Arcturus (the Bear-Watcher).
2.	The brightest star in the Lyre	Vega, in Lyra. (Falling Eagle, Arabic).
3.	Aichs (the Goat)	Capella (Little Goat, Latin).
4.	The brightest star of the Hyades, fire-coloured	Aldebaran. (The follower, because following the Pleiades, Arabic).
5.	Basiliskos (the royal), the heart (of Leo).	Regulus (Latin equivalent).
6.	The tip of the tail (of Leo)	Denebola (Lion's Tail, Arabic).
7.	Stachys (Ear-of-Corn)	Spica (Latin equivalent).
8.	The last of the Water and Mouth of the Southern Fish.	Fomalhaut. (Mouth of the Fish, Arabic).
9.	The fire-coloured bright star on the shoulder of Orion.	Betelgueux. (Shoulder of Giant, Arabic).
10.	On the left foot of Orion, common to the Water.	Rigel. (Foot of the Giant, Arabic).
11.	The last of the River	Achernar (Arabic equivalent).
12.	The very brilliant star, fire-coloured, at the mouth of the	Sirius (Greek *Seirios*, or *Sothis*, from the Egyptian

	Name and Description	*Modern Name and Meaning. in the Catalogue.*
	Dog, called the Dog.	*Sept*), in Canis Major.[56]
13.	Prokuon (the preceding Dog), on the thigh of Prokuon	Procyon (in Canis Minor), which rises before Canis Major.
14.	Canopos, on the rudder (of Argo)	Canopus (an Egyptian god, and a town on the Nile delta).
15.	The tip of the right forefoot (of the Centaur).	Alpha Centauri.

Altair and Antares were counted as second magnitude, though we now class them among the first.

It will be seen from the above that only two of these fifteen brightest stars of Ptolemy's still bear their Greek names, Arcturus, and Procyon; but most of the other modern names are direct translations, either into Latin or Arabic, of Ptolemy's description or name for the star, while Vega and Aldebaran have preserved their original Arabic names (much corrupted), and Canopus and Sirius are derived from the Egyptian. Regulus-Basiliskos apparently comes from Babylon, for the name of this star on tablets of the second century B.C. was Shar-ru, which means "royal."

Ptolemy's "Last of the River" was taken by his Arab and other commentators to be the first-magnitude star which consequently they named Achernar, but this was too far south to be visible at Alexandria, though possibly he heard of it or saw it himself at Syene, where it rose over the horizon in his times. Delambre thinks it is the same as Fomalhaut, which already belongs to two other constellations in the star catalogue; Brown suggests it was Theta Eridani, which may have been brighter sixteen centuries ago. Perhaps it was the star we now call Alpha in the Phoenix, for this lies between the last bend of the River and the Water of Aquarius, and its magnitude is between first and second. The stars of the Southern Cross are included by Ptolemy in the stars about the hind feet of the Centaur, though it is difficult to identify each one with certainty: the positions are not very accurately given.

Ptolemy wrote also on Astrology, or as it was then called Judicial Astronomy, and his work on the subject, which is in four books, received the name of the *Tetrabiblos*. It seems that astrology was not very much believed in by the Greeks, for he protests that the influences of the stars are real, and can be known and predicted, but allows that this part of astronomy is much more difficult than the mathematical part, with which he had dealt in the *Syntax*, and that it has not yet been perfected, and therefore it is sometimes slandered as untrue. Astrologers sometimes make mistakes, like doctors, yet both astrology and medicine are useful arts. He mentions especially the Egyptians as practising it.

Only second in importance to the *Almagest*, and even better known in old days, was Ptolemy's work on Geography. In this he shows how the size of the earth and the latitudes and longitudes of places on Earth, can be discovered by observations of the heavens. For Earth's diameter he adopts the value found by Poseidonius. His terrestrial globe had a moveable half-circle attached to the poles, which was divided into 90° from the equator in both directions, so that by placing this against any spot on the globe, its latitude north or south of the equator could be immediately read off. Only half the equator was marked on this globe, and divided into 180°, for the known part of the earth all lay within these limits. The parallels of latitude marked were not the same as we use, though the most northerly fell very close to the Arctic Circle: it showed the latitude of the Island of Thule, far away "towards the Bears," as Ptolemy expresses it, meaning that it lay under the constellations near the North Pole. The most southerly was not far below the equator, and others were marked between these which divided the known Earth into "climates," according to the height of the Pole and the length of the longest day in each. Among the many towns which figured on this globe we may mention Alexandria, Rome, Athens, Jerusalem, Florence, Cadiz, Paris, Strasburg, London, Bath. Meridians were marked at every 5 degrees. From this globe Ptolemy shows how a plane map may be constructed. He took Rhodes as the central meridian, because it had a nearly central position among the "climates," and was the place in which Hipparchus had observed.

Ptolemy also wrote books on Optics, on the theory and construction of Dials, and on Music!

V

RETROSPECT.

The *Almagest* is the last word of Greek astronomy. We have seen how from the dawn of Greek history, as far as we can trace it back in literature, the Greeks were familiar with the skies, and how the most ancient of their philosophers had tried, chiefly by abstract reasoning, to discover the cause of the circling motions—never changing, never ceasing—of sun and stars. When the seemingly unruly movements of the planets were known, they still felt sure that there must be some underlying principle which would bring all into harmony. Inspired by Plato, and helped by the Egyptians, Eudoxus took the first step towards finding this principle by careful study of the planetary motions, and at last, after many generations of observers and mathematicians, Ptolemy was able to describe these motions as accurately as was possible with the methods available.

At first the heavenly bodies had been thought of as gods, then as worlds like our own, then as spheres of ethereal fire. They were swept round by a mighty wind, they ran on wheels, they floated in the ether, they were set in crystal spheres; the controlling force was the principle of number or harmony, an all-pervading World-Soul, a host of immaterial intellectual Beings subject to one eternal First Mover. It was also suggested that the greater part of the motions were apparent only, that Earth was really in motion, spinning on her axis, revolving round a Central Fire, or revolving round the Sun. But these ideas had not enough evidence to support them when suggested. Had a great imaginative thinker, a Pythagoras or an Aristarchus, arisen after Ptolemy, he could have shown in detail how by assuming these two motions the phenomena could be more simply accounted for, and he could have made out a very good case for their probability.[57] But the time was past then for such bold originality, and the explanation of Aristotle was universally adopted. He, as we saw, placed the abode of the gods, the rulers of the universe, beyond the outermost sphere, and found the principle for which all were seeking, which should be the key to all celestial motions, in the law of circular motion.

The Greeks had at first thought that the earth was a disc, under a tent-like sky; then it was a cylinder, with the sky in tiers above; then a huge hemisphere filling half the universe. But they discovered that it was a sphere, surrounded on every side by the heavens; and they found its true size, and that the portion of it which was known to them was less than a quarter of its whole extent. They explained the marvel of Earth's remaining unsupported in space by the known facts of gravity, arguing that the falling of every particle of earth towards Earth's centre proved that it was also the centre of the Universe, and that every heavy thing tended thither by a law of nature.

They also knew the size and the distance of the moon; they realized that the planets were all immeasurably remote, and the stars vastly more distant still. They believed that there were great intervals between the planets, corresponding to the differences in their periods of revolution, but the stars were always thought of as set in a great sphere, and therefore all at the same distance from Earth at its centre. The relative positions of the stars on this great sphere, and times of rising and setting of many of them had been known in a rough way for many ages, through the familiar appearance of the constellations, named we know not by whom; but Hipparchus measured their positions in degrees, and was able to foretell where any star would be in the sky for any place or time. He also discovered the Precession of the Equinoxes, and this was thought to prove a slow rotation of the star sphere.

Astrology was learned by the Greeks in Egypt and the East, but was never practised with the enthusiasm shown by their teachers.

As regards the division of time, the Greeks adopted from the Babylonians the twenty-four hour day which we still use, and their months began with the young moon. Their year therefore had to contain a whole number of months, and it had sometimes twelve, and sometimes thirteen, as with the ancient Babylonians, but the system by which this was arranged was quite different, as they did not depend upon observations of the stars, but counted the number of days between equinox and equinox by means of gnomons. Great difficulties were found in trying to reconcile the lunar and solar periods. The ancient 12-month year, used in the time of Hesiod, was found to be much too short, and alternate years of 12 and 13 months made the period too long; then an eight-year cycle was invented, but this had to be constantly corrected, which led to great confusion. As we have seen how carefully Meton and Euctemon had determined the length of the tropical year, even before the days of Eudoxus, surprise may be felt that a calendar year was not fixed to correspond accurately with the movements of the sun, ignoring the irreconcileable movements of the moon, but it is difficult for us to realize in these days how wrong and strange it seemed if a new moon occurred in the middle or at the end of a month, instead of at the beginning. In Aristophanes' play of *The Clouds*, which was acted in B.C. 423, the moon was said to grumble because men would not keep the months as she showed them:—

"Yet you will not mark your days As she bids you, but confuse them, jumbling them all sorts of ways, And she says the gods in chorus shower reproaches on her head, When in bitter disappointment they go supperless to bed, Not obtaining festal banquets duly on the festal day."

Then Meton made his celebrated discovery that nineteen tropical years correspond almost exactly with 235 synodic months (the difference is in fact only a few hours), and a cycle of 19 years was arranged, which was adopted by all Greek states and dependencies. Some of the years had twelve and some thirteen months, and some of the months 29 and others 30 days, but all followed in a regular order, and when one cycle was completed another was begun. The total number of days in each cycle was 6940, and as this is only 9½ hours longer than 19 true tropical years, it follows that the average year in the Metonic cycle was only half an hour longer than it should have been.[58] The average month was not quite two minutes longer than the true synodic month.

An improvement even on Meton's cycle was made by Calippus, who proposed to correct its too great length by quadrupling the period, and then deducting one day from the whole. This would have given a cycle of 76 calendar years, in which the average year was 365¼ days, or only 11¼ minutes too long. It does not seem, however, to have been ever brought into actual use as a calendar, but Ptolemy often refers to the Calippic epoch as a date from which to calculate celestial phenomena.

Theon of Alexandria *c.* 380 A.D.

After Ptolemy, the Alexandrian school of astronomy produced only copyists and commentators, the last of whom was Theon, who saw his daughter Hypatia murdered and the library burned by fanatical mobs. It only remains for us to see how the Greek system of astronomy, brought to so great perfection in the *Almagest*, was neglected for many centuries, and by whom it was at length rediscovered and made to live again.

VI.

ASTRONOMY UNDER IMPERIAL AND CHRISTIAN ROME.

B.C. 46 TO A.D. 1000.

Vos O clarissima mundi Lumina labentem cœlo quæ ducitis annum. Georgic I.

No new school of astronomy arose under the Roman Empire, nor do we even know of one Roman who devoted his life to the science. The genius of this people lay in other directions, and Dante truly says:—

"'Nature ordained in the world a place and a people for universal command ... to wit, Rome and her citizens or people. The which our poet too has touched upon right subtly in the sixth [Virgil, in the sixth book of the Æneid], introducing Anchises admonishing Æneas, the father of the Romans, thus:—

"Others shall beat out the breathing bronze more softly, I do well believe it! And shall draw the living features from the marble; shall plead causes better, and trace with the rod the movements of the sky, and tell of the rising stars. Roman! do thou be mindful how to sway the peoples with command. These be thy arts: to lay upon them the custom of peace, to spare the subject and fight down the proud.'"[59]

Yet there were some enthusiastic amateur astronomers in Rome, for Cicero tells us of one who had felt old age to be no burden because he was so eager over his astronomical studies, sitting up sometimes all night to finish his calculations, and delighted when an eclipse he had foretold came to pass.[60] This ardent amateur, Sulpicius Gallus by name, had found such knowledge of practical value in his younger days, when he was with the legions in Macedonia, for he had been able to persuade the troops not to be alarmed by an eclipse of the moon which was about to happen, explaining how it was due to natural causes; and while the soldiers in the opposite camp were shrieking and moaning, believing that the eclipse portended the death of their king, the Roman soldiers remained quite calm. This was on the eve of the battle of Pydna, in B.C. 168, a little before the time of Hipparchus.

Ovid B.C. 43-A.D. 17.

Virgil B.C. 70-19.

Manilius *c.* A.D. 10.

Cicero B.C. 106-43.

In Cicero's own life-time, and throughout the early days of the Empire, it was the fashion to have at least a smattering of Greek astronomy. Ovid tells legends of the constellations; Virgil, at his farm where he lingered happily among his vines, his cattle, and his bees, studied the varying aspects of the constellations in connection with the seasons and the weather; Manilius wrote a long poem in five books on astronomy and astrology; Cicero himself was quite learned in the subject, and made a translation of Aratus, which had a great vogue.

Very popular also, both in classical and mediæval times, was his *Dream of Scipio*, which was an imitation of Plato's fable of the vision of Er, in the *Republic*. The moral is that earthly fame is valueless, since Earth itself is insignificant compared with the starry heavens, but that those who practice virtue for its own sake shall return to the stars whence their souls originally came. The youthful Scipio is transported to the skies in a dream, and meets the souls of his father and of the elder Scipio Africanus in "a radiant circle of dazzling whiteness, which you have learned from the Greeks to call the Milky Way." There he sees "stars which we never saw from this place, and their magnitudes were such as we never imagined, the smallest of which was that which, placed upon the extremity of the heavens, but nearest to the earth, shone with borrowed light." But the shining globes of the stars are so great that Earth seems to have contracted to a point, and Scipio gazes at it, grieved.

"How long will your gaze be fixed on Earth?" cries Africanus. "Do you not see into what temples you have entered?" and he points out nine spheres which compose the whole universe. The outermost, in which the stars are fixed, is most divine, and within this are seven, one of which contains the planet called Saturn upon earth, the next the glorious Jupiter, friendly and helpful to mankind, then Mars, ruddy and terrible, and the next place in the middle region is held by the sun, the leader, prince, and governor of all other luminaries, the soul of the world, filling all things with his light. Venus and Mercury follow him in their courses, like attendants, and in the lowest sphere rolls the moon, kindled by his rays. Below this, all is mortal and transitory, except the souls given to the human race by the grace of the gods; above the moon all is eternal. Earth, which is at the centre and forms the ninth sphere, is immoveable and below all the rest; and all weights, by their natural gravitation, fall towards her.

Scipio then asks what is the sound which fills his ears, so loud and yet so sweet? and he is told that he hears the music of the spheres, which is too great for mortal ears, just as the sun is too bright for human eyes to look upon. Yet those who make music upon Earth, with strings or voice, like all others who follow heavenly pursuits, are opening for themselves a path by which to return to the stars, the true home of the soul.

Strabo born *c.* B.C. 63.

Seneca B.C. 3-A.D. 65.

Pliny *c.* A.D. 23-79.

Proclus died *c.* 480 A.D.

Martianus Capella, 5th century A.D.

Simplicius 5th century A.D.

Other famous Latin authors who wrote on astronomy were Strabo, Seneca, and Pliny, who quote Eudoxus and Aristotle, Poseidonius, and Hipparchus. When Ptolemy's work was done there was no great writer to popularize it. Three hundred years later Proclus writes a commentary on it, and Martianus Capella mentions it, but Simplicius, in his commentary on Aristotle's *De Cœlo*, though he speaks of "the admirable Ptolemy," is evidently unacquainted with his work.

No Roman added anything new to astronomy, and the most precious parts of their writings for the history of astronomy are some fragmentary notes of early Greek astronomers whose original works are lost.

The practical use of astronomy for measuring time appealed, however, to the Roman people. The most ancient Roman year (said to have been introduced by Romulus) had only ten months, March being the first, which explains why the ninth to the twelfth of our present months have the names of September, October, November and December, as if they came in the order of seventh to tenth. Two more months were added later, and at some unknown date the old Octennial or 8-year Cycle was adopted from the Greeks. This involved the use of intercalary months of varying length, and the priests were entrusted with the business of arranging them. But the priests thought it much more important that the length of the year should suit their convenience than that it should conform to the celestial movements, so they made it long or short according as they approved or not of the persons holding office at the time; and by the time of Julius Cæsar the calendar had fallen into such confusion that March 25, which was supposed to be the date of the spring equinox, came in the middle of winter!

Julius Cæsar and
Sosigenes B.C. 46.

A drastic reform was necessary. Julius Cæsar called in the Alexandrian astronomer Sosigenes, and gave the Empire the calendar which, with the exception of one small reform made since, we still use. The moon was thrown over altogether: there were to be no more intercalary months, and every year was to be exactly like the last, except for the addition of one day in every fourth year, so as to make the average year equal to 365¼ days. Since the tropical solar year is only 11 minutes, 14 seconds shorter than this, many centuries would elapse before the months of this calendar would depart from their proper seasons. March 25 was restored to the time of the vernal equinox, but the first day of the new year was to be January 1, and Julius Cæsar gave his own name to the old fifth month of Quintilis. Each month was to be alternately of 31 and 30 days, except February, which would only have its full complement of 30 days in the fourth year (Leap Year), and 29 every other year. This reform was regarded by some as an unwarrantable interference by a despot. Cicero, when some one mentioned that the constellation Lyra would rise at a certain hour, answered bitterly, "Yes, if the edict allows it!"

Unfortunately, the simplicity of the scheme was a good deal spoiled by the folly of Augustus, who could not bear that the month of his predecessor should have 31 days while the next, the old Sextilis, which he turned into August, named after himself, should have only 30. So he made two months of 31 days come together, and took away a day from February. Afterwards, Nero gave his name to April, and Domitian his to October, but this was more than a long-suffering world could bear, and the new names were gladly forgotten as soon as the tyrants were dead.

Although, therefore, Rome was obliged to apply to Alexandria, that is to Greek astronomy, to carry out the project, it was Rome that gave us the most correct and convenient calendar which exists, better in both respects than that which had been used in Greece itself.[61]

But alas! the celestial science, a willing servant as time-measurer for the daily uses of humanity, a docile captive to adorn the triumph of literature, was doomed to a baser servitude. It was in the early days of the Empire, and all through the Middle Ages, that the pseudo-science of Astrology was pursued with passion, and men spent their lives in studying the paths of planets and positions of the stars, urged solely by the delusive hope of being able therefrom to read the book of fate and guide the lives of their superstitious clients. From Chaldea, its ancient home, came the most famous astrologers, but their art was soon learned in every country of Europe, and its professors were sought after by peasants and kings—now reviled and banished as impious and leagued with devils, now loaded with honours and rewards, revered, hated, welcomed, forbidden, but always believed in. Although when Christianity was established the seven planets could no longer be regarded as great gods ruling over the lesser gods of the stars, they were still thought to be mighty revealers of fate. Each had his special attributes and influence over man: the fiery colour of Mars no doubt suggested the warlike and hostile spirit ascribed from time immemorial to this planet-god; the slow motion of Saturn in his distant sphere gave an impression of a mournful morose being, the "frigida Saturni stella" of Virgil; Venus was the planet of love; the sun, of honour and power, and so forth. Each planet was also mysteriously connected with a colour, and, the alchemists said, with a metal; the sun with gold, the moon with silver, Saturn with pale heavy lead, etc. Each also influenced a special part of the body: thus, if Mercury were unfavourably placed at the moment of a child's birth it would be liable to suffer from lung-disease; the moon's position affected the brain; the sun's the heart, etc.

As the planets had distinct and often contrary influences in different positions, it was necessary, in determining a man's fate, to consider the aspect of the whole heavens, especially at the moment of his birth, but horoscopes were also cast for any period in his career, past, present, or to come, enabling him to guard against threatening evils or bad tendencies, and to seize favourable opportunities. The method was as follows: the sky-sphere, as it appeared at the given time and place, was divided into twelve "houses," by drawing meridians (called "circles of position") 30° apart. The house just about to rise on the eastern horizon was called the "ascendant," and was the first and most important, planets situated there having more power than anywhere else; but each house had its special significance, the second (just above the eastern horizon) being the House of Riches, the seventh of Marriage, the twelfth of Enemies, etc. The kind and the strength of each planet's influence depended mainly upon the house in which it happened to be, and was strongest when the planet was in its own house and also in its favourite zodiacal sign. The sun was considered to be most at home in Leo, the moon in Cancer, and each of the other five planets owned two of the remaining signs. Another important point was the "aspect" of the planets with regard to one another, that is, their angular distance apart on the sky-sphere. If Mars and Jupiter, for instance, were in "opposition," *i.e.* 180° apart, the portent was unfavourable, but in "trine" or "sextile" aspect (120° or 60° apart), favourable.

It is evident that for casting horoscopes it was necessary to be able to calculate for any given time or place the positions of the heavenly bodies; and for this the skies must be watched, and the movements known of the stars, of sun and moon, and of planets. To this extent, therefore, an interest in genuine astronomy was kept alive; but on the other hand, the system fostered belief in the overwhelming importance of Earth in the Universe, and the existence of the heavenly bodies for the sole purpose of ruling and foretelling human destinies: no one cared to inquire what were the underlying laws, and what the real nature, of the heavenly phenomena.

Closely allied to this superstitious belief in planetary influences, was the dread of comets, meteors, and eclipses, which were everywhere regarded as omens. It was in vain that Seneca urged the greater importance of investigating the nature of the heavenly bodies about which something was already known, and of trying to solve the problem worthy of highest consideration, viz. whether the earth, as some had asserted, was turning rapidly, or was stationary in a turning World. His very protest, as well as his lengthy dissertation on comets, shows how far less interesting, alike to philosophers and public, were the ordered courses of stars and planets, than the startling apparition of such rare objects as that great "hairy star" which, appearing suddenly during the games instituted by Augustus after the assassination of Julius Cæsar, was thought to be the soul of the dead Emperor. Augustus erected a temple in its honour.

Rome, considered as the metropolis of the dominant temporal power, failed to encourage astronomy; Rome as the centre of the spiritual power directly discouraged it.

Cosmas c. 540 A.D.

Augustine 354-430 A.D.

In the fourth and fifth centuries after Christ "the old heathen theory" that Earth is a sphere was opposed by some of the Fathers, as inconsistent with certain expressions in the Bible; and in the sixth an Egyptian monk, Cosmas Indicopleustes, formulated a scheme of the Universe according to which the Jewish Tabernacle was a type and pattern of the World. The earth is the flat oblong floor, surrounded by four seas; these are enclosed by four massive walls which support a roof (the firmament or sky), and above this live the angels, who move the sun, moon, and stars across the firmament, and let down rain through its window. This childish cosmogony, supposed to be in entire accordance with Genesis, Isaiah, and the Psalms, bears a curious similarity to one of the oldest "heathen theories" of all, born in the land where Cosmas lived.[62] Saint Augustine, however, author of the paralysing doctrine: "Nothing is to be accepted save on the authority of Scripture," does not seem to have considered belief in a spherical Earth forbidden, but it was to him a matter of perfect indifference, while he upheld as an article of faith that in no case could any antipodean inhabitants exist. For they could not be descendants of Adam, nor ever hear the Gospel, since everyone knew the torrid zone to be an impossible barrier between north and south. Another favourite Church doctrine was that Jerusalem was the centre of the earth: this idea, which it will be remembered has a place in Dante's cosmogony, was based on the words in Ezekiel:[63] "This is Jerusalem: I have set it in the midst of the nations and countries that are round about her." Good bishop Arculf, and other pilgrims, were shown a pillar "on the north side of the holy places, and in the middle of the city," which marked the exact spot, and were told in proof of the assertion that at midday at the summer solstice this pillar cast no shadow! How this proved its central position is a mystery, and if true the pillar must have been deplorably crooked, for the sun can never pass overhead in Jerusalem, in a latitude of nearly 32° north.

The Church, like the State, saw that astronomy had one use, and applied, like the State, to Greek astronomy for a calendar. It was necessary that the ecclesiastical calendar should be luni-solar, since Easter, which corresponds with the Jewish passover, must fall on the Sunday following the first full moon after the Vernal Equinox. By 325 A.D., when the Council of Nice was held, at which this question was settled, the Vernal Equinox fell on March 21 instead of March 25, owing to the neglected eleven minutes in the Julian year. March 21 was therefore adopted by the Church as the date of the equinox, which was assumed to remain constant; and the old luni-solar cycle of Meton was used, and still is used in all churches which celebrate Easter, as a basis for the ecclesiastical calendar.

The custom of reckoning years forwards and backwards from the birth of Christ was first introduced by a Roman abbot, Dionysius Exiguus, in the sixth century, but it did not become general in Christian countries until the ninth century. Dionysius adopted as the first day of the epoch, not January 1, but March 25, the old Roman date of the vernal equinox. This was because it was Annunciation Day, and it was a belief of the Middle Ages that the Annunciation and also the Crucifixion actually took place on this day, and also that the Creation began on the same date.

| Martianus Capella 5th cent. A.D. |
| Cassiodorus *c.* 530 A.D. |
| Boëthius died 525. |

At the break-up of the Roman Empire some fragments of classical learning were saved from the wreck, mainly in the text-books of the "heathen" writers, Capella, Cassiodorus, and Boëthius. These were preserved by the Church, now the only repository of learning. The secular instruction given to churchmen included astronomy, for while the "Trivium" comprised the three elementary sciences of Grammar, Rhetoric, and Dialectic, the "Quadrivium" comprised the four advanced sciences of Arithmetic, Astronomy, Geometry, and Music. But only a mere smattering of the "Quadrivium" was taught in this period, and scarcely more of astronomy than was necessary for determining the date of Easter. The intimate knowledge and ingenious theories of the Greeks concerning the celestial motions interested no one any more.

| Charlemagne 732-814. |
| Isidore died 636. |
| Bede c. 673-735 |
| Fergil, 745. |
| Dicuil, 825 |

From the seventh century, however, the ignorance began to be less dense. Charlemagne established many schools, and there were enlightened monks here and there—Saint Isidore of Seville, the Venerable Bede, and Irish scholars like Fergil and Dicuil—whose teachings show that the elements of astronomy as taught by the Greeks were not totally forgotten everywhere. From the beginning of the ninth century all famous monasteries had schools for laymen as well as for monks.

In Italy the darkness was never quite so deep as in northern Europe, for traditions of classical culture never quite died out. And although throughout the long period from Ptolemy to the end of the tenth century, astronomy had been almost completely neglected in Europe, the way was slowly being made plain for a great revival. The ideal Empire, governing the whole world from the Eternal City, and the ideal Church making all men brothers, though neither has ever existed in fact, did so in men's minds, as we see clearly in Dante, and both exercised a powerful influence. The Roman Empire and the Roman Church did impress a kind of unity on Europe as it grew: there was one civilization, one religion, one language in which new thoughts could be conveyed to all. Thus the ground was prepared, and whenever a new school of astronomy should arise or be imported, it would not remain the property of one nation surrounded by barbarians, but might at once be shared in and advanced by the whole of Europe.

VII.
ARAB ASTRONOMY.

A.D. 750 TO 1250.

Once more the scene changes. While European science is at low ebb, if we look to the banks of the Tigris, not many miles from the ruins of ancient Babylon, we find the centre of a new and famous school of astronomy.

Al Mansur 753-775 A.D.

From the deserts of Arabia an immense empire had arisen, which in less than a century spread eastwards as far as India, and westwards to Morocco and Spain. Its first capital was Damascus, but in 755, after the defeat of the Omeyyad dynasty, the new Caliph fixed his capital at Baghdad, the wondrous city of the Thousand and one Nights. This Caliph was the renowned Al Mansur. To his court there came one day a scholar from India, who was skilled in the knowledge of the stars, and he laid before the Caliph a book which treated of things celestial and showed how to foretell eclipses. Al Mansur was profoundly interested, and ordered a translation of the book to be made into Arabic. The astronomical system of the Hindus was at that time very similar to the Greek, and there can be no doubt that Greek astronomy had found its way to India several centuries before this.

Haroun al Raschid 786-809 A.D.

Shortly after this the writings of Greek philosophers and astronomers were brought to the court at Baghdad. They had been carefully preserved, copied, and translated into Syriac, by Nestorian monks in some of the many monasteries which were founded in Persia and other countries of the East when these heretic Christians were driven out of Europe in the fifth century; and many of the Court physicians of Baghdad came from a Nestorian school of medicine. Haroun al Raschid, son of Al Mansur, gave orders for Ptolemy's *Syntax* to be translated into Arabic, and it now received its name of *Almagest*: several other translations were made later, and Aristotle was eagerly studied. This Caliph sent an embassy to Charlemagne, and among the presents sent by the East to the West were an elephant and a clepsydra.

Al Mamun 813-833 A.D.

Al Mamun, son and successor of Haroun Al Raschid, is said to have learned astronomy under a Persian teacher. He also added to his father's library, and one of the terms of a treaty he made with Michael, the Greek emperor, was that a collection of Greek writings should be made throughout the empire, and forwarded (originals or copies) to Baghdad. Moreover, he was not content with merely reading astronomy, Greek, Persian, or Hindu: he ordered Ptolemy's estimate of the size of the earth to be tested, by measuring an arc of a meridian in his own country, and he founded a splendid observatory in the province of Baghdad. The instruments were of the same kinds as the Alexandrian, but they were larger, and better made, and the circles were more accurately divided. The Arab astronomers were good observers, and among them for the first time we hear of astronomers winning fame by skill in instrument making. Their dials were superior to those of any other race, and they made some important improvements in mathematics, which were immensely useful to astronomers. One great service was the introduction of the decimal notation, which they learned from the Hindus.

Astrology is forbidden by the Koran, but it was practised eagerly, nevertheless, by the Arabs, who constructed tables for this purpose, and made improvements in the methods used.

Traces of Arab contributions to astronomy survive in our words "zenith" and "nadir," and "almanac"; our word for a "cipher" is the Arabic "zifra," and indicates the main advantage of the decimal notation in arithmetic; while "sine" is the Latin translation of an Arabic word, and reminds us of the great improvements made in trigonometry.[64]

Alfraganus *c.* 840 A.D.

Albategnius *c.* 900 A.D.

Abul Wefa 940-998.

Among the many names of Ptolemy's successors at Baghdad, strange and uncouth to our ears, the three most famous are Mohammed ebn Ketir of Fargana, Mohammed ben Geber Albatani, and Mohammed Abul Wefa al Buzjani, which became known to the West under the shortened forms of Alfraganus, Albategnius, and Abul Wefa.

Ebn Jounis died 1009.

Egypt too had her Arab school of astronomy, as she had had her Greek. A little later than Abul Wefa, Ebn Jounis drew up the famous Hakemite Tables of the sun, moon, and planets, under the patronage of the Caliph Hakem of Cairo.

At the western end of the Arab dominions there were centres of intellectual activity in Morocco and southern Spain. Cordova, the city of the marvellous mosque, had also, in the tenth century, an Academy, and a library which rivalled that of Baghdad; and here, in the midst of almost ceaseless public strife and agitation, in a strangely mingled atmosphere of cultured refinement unknown to the rest of Europe, and of ferocious barbarism, of tyranny and tolerance, heroic deeds of chivalry and treacherous intrigues, there lived and dreamed and worked men of science and philosophers, poets, and artists.

| Arzachel *c.* 1080. |
| Averroës 1126-1198. |
| Al Betrugi *c.* 1150. |
| Abul Hazan *c.* 1200. |

The best known of the Spanish Arabs whose names are connected with astronomy are Arzachel, who drew up the Tables of Toledo; Averroës, the great philosopher, author of *De Substantia Orbis*, and a commentary on Aristotle's Metaphysics, who saw "a black spot on the sun" on the day he had predicted a transit of Mercury; and Al Betrugi (or Alpetragius) who wrote on the Spheres. There was also a certain Abul Hazan, a renowned geographer, who travelled across North Africa, and made a catalogue of 240 stars, including some not given by Ptolemy.

Of all the astronomical writings of the Arabs, one of those which became earliest and best known in Europe was the *Elements of Astronomy and Chronology* of Alfraganus. This was actually used by Dante as his favourite text-book, and he mentions it in the *Convivio*. I shall therefore give a short account of its contents, following the edition of Golius, printed in Arabic and Latin at Amsterdam in 1669.

In the first chapter Alfraganus gives an account of the calendars used by different nations—Arabs and Berbers, Syrians, Romans, Persians, and Egyptians. After this he plunges at once into a description of the Universe, as portrayed by Ptolemy, and follows his master so closely that his book is almost a much-abridged and simplified *Almagest*, with a few additions, and with all the mathematics left out.

It is accepted almost without dispute among learned men, says Alfraganus, that the sky is spherical, and rotates on two fixed poles, one north, one south. This is proved by the observed movements of the stars. Equally undisputed among the learned is the fact that Earth and water together form a globe, which is surrounded by air. The spherical form of Earth is proved by the fact that phenomena such as lunar eclipses and shooting stars are seen at a later hour by observers in the East, and by the changing height of stars above the horizon as one travels north or south.

Earth is at the centre of the universe, and is but a point compared with the heavens.

There are two principal celestial motions: (1) the "prime motion" which causes the whole sky to revolve with every celestial body, and produces day and night; (2) the proper motions of the sun and other stars in the opposite direction and round other poles. The great circle of the first motion Alfraganus calls the Equator of the Day; that of the second, the Star-bearing Circle, *i.e.* the zodiac, or more precisely, the ecliptic.

The twelve zodiacal signs are then described, with the division of each into degrees and minutes, the positions of the equinoxes at the beginning of Aries and Libra, and of the solstices at the beginning of Cancer and Capricorn. The Colure is described as a great circle cutting the zodiac (*i.e.* ecliptic) and equator at the points where they are furthest apart (*i.e.* at the solstices). This greatest difference was found by Ptolemy to be 23° 51' but according to the measurement ordered to be made by Al Mamun of pious memory, which was carried out by a number of experts, it is 23° 35'. This value is adopted and quoted subsequently throughout the book of Alfraganus.

Alfraganus then proceeds to explain very clearly how the movements of sun and stars appear from different latitudes on Earth—on the equator, at stations further north, and finally at the pole. He shows how it happens that on the equator day and night are always of equal length, and the sun passes exactly overhead twice a year; whereas day and night vary more and more in length, according to season, as one travels further north, and the sun is lower; until at last at the pole the year consists of one long day and one equally long night, the celestial pole is in the zenith and the celestial equator on the horizon, so that the sky revolves like a mill-stone (*i.e.* the stars do not rise and set, but trace out horizontal circles, like a wheel which is not upright but flat on the ground).

The circumference of Earth, as determined by Al Mamun of glorious memory, is 20,400 miles, and the diameter, therefore, is nearly 6500.[65] Alfraganus deduces from this the area of the whole Earth and also of the habitable portion of Earth. The latter extends only from the equator to 66° 25' North, and its longitude at the equator is equal to 180° or 10,200 miles, at the northern limit to 4080 miles. This is divided into seven "Climates," as in Ptolemy's Geography, the first lying a little north of the equator. Alfraganus gives for each the length of the longest day, the height of the pole above the horizon, the extent of territory, and the principal regions and towns comprised. He admits that south of the first climate as far as 0° is some land, surrounded by sea, and sparsely inhabited, and north of the seventh climate are a few towns, but these are of no account.

Our author treats next of the risings and settings of the zodiacal signs, and of the division of the day into 24 equal or 24 "temporary" hours (see p. 26).

After this, the unanimous opinion of wise and learned men concerning the spheres is duly set forth in seven chapters; how there are eight great Orbs, the smaller enclosed within the greater, the star sphere being the outermost and largest of all; how epicycles are fixed in these; how only the star sphere has its centre exactly in Earth, the others being slightly eccentric; what are the positions of the poles and centres of the great spheres and the small epicycles, their relative sizes, and their different velocities as they turn; finally, how well the system represents the movements of sun, moon, stars, and planets. The moon's mean daily motion, resulting from a wonderful combination of five circular motions, amounts to about 13° 11'; the sun's is 59', and he completes a revolution in 365¼ days "less an insignificant fraction." (This being a popular treatise Alfraganus apparently thinks it unnecessary to state the length of the year more precisely). The sluggish motion ("motus tardissimus") of the star sphere, which is communicated to all the rest in addition to their own motions, is 1° in a century, according to Ptolemy, so that it completes a revolution in 36,000 years.

Coming now to the fixed stars, their number, and brightness, Alfraganus does not copy Ptolemy's great catalogue, but informs us that learned men ("sapientes") did number all the fixed stars as far south as they could see in the 3rd climate, and divided them according to magnitude into six classes. "To the first class they assigned the bright and shining stars such as Canis (Sirius) and Procyon, Vultur Cadens (Vega) and Cor Leonis (Regulus). Stars a little less bright they called second magnitude: such are Alfarcatein and Benet Naax," Arab constellations which the Latin version describes as the two bright stars of Ursa Minor, and those brilliant ones in the tail of Ursa Major. Thus they proceeded with the other magnitudes, the smallest measured being of the sixth magnitude. The number of stars in each class is given, and the total of 1022;[66] then a list of the 15 first-magnitude stars, which are the same as Ptolemy's (See p. 155). This is followed by a list of the Arab "Mansions of the moon."

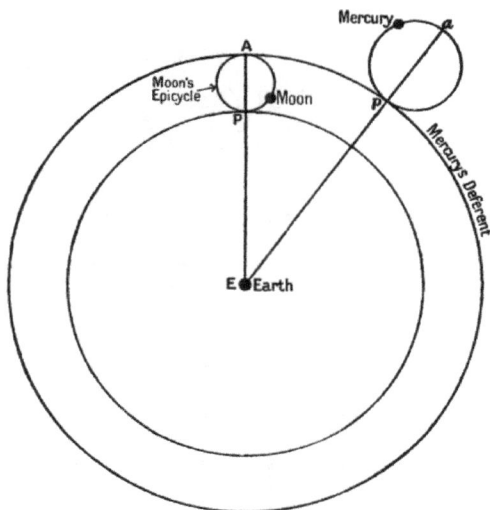

Fig. 36. Method (erroneous) of estimating planetary distances, described by
Alfraganus.

The distance $E\,A$ being known, and $E\,p$
being assumed equal to it, the distance $E\,a$ can
be calculated from the known ratio $E\,a\!:p\,a$.

So far (with the exception of the mansions of the moon) the Arab
writer has followed the Greek, but we have now reached a point where he
diverges. Ptolemy, he says, only tells us the distances and sizes of the sun
and moon, and said nothing about the other heavenly bodies; but if we
suppose the greatest distance of the moon to be the same as the least
distance of Mercury, and from this calculate his greatest distance (for the
ratio is known), and if we proceed in the same way with Mercury and
Venus, we shall find that the greatest distance of Venus equals the least
distance of the sun as given by Ptolemy. Ptolemy's least distance for the
sun, which was totally wrong, was 1160 Earth-radii: the greatest distance of
Venus, calculated in this way from Ptolemy's figures, was 1150. Alfraganus
takes this unlucky coincidence as an indication that there is only just
sufficient space between each sphere and the next to allow their respective
epicycles to pass one another, and upon this entirely erroneous assumption
he proceeds to lay down the distances of each planet from the earth, and
finally of the stars, which are all supposed to be at the same distance, equal
to the greatest distance of Saturn.

Who first suggested this method of estimating distances we do not know: the first mention of it occurs in Europe in the fifth century A.D. The following table shows the distances obtained in this way:—

GREATEST DISTANCE.

	In Semi-Diameters of Earth.
Moon	64⅛
Mercury	167
Venus	1120
Sun	1220
Mars	8876
Jupiter	14,405
Saturn and Stars	20,110

In this, the moon's distance is approximately correct, but the sun's is not much more than one-twentieth of its true value. To set the stars at a distance of only twenty thousand times Earth's semi-diameter seems to us to bring them very close,[67] but they would still be beyond measurement by naked eye methods, so it is no contradiction to Alfraganus' earlier statement that Earth is a point compared with the heavens.

The Arabs also believed that they had succeeded in measuring the apparent diameters of the planets and even of the points of light which are all we can see of stars, so Alfraganus gives the accepted sizes of all. I give them below in descending order of size. The modern values in the third column show how false were the results obtained by this mistaken method.

DIAMETER: EARTH = 1.

	Alfraganus.	Modern Values.
Sun	5½	109½
The 15 first-magnitude stars	4¾	Arcturus, Sirius, Spica, and others, much larger than the sun.
Jupiter	$4^9/_{16}$	11
Saturn	4½	9

	Alfraganus.	*Modern Values.*
Other stars, in order of magnitude, 2nd to 6th ...		Various. Some certainly larger than the sun.
Mars	$1\frac{1}{8}$	$\frac{1}{2}$
Earth	1	1
Venus	$\frac{3}{10}$	$\frac{9}{10}$
Moon	$\frac{5}{17}$	$\frac{1}{4}$
Mercury	$\frac{1}{18}$	$\frac{1}{3}$

In the above table the size of the moon (whose parallax had been found by the Greeks) is the only one which is nearly right. The sun is far too small, and so are the stars. We cannot yet know with certainty the diameter of any star, but they are all comparable with the sun, and many are enormously larger.[68] As their distances are all different, some of the brightest may be comparatively small, and some of the faintest the largest of all.

The Arabs took a backward step in adopting these imaginary measurements, for Hipparchus had recognized that only the moon was near enough to measure, and although Ptolemy accepted Aristarchus' value for the sun, he distinctly stated that the planets had no parallaxes and he could not tell their distances.

The next four chapters describe briefly the risings, settings, and meridian transits of stars as seen from different latitudes on Earth; the heliacal risings and settings and the conjunctions with the sun of planets, stars, and moon: the phases of the moon and the direction of her horns at different times of the year. Parallax is then clearly defined and discussed.

A description follows of the earth's shadow, thrown by the sun into space—its tapering form, its position, always pointing away from the sun, its width at the distance of the moon, and its length according to Ptolemy. This is stated to be 268 times Earth's semi-diameter, which is nearly correct,[69] for although Alfraganus believed (from Ptolemy's erroneous parallax) that the sun, was much nearer than it really is, it followed from this—since the size was deduced from the distance—that he also thought it much smaller, and the length of a shadow thrown by any dark body is longer the nearer it is to the light-source, but shorter, the smaller is the light-source.

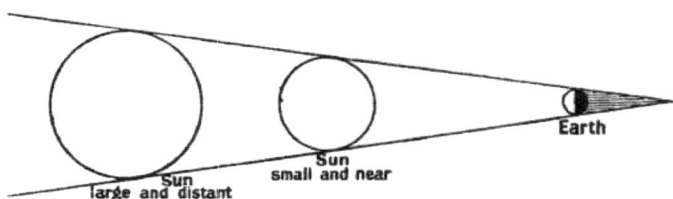

Fig. 37. Earth's Shadow.

The two last chapters are devoted to eclipses, lunar and solar, and Alfraganus points out that, unlike lunar eclipses, eclipses of the sun vary in duration and magnitude according to the place on Earth from which they are viewed.

The book concludes: "Enough having now been said concerning the eclipses of sun and moon, by the goodness of God we have been enabled to bring this writing to an end; and for this *Deo Laus et Gloria*."

The time at which Alfraganus lived is not precisely known, but it seems to have been in the first half of the 9th century, since from internal evidence he wrote after, but probably not much after, the death of Al Mamun. His name indicates that he was a native of the beautiful and fertile country, shut in by lofty mountains, which lies on either side of the ancient river Jaxartes. He was surnamed the Calculator, and wrote on sun dials and the astrolabe, but we do not know of any observational work of his.

Later Arab writers although they all continued to base their work on that of Ptolemy, improved on some of his estimates. Albategnius began his book "On the Number and Motions of the Stars" by saying that having studied Ptolemy's *Syntax* and mastered the Greek methods, and having noticed some errors in the positions of the stars, he felt impelled to add to Ptolemy's observations, as the latter had added to those of Abrachis (Hipparchus), for it is not given to man to attain perfection. He gives a much more accurate value for precession than Alfraganus had done, who merely copied Ptolemy, for he makes it 54½ seconds yearly, or one degree in 66 instead of in 100 years. His tropical year, too, is only two minutes shorter than the modern value, so that he improved upon Hipparchus in this respect; and he made the discovery which Ptolemy missed, the motion of the sun's apogee. He merely notes, however, that the position found by himself differed from that given in the *Almagest*, so it is doubtful whether he realized his discovery, or merely thought that a large error had been made.

Although some of Ptolemy's values were thus corrected by the Baghdad astronomers, no change was made in his theory of epicycles and eccentrics, which all accepted as having a concrete existence, partly because at first they did not distinguish between these and the spheres discussed by Aristotle, which he had described as formed of the same material as the planets. As Ptolemy does not explicitly state that his circles were only symbols, this is not surprising. They are formed, say the Arabs, of the fifth essence al-acir (the æther); and they conceived the epicycle as gliding over the outer surface of the deferent, like a small soap-bubble on the back of a big one. This notion, coupled with Ptolemy's erroneous distance of the sun, misled them, as we have seen, into their imaginary discovery of planetary distances.

Tabit ben Korra 826-901.

It was also partly due to Ptolemy's unfortunate habit of adopting the doubtful or merely provisional results of his predecessors, and representing them as well established and confirmed by himself, that they fell into another error. Tabit ben Korra, noting the discrepancy between the Greek and the Arab value of precession, investigated the question, and put forward a theory that the motion varies both in amount and direction. In his book "On the Motion of the Eighth Sphere" he describes an elaborate apparatus which he had invented to account for this variation, but he is very modest about it, and after narrating the results obtained by others, and how they had left them to be judged by posterity, he adds "And this is what we have done, with God's blessing." Then follow his figures and tables. This imaginary discovery was accepted by some of the Arab school, and it appears in many mediæval tables, instead of precession, under the name of the "trepidation."

The belief in the material reality of the spheres caused the Arabs to add a ninth sphere to the eighth of Ptolemy and Alfraganus, for they thought it was enough to demand of the eighth that it should carry all the stars and give them their slow movement of precession (or trepidation). This ninth sphere, therefore, enveloped the whole universe: upon it were fixed no epicycles, no stars, no planets, but it originated the "prime motion" by turning once in a day and night, and communicating this revolution to all the inner spheres. It became known in mediæval astronomy as the Primum Mobile or first moving.

Nasir-ed-din 1201-1274.

Alfonso 1223-1284.

The Baghdad school of astronomy came to an end with Abul Wefa in 998, and the Spanish schools died out when Seville and Cordova were captured by the Christians in the thirteenth century; but the impetus given to the study of Greek astronomy and astronomical observation was carried on by other nations. In Persia a fine observatory was founded by Nasir-ed-din; in Spain the Christian king, Alfonso X, ordered tables to be drawn up to replace those of Arzachel, and the *Libros de Saber* to be compiled. The movement in Persia was short-lived, but in Europe the revival of astronomy had begun.

VIII.

THE RETURN OF GREEK ASTRONOMY
TO EUROPE.

A.D. 1000 TO 1300.

"He hath made everything beautiful in His time; also He hath set the infinite in their heart."

In the whole cycle of the changing year there is no moment so wonderful in northern climes as that which comes in early February, when winter is not yet past, but for the first time the promise of spring is felt in the air. Not a leaf has unfolded its green, but the swelling buds on the trees make a purple flush all over the woods, the blackbird sings an exultant strain, and in some sheltered copse you may find a delicate daring primrose already in bloom.

Such a moment in the history of Europe was the year 1000 Anno Domini. After the apathy, the ignorance, the despair of the Dark Ages, a new spirit began to breathe hope into the hearts of men. A love of beauty, a new religious fervour, a passionate desire for knowledge took possession of them. Yet it was nearly a hundred years before the great universities which were one expression of this new spirit sprang up in Paris, Bologna, Oxford, to be followed later by similar centres of intellectual activity in all parts of Europe.

By the end of the twelfth century it is said that there were 100,000 students in Bologna. A large number were foreigners from many lands, for as Latin was the universal tongue in education, all nations could understand each other, and scholars often wandered from one university to another, attracted by the fame of some great master. Similarly the doctors would teach first in one town and then another. Men of all ages and classes met together, for among the students were young boys and elderly ecclesiastics, poor scholars who begged their bread, and rich nobles who came with a tutor, a chaplain, and a whole suite of servants. There were no colleges or even lecture-halls: the students joined together in small groups to take a house and share expenses, and the professor lectured in his own house, or in a hired room, or, if the audience was large, in a city square, speaking from an open-air pulpit. All were united in the ardent pursuit of learning, and none complained if the floors were merely covered with straw, and the lectures, which often lasted three hours, began before sunrise on winter mornings in rooms which had no light and no fire. Was it not enough that when leaving at the end of university life one was technically said to be "going home a wise man"?

One cause of this intellectual fervour was the influence of Arab culture, with which Europe came into contact through the crusades, and through the Saracens in Sicily, and the Moors in Spain. For this reason astronomy and astrology took a high rank among the new studies. To distinguish between the two is quite a modern idea, and in mediæval times either name was used indifferently to cover both subjects. In Bologna university in the thirteenth century an important school of medicine and arts arose, through Arab influence, and the Arab doctors of medicine introduced the system of astronomy which they had learned from the Greeks. "A doctor without astrology," it was said, "is like an eye that cannot see;" and before prescribing for a patient it was thought quite as important to determine the positions of the planets, as the nature of the disease. By the beginning of the fourteenth century there were salaried professors of astrology in Bologna, and they were more highly esteemed than any other professors except those of philosophy.

[*To face p. 200.*

A MEDIÆVAL ASTRONOMER.

From a painting by Gerard Dow.

Cecco d'Ascoli *d.* 1327.

One of their duties was to provide "judgments" (*i.e.* to cast the horoscope) gratis for students. But the dignity was a perilous one. Legitimate prediction by astrology bordered close on necromancy, which was banned by the Church, and one of Bologna's most famous professors in astrology, the learned Cecco d'Ascoli, was burned at the stake in Florence in 1327 for the crime of sorcery.

Astronomy, like other subjects, was taught chiefly by lectures and "repetitions," or classes for catechizing the students on what they had already heard. Books could also be had, though they were dear, on hire or purchase, from the university "stationers" or librarians.

The first books on Greek astronomy which found their way into European universities were Latin translations of Arabic commentaries and paraphrases of Aristotle, which travelled from Moorish academies in Spain to Paris. The astronomical treatises with strange technical terms in Arabic were hard work to translate, especially when they had already passed through several languages, as often happened. Thus it was possible to possess a work of Aristotle which was a Latin translation of a Hebrew translation of a commentary upon an Arabic translation of a Syriac translation of the original Greek text!

Urban IV., pope 1261-1264.

But meanwhile some of Aristotle's works in the Greek entered Italy from the East, as a result of the crusading conquest of Constantinople in 1204. The first translations of these into Latin were very poor, but later on St. Thomas Aquinas, with the help of Pope Urban IV., had a better version made.

Frederick II. 1194-1250.

Sacrobosco died *c.* 1256.

Roger Bacon *c.* 1214 to *c.* 1294.

It was long before a good Latin version of Ptolemy's *Almagest* could be obtained. A translation was made from the Arabic in 1230, at the bidding of the Emperor Frederick II., who did much, at his Sicilian court, to encourage Arab literature, but this translation was not much known or used.[70] The teachings of the *Almagest* became known chiefly through popular expositions such as those of Alfraganus, Albategnius, and John Halifax of Holywood, an English monk who became famous under his Latinized name of Sacrobosco. He was not an astronomer but had studied Greek and Arab writings, and finding that the study of astronomy was neglected because books on the subject were difficult both to procure and to understand, he compiled a useful handbook, which became widely popular and remained so for several centuries. Several other writers, notably Roger Bacon, wrote on the spheres, on the use of astrolabes, and on astrology. The books prescribed in Bologna for the course in Astrology and Mathematics were as follows:—

A work on Arithmetic or Algebra.

Euclid, with a thirteenth century commentary.

The "Theorica Planetarum," which was either a free translation

of Ptolemy's *Almagest* or an exposition of its principles.

The Alfonsine Tables.

The Canons of De Lignières (of Amiens, 1330 A.D.),

i.e. rules for the use of astronomical tables to determine

the motions of the heavenly bodies.

Portions of the Canon of Avicenna (the Arab philosopher).

A treatise on the Astrolabe by a Jewish astrologer of the ninth century.

A treatise on the Quadrant.

The astrological works of Ptolemy, with a commentary.

A book by Alchabicius (fl. *c.* 850 A.D.), probably his work on astrology.

A book on astrological medicine.

Alfonso X. 1223-1284.

From the above we gather that a past master in Astrology would understand the elements of mathematics, and all the astronomy that Ptolemy's translator or commentator could teach; that he had learned—at least in theory—the use of astronomical instruments and tables, and a good deal of astrology, including its use in medical practice. The tables were intended mainly for astrological predictions. The standard Alfonsine Tables had been drawn up in 1252 by Christians, Jews, and Moors, under the direction of Alfonso X. king of Castile. They contained lists of positions of the planets, dates of Easter moons, "golden numbers" and "dominical letters" of the ecclesiastical calendar, times and other details of eclipses, together with methods for finding the places of planets, and for casting horoscopes. This is the Alfonso who was so much shocked at the complexity of Ptolemy's multitudinous circles that "the ointment of his name is marred," says Fuller, "with the dead fly of his atheisticall speech": "If only the Creator had consulted me, when He made the world, I would have given some good advice!"

Sylvester II. (Gerbert) Pope, 999-1003.

Not one of all these books pretended to add any new discovery to astronomy: all intellectual energy was absorbed in eagerly assimilating the knowledge stored by Greeks and Arabs. Nor were any great observatories founded in Europe yet, in imitation of Alexandria, Rhodes, or Baghdad. The instruments in use were celestial globes and small portable astrolabes and quadrants for determining positions and angular distances between the heavenly bodies. The learned pope, Sylvester II., who had studied astronomy among the Moors in Spain, was so skilful in making astrolabes that some accused him of gaining the art by selling his soul to the devil!

Albertus Magnus 1193-1280.

Aquinas 1225-1274.

When Aristotle first came to Paris (about 1200 A.D.), in Oriental dress, and accompanied by Moslem authors, the Council of Paris denounced him as an infidel; yet in less than fifty years all his works were placed on the list of books prescribed in the university course. This was brought about by the Dominicans. In six of his twenty-one ponderous volumes the German friar Albertus Magnus paraphrased the whole of Aristotle's works, and stripping his philosophy of the pantheistic and materialistic garb in which the Spanish Arab Averroës had clothed it, presented the Greek philosopher as an ally of Christianity. The Italian saint, Thomas Aquinas, pupil of Albertus, in his much more readable commentaries and treatises, popularized this idea so successfully that Aristotle—*the* Philosopher, as he was called—speedily became as great an authority on every other subject as he had always been on logic.

It was a mutual victory. Aquinas captured the Greek for the Christian faith; Aristotle won the western world to accept his theories. No longer was the doctrine of a spherical Earth called "an old heathen theory": it was almost an integral part of the Catholic faith. Aristotle's demonstration that there must be a First Mover, himself unmoved, became an argument for the existence of the Christian Deity; the intelligences which preside over the celestial movements were interpreted as the nine hierarchies of angels whose existence was taught by the Church. To the eight spheres of the Greeks and the Primum Mobile of the Arabs, the thirteenth-century Christians added the all-embracing heaven of heavens, the Empyrean, to be the abode of the Creator and blessed spirits. Within the central immoveable earth they placed Purgatory and the fires of Hell. They accepted the limits of the habitable earth as laid down by Ptolemy, but kept Jerusalem as the centre by asserting that it was situated 90° from the Pillars of Hercules, and 90° from the mouths of the Ganges. Eden, the earthly paradise of Adam and Eve, was represented on contemporary maps as in the extreme East, separated by sea from the eastern boundary of the inhabited earth.

Thus theology and science supported one another. All learning was sacred, and all that man's mind is capable of understanding he might aspire to know, for the search if rightly pursued would lead at last to the perfect bliss of beholding with unveiled eyes the Source of all Truth.

SECOND PART.

THE ASTRONOMY OF DANTE.

I.
POPULAR ASTRONOMY IN ITALY
IN DANTE'S TIME.

In the first part of this book we have sketched the story of man's thoughts about the stars, from primitive days until the thirteenth century of our era. We have seen what a wealth of imagination and invention the Greeks brought to bear on the purely empirical science of Egypt and Babylon and Homeric Greece, and how out of all the systems devised by them between 600 B.C. and 100 B.C. one survived, which was completed and expounded by Ptolemy in the second century after Christ in his great *Syntax* or *Almagest*. Its fundamental principles were that Earth is a sphere, at rest in the centre of the Universe, surrounded on all sides by spherical heavens, and that the movements of all the heavenly bodies are explained by a combination of uniform circular movements.

In the succeeding centuries there were few who could appreciate his work, and during the Dark Ages it was scarcely known in Europe, but was preserved by Nestorian Christians in schools and monasteries of Persia, whence it was unearthed by Mahomedan princes, five centuries after Ptolemy's death; and when another five centuries had passed it was brought back to Europe, tinged with Oriental thought, and almost immediately became immensely popular among scholars.

Then there arose one of the world's greatest poets, and, a thousand years after Ptolemy's death, immortalized his work, writing in a tongue unknown to Ptolemy, and for nations which in his day were only just struggling into existence. As Homer reflects to us man's primitive conceptions of the Universe, so Dante reflects the ideas of Ptolemy and his school.

And because he lived just at this time he was able to write with perfect confidence, quoting Ptolemy and the Catholic Faith side by side as infallible authorities in astronomy. Had he lived in the early Christian centuries he would have been obliged to choose between classical and orthodox views; had he been born three centuries later, he would have found Copernicus and Galileo ranged against Ptolemy and the Church. Even Milton writing a hundred years after the death of Copernicus, could not make up his mind which system to adopt, and the astronomy of *Paradise Lost* is a curious jumble of ancient, modern, and transitional ideas. He describes the Primum Mobile as a "firm opacous globe"[71] on which Satan alights and walks about, yet later on tells us that we need not believe in its existence if Earth is turning on her axis;[72] the archangel Raphæl describes to Adam the Creation, at which he was present, yet declares that he himself does not know whether the sun circles round Earth, or Earth round the sun. This the great Architect had wisely concealed from man and angel,

"perhaps to move His laughter at their quaint opinions wide, Hereafter when they come to model heaven, And calculate the stars, how they will wield The mighty frame, how build, unbuild, contrive, To save appearances; how gird the sphere With centric and eccentric scribbled o'er, Cycle and epicycle, orb in orb."[73]

Such inconsistencies are not found in Dante's work, and nothing could have been further from his thoughts than to imagine the Creator mocking at man's mistakes and ignorance. Like the angelic doctor, Thomas Aquinas, he considered man's desire for knowledge as one of his highest attributes, and believed that it had been given to him in order to be satisfied.

But before we examine Dante's writings, it will be interesting to form an idea of the spirit in which astronomy was generally regarded by his fellow-countrymen, and what were his opportunities of studying it.

Dante Alighieri was born in Florence in or about the year 1265, and died in 1321. There can be no doubt that the part of star-lore which appealed most to the general public in Italy at this time, educated and ignorant alike, was the art of the astrologers. The movements of the heavenly bodies were regarded not merely as omens, but as the actual instruments by which every event on Earth was brought to pass. Every class of plants, every race of animals, was thought to be under the protection of some planet or constellation; so if it was a bad year for certain fruits, or if an epidemic broke out among the cattle, this was because the guardian planet was unfavourably placed, or an evil planet was in the protecting constellation. Floods and drought, prosperity and death, properties of minerals such as the lodestone's attraction for iron or the emerald's alleged power of blinding serpents, the instincts of animals and the impulses of men, all were subject to the influences of the stars; and naturally the men who understood and interpreted their movements were held in great repute.

Some of them were unscrupulous quacks, like the one-eyed prophet of Brescia, mentioned by the gossiping friar Salimbene.[74] He "called himself an astrologer and diviner," and received daily "ten great pennies of silver, and nightly three great Genoese candles of the purest wax" from a political party in Modena, as a recompense for advising them how to act. On one occasion he prophesied a victory for them, but he had little faith in his own words, for being threatened with violence if his prophecy should fail, he "carried off all that he had gained and went his way without saluting his hosts." "Then," adds the friar, "the men of Sassuolo began to mock them, as men who sacrifice to devils and not to God, as it is written in Deuteronomy."

But most of these men, like Asdente of Parma, sincerely believed in their own ability to foretell events, and they usually combined some other favourite forms of soothsaying, as well as a little alchemy, with astrology. Asdente is described by Salimbene as "a poor working cobbler, pure and simple, and fearing God, and courteous and urbane; illiterate, but with great illumination of mind." His proper name was Master Benvenuto, but he was "commonly called Asdente, that is, toothless, by way of contrary, for he hath great and disordered teeth and an impediment in his speech, yet he understands and is understood well. He dwells at the bridge-head of Parma, hard by the city moat and the well, along the street which goes to Borgo San Donnino." This humble prophet was said by Dante to be the best known citizen of all Parma (*Conv.*, IV. xvi. 65-71): he was asked to dinner by a bishop, and consulted by the warring factions of Reggio and Parma. He was said to have foretold the death of two popes, and a naval defeat of Pisa by Genoa.

The greatest generals of the day governed their tactics by the advice of astrologers who regularly accompanied them to the field and the camp. The famous Ghibelline, Guido of Montefeltro, who is called by Villani the cleverest soldier of his times[75] retained Guido Bonatti[76] in his service and was believed to have gained his great victory at Forli (in 1282) through the advice of this astrologer. Bonatti is diversely described as a tiler and a lawyer, but whatever his original occupation may have been he found that the position of private astrologer brought him both more fame and more money. He wrote a book on Judicial Astronomy, and Vincent de Beauvais describes him as celebrated throughout the western world for his knowledge of the art.

But Bonatti's fame was faint and fugitive compared with that of the wizard Michael Scot. He is one of the picturesque figures of the thirteenth century, round whom so many legends have gathered that the facts of his life are difficult to glean. It seems that he was born in Fifeshire of a noble Scottish family, at the end of the twelfth or beginning of the thirteenth century, that he studied in Oxford and Paris, and then spent some time in Toledo. Here he learned Arabic, and probably also astrology, for it was so commonly practised there, especially among the Arabs and the Jews, that it was sometimes called the Toletan art. Afterwards he went to Germany, and was discovered by Frederick II., who took him to Italy. His great learning earned the admiration of Pope Gregory IX., who speaks of him quite affectionately in a letter to the Archbishop of Canterbury; and it is said that

Honorius II. would have liked to make him an archbishop. But Sir Michael the Scot found Frederick's court more congenial. The Emperor, who was himself a poet, was a munificent patron of literature and art, and attracted to himself men of talent from all parts of the world. The culture of both East and West met at that brilliant Sicilian court[77] which was his for fifteen years before the title of Emperor was added to that of King of Naples and Sicily. Indeed it was his "fellowship with Saracens"[78] which was one great reason for the accusation of heresy on account of which Dante placed him among the Epicureans in the *Inferno*. He knew Arabic, as well as French, German, Italian, Latin, and Greek. Michael Scot's acquaintance with Moorish literature and language was a bond of sympathy between them; he became astrologer to Frederick, and at the Emperor's wish he superintended a new translation of Aristotle's works from Arabic into Latin. His taste for astronomy is evidenced by the fact that out of these he chose to translate the *De Cælo* himself.

An absurd story is told by Salimbene about Frederick and Michael Scot, which, however, shows what was believed of his capabilities as an astronomer. The Emperor one day asked him, when they were in the palace together, how far they were from the sky, and the astrologer told him the distance. They then took a long journey together, during which the palace was secretly lowered, and on their return Frederick asked casually whether the sky could really be so distant as Michael had said. "Whereupon he made his calculations, and made answer that certainly either the sky had been raised or the earth lowered; and then the Emperor knew that he spake truth."

Michael Scot is said to have warned Frederick that he would die in Florence, for which reason the Emperor would not enter that city; but having thoughtlessly gone to a town called Florentiola he died there; "for this," adds the historian, "is almost always the way, the devil tricks one by a play upon words." It is curious to contrast this remark, attributing Michael's prophecy to the evil powers, with Salimbene's quotation from him, in exactly the same spirit as if he were quoting from an Old Testament prophet:—"that the word of Michael Scot may be fulfilled in them, which he wrote in his verses wherein he predicted the future, 'And the factions at Reggio shall hold ill words together.'" The same author brackets him with others who have foretold the future, in a list which reads curiously to us— "Abbot Joachim, Merlin, Methodius and the Sybil, Isaiah, Jeremiah, Hosea, Daniel, and the Apocalypse, and Michael Scot who was astrologer to the deposed Emperor Frederick II."

Besides these prophetic verses, Michael wrote several books which treat almost exclusively of astrology, alchemy, and other occult "sciences," and even in the fifteenth century it was said that his magic books could not be opened without danger, because of the fiends who were thereby invoked! He seems to have returned to his native Scotland to die, but the date is very uncertain. We do not know whether Dante's picture of him is drawn from memory, or hearsay of some who had seen the lanky Scotsman among the southerners, by nature taller and thinner than they, and worn by his prolonged studies.

"Quell' altro che ne' fianchi è così poco Michele Scotto fu, che veramente Delle magiche frode seppe il gioco."[79] (*Inf.* xx. 115-117).

It was perhaps the winning personality or the prudence of the canny Scot which enabled him to bring a brilliant career to a peaceful close, favoured by the Church as well as by the excommunicated Emperor, although his studies were of so dubious a nature, and he was intimate with heretic Mahomedans and Jews. Cecco d'Ascoli was not so fortunate. This learned Italian received the high and honoured post of professor of astrology in Bologna, where he lectured and cast horoscopes for his students. He was well versed in natural science, but the shady side of astrology had a fatal attraction for him: he fell under suspicion as a sorcerer, was condemned, and burned at the stake in Florence in 1327 (six years after Dante's death).

The reputations achieved by these and other thirteenth-century astrologers in Italy, belonging to such different ranks of life, show what an immense importance was attached to their art by the general public. Yet we shall be greatly mistaken if we think that it was only to acquire skill in this fascinating pursuit that men thronged to hear Cecco lecture, or pored over Latin manuscripts. The intense ardour for knowledge which marks this period made them eager to understand the world around them and the sky above their heads.

ASTRONOMY.

From a fresco in the Spanish Chapel of Santa Maria Novella, Florence.
[To face p. 217.}

It would be a shame, writes one, to live in a house and not know how it is built or what shape it has, never to examine the walls, and floors, and ceilings, nor to consider the use of the wooden beams used in its construction. In like manner we should not be content without understanding the form and structure of the Universe in which we live. Man, with his upright attitude and his head held high, unlike the animals, was designed by his Creator to look and listen, to know and comprehend this marvellous Universe, and especially that noblest part of it above him, the heavens and their wonderful movements. For thus alone can he learn to know God Himself, the great Architect of the World.

In writing thus, Ristoro, the monk of Arezzo, was not only echoing the thoughts of Plato and Cicero, he was expressing the feeling for astronomy as a noble and elevating study which was general among thoughtful men of his time. It was expressed also in contemporary art. Visitors to the "Spanish Chapel" in the cloisters of a Florentine church will remember seeing on the frescoed walls the figure of Astronomy as she was personified in Tuscany in the fourteenth century. She sits among her peers, the sciences of the Trivium and Quadrivium, the only one who wears a crown; her fair hair frames a spiritual face, one hand is lifted heavenwards, the other holds a celestial sphere, on which the broad band of the zodiac crosses the "equator of the day." At her feet sits a kingly figure in flowing robes, also crowned, and with a face of singular beauty and refinement; he gazes up into the skies with a rapt expression, and on his knee is a book in which he writes what he sees.

This nameless figure was identified doubtfully by Ruskin as Zoroaster, who was considered by many as the inventor of astrology, but surely it can be no other than Ptolemy with his *Almagest*. For Ptolemy, the prince of astronomers, was often and naturally confused with the royal race who had patronized astronomy at Alexandria; as for instance by Omons, a thirteenth-century writer, who says in his *Image du Monde* that "Ptolemy king of Egypt" wrote the *Almagest*. The curious mode of dressing the hair and beard may have been thought by the artist to represent an ancient Egyptian fashion.

Ptolemy, we know, was universely acknowledged at the time to be "Master of Astronomy," as Brunetto Latini calls him. His *Almagest* was only known indirectly, but it was believed to contain all that could be known about the movements and the nature of the heavens. Some minor additions and corrections had been made, as we have seen, by the Arab astronomers, but the system was accepted as a complete and satisfactory explanation of all celestial phenomena. Hence no professional astronomer was expected to make discoveries; he was simply well versed in the work of those who went before him, skilful in the use of a few simple instruments and tables, and practised in applying the principles of astrology.

A general notion of the Ptolemaic system was widely diffused. For those who could not read Latin there were encyclopædic works written in the vernacular and in a popular style, such as the *Trésor* of Brunetto Latini, and these always contained a section on astronomy. The average educated man probably had only vague ideas about epicycles and eccentrics, and perhaps had never heard of the Arab estimates of the sizes of the planets; but he would know that astronomy taught that Earth is a globe, motionless at the centre of the universe, and smaller than any of the stars; he would know the names of the seven planets (including among these the sun and moon), and probably also their colours, their periods, and their astrological significance; the zodiacal constellations would be familiar, especially as they were often used decoratively; and he would believe that stars and planets are set in crystalline transparent spheres.

Moreover, he would often be more of an astronomer than he knew, for he would learn almost unconsciously many things of which modern men are ignorant. The ill-lighted streets and the dangers of night journeys would force him to be better acquainted with the motions and phases of the moon than most of us are to-day; he would know when and where to look for different stars; and the want of a watch would make it necessary for him to be able to take his time from the sun at any season of the year. He could, however, sometimes consult a sundial on a church wall or in a private garden, and the church chimes rang out at tierce, and nones, and vespers. These were heard at intervals which were much longer in summer than in winter, for the system of "temporary hours" was used by the Church, and the service of tierce was held halfway between sunrise and noon (or nones), and vespers was halfway between noon and sunset.

II.

DANTE'S STUDIES.

Dante was far above the level of the average educated man. Not that his scientific ideas were in advance of his age: on the contrary, one special interest that they have for us is that they illustrate, like his political and religious views, the beliefs and feelings of the period. His authorities were the authorities of all, but he had studied them and made their thoughts his own, as few others did, except some churchmen and professed scholars. The extent and depth of his reading is evident from his own writings, and his great learning is noted with admiration by all his biographers. Giovanni Villani, in the earliest account we possess of Dante, says that he was "a great scholar in almost every branch of learning, although he was a layman." Boccaccio would have us believe that while still a child, so young that he might be expected to spend his time playing with other children or sitting on his mother's knee, he gave the whole of his time to reading and learning. Lionardo Bruni, however, assures us that though he was an ardent student, and showed unusual powers at an early age, he by no means tried to "sever himself from the world, but living and moving about amongst other young men of his age, he approved himself gracious and skilful in every youthful exercise." It was wonderful, he says, how Dante maintained all his social and civic intercourse while he pursued his studies so fervently.

In truth, the poet's troubled life was far removed from that life of calm retirement which one thinks suitable for a scholar. In his early youth he experienced a passionate love and sorrow; a year before the death of Beatrice he was fighting for Florence in the great battle of Campaldino, nor was it the first time he had borne arms; in 1296 he spoke in the council of the Hundred; in 1300 he was ambassador for the Tuscan League to San Gemignano, and was elected to the highest office a citizen could hold in his native city, that of Prior; in 1301 he was ambassador to the Pope in Rome, and in the year following he was exiled. After this he was always wandering, often in great poverty, dependent on first one patron and then another, always hoping that some turn of affairs would restore him to Florence, always taking a keen and active interest in Italian politics, until he died, still in exile, at Ravenna. Add to this the difficulties common to all scholars of his day, viz. absence of printed books, public libraries, and journals, etc., and we must marvel how he ever found the opportunities and the serenity of mind for his prolonged studies.

Boccaccio adds another obstacle—his wife! To console him for the death of Beatrice, his friends and relatives persuaded him to marry a wife of their choosing with melancholy results:—

"Dante formerly had been used to spend his time over his precious studies whenever he was inclined, and would converse with kings and princes, dispute with philosophers, and frequent the company of poets.... Now, whenever it pleased his new mistress he must at her bidding quit this distinguished company, and bear with the talk of women, and to avoid a worse vexation must not only assent to their opinions, but against his inclination must even approve them. He who, whenever the presence of the vulgar herd annoyed him, had been accustomed to retire to some solitary spot, and there to speculate on the motions of the heavens, or the source of animal life, or the beginnings of created things, or may be to indulge some strange fancy, or to compose somewhat which after his death should make his name live into future ages, he now, as often as the whim took his new mistress, must abandon all such sweet contemplation, and go in company with those who had little mind for such things."[80]

However, after a long tirade against wives, Boccaccio owns that as far as concerns Dante the picture is entirely imaginary. He believes the marriage was an unhappy one, because after his exile Dante and his wife never met again, but we have no evidence whatever that while they were together their life was as Boccaccio depicts it. Doubtless family life and the care of four children interfered to some extent with Dante's studies, but the one detail Boccaccio affirms as fact among his fancies, namely that the poet refused for nineteen years to see his wife, seems rather to imply that one thing poor Gemma did know how to do was to leave her husband in peace when he did not desire her company.

When and where Dante began to study astronomy seriously it is not easy to say. His favourite study in his youth was poetry, but his parents (says Bruni) gave him a good general education, engaging such teachers as could be found in Florence, where as yet there was no university. But seeing how clever the boy was, his relatives and friends helped and encouraged him, and among the latter Bruni mentions Brunetto Latini.

This famous Florentine was a very learned lawyer, who during Dante's infancy and boyhood held high offices of state in the city. It is not possible, therefore, that he should have had pupils at this time, and he cannot have been Dante's master in any strict sense of the word, as Vasari affirms, and as many commentators have assumed on the strength of the moving interview between these two in the *Inferno*.[81] Dante's affectionate greeting mingled with reverence, and the fatherly solicitude of Brunetto, suggest just such a connection as Bruni indicates. The elderly man had taken an interest in the budding genius of the boy, and had held inspiring conversations with him from time to time on serious subjects, and by his own example had encouraged the youth to win fame through his pen. For Dante was no doubt well acquainted with Brunetto's *Tesoretto*, and his more ambitious and voluminous *Trésor*, which Brunetto specially commends to his care.[82] This was a compendium of knowledge, a small part of which was devoted to the elements of astronomy, and it may well be that this short epitome was Dante's first introduction to the science.

Some have thought that Brunetto's advice to Dante to follow his star "Se tu segui tua stella ...,"[83] indicates that he had cast the poet's horoscope; but there is no evidence that Brunetto had ever practised astrology, or even that he took special interest in it above other branches of learning.

Villani and other biographers tell us that after his banishment from Florence, therefore in middle age, Dante went to the University at Bologna, but they seem to have made this statement rather because of its probability than because they knew it for fact. It is, however, quite possible that he spent a couple of years there, between 1304 and 1306. He probably left his "primo rifugio"[84] at Verona when Bartolommeo della Scala[85] died in March 1304, and we know nothing of his whereabouts till August 1306, when he was in Padua, as is proved by a fifteenth century document.[86] Now in 1306 a number of Ghibelline students and professors left Bologna for Padua; and we know that Dante had friends among the Bolognese professors, for later on he seems to have corresponded with Cecco d' Ascoli,[87] and Del Virgilio entreated him to come and receive the laurel crown, an invitation which the poet refused in one of his Latin Eclogues. Bologna and the Bolognese are referred to several times in Dante's works, and the friar Catalano is no doubt alluding to the school of theology at the University when he says in *Inf.* xxiii. 142-144 that he used to hear at Bologna much about the wickedness and the lies of the Devil. One vivid passage can hardly be anything but a personal reminiscence of the Carisenda, the leaning tower at Bologna. The poet compares his fear when he saw the giant Antæus stooping over Virgil and himself to the sensation of looking up at the Carisenda from beneath the leaning side, when a cloud is passing over it, and the spectator feels that the tower is about to fall upon him.[88] (The tower was much higher in Dante's day than it is now, part having been pulled down in the middle of the fourteenth century).

The University at Padua was also famous, and as it was originally an offshoot from Bologna and modelled on the same plan, it is sure to have included astronomy in its curriculum. But Dante does not seem to have stayed here more than a few months.

There is also a tradition that he went to Paris, the greatest intellectual centre of his times, and that he himself heard those lectures of Sigieri, the philosopher, in the Street of Straw, of which he makes mention in *Par.* x.

137. This is possible, but quite uncertain; the allusion to Sigieri is no proof, since it has lately been discovered that he died in Italy, and his story was well known there.

After sifting all the evidence available, we can only echo Boccaccio, and say of Dante that "As it was at divers ages that he studied and learned the divers sciences, so likewise it was at divers places of study that he mastered them under divers teachers."

III.

BOOKS ON ASTRONOMY USED BY DANTE.

We shall probably be right if we conclude that Dante's knowledge of astronomy was principally gained from independent reading, and from conversation and discussion with learned men. Like a true scholar, he was learning all his life, for in the *Paradiso* he corrects the opinion expressed earlier in the *Convivio* about the origin of the markings in the moon.

Boccaccio's description of his planning out a course of study and attacking some subjects alone in his youth agrees with his own account of his lonely struggles with Latin authors;[89] and it is quite credible that he suffered heat and cold, and went without food and sleep in his eagerness to learn, since he himself mentions quite casually that at one time he injured his sight by constant reading, so that the stars appeared blurred to him, until by rest and bathing with cold water it became strong again.[90]

His diligence must have been great, and his memory wonderful, judging from the numerous quotations from classical and contemporary authors to be found in his writings. They seem to be chiefly from memory, and it is not likely that he can often have had many books by him when writing. If he gives a reference it is often a vague one, and occasionally wrong: his authorities on astronomical questions are sometimes "the mathematicians," "the astrologers," or "the sages of Egypt," by whom he means the Alexandrian astronomers.

One does not, of course, expect mediæval writers to verify their references, or to have our modern scruples about quoting other authors without acknowledgement: in those days they were only too glad to get their information wherever they could, often at second or third hand, and to make it their own by storing it in their memories. Nevertheless, from a careful study of Dante's direct quotations, and his allusions and reminiscences, which are sometimes unconscious, a good idea may be gained of his range of reading, and of the books and authors on whom he chiefly relied for his astronomical data.

His supreme authority was of course Ptolemy. But it would only be by a rare chance that he could see the *Almagest*, even in a translation, and all the evidence that we can find in his own writings points to its being entirely unknown to him. On each of the three occasions that he quotes Ptolemy's opinion on subjects dealt with in the *Almagest* he is wrong: once it is Ptolemy's view on the physical nature of the Galaxy,[91] as to which none had been expressed, although its appearance was carefully described; in another case he implies that Ptolemy discovered precession, and says that he added a ninth heaven to account for it,[92] and this double mistake was apparently copied from Albertus Magnus or Averroës.[93] As we know, it was Hipparchus who discovered precession, and Ptolemy gives him the credit for in the *Almagest*; it was the Arabian astronomers who added the ninth sphere.

On the other hand Dante quotes twice quite correctly from the *Tetrabiblios*, which was more widely known.[94] He may, therefore, possibly have read it, and he alludes to it in the *Convivio*, though by a curious slip he does not give its name, thinking he has done so already: "Tolommeo dice nello allegato libro...."[95] This is in the fourteenth chapter of the second treatise, but so far from having just quoted any special work by Ptolemy he has said nothing at all about him since the third chapter, and then it was only his opinion and name that were mentioned. I have not been able to trace the quotation given in the *Quæstio de Aqua et Terra* (xxi. 29-31), where Ptolemy is said to have asserted that things on earth resemble things in heaven, but this probably comes also (or was supposed to come), from the "judicial astronomy" of the *Tetrabiblios*. It is not taken from the *Almagest*.

The elements of Ptolemy's system, however, could be learned indirectly, and it seems that Dante had recourse to the excellent epitome of the Arab astronomer Alfraganus. It is true that his name is only mentioned once, and his book once; but nearly all Dante's astronomical data appear to have been taken from him, and his very expressions are sometimes repeated.

One of the most striking examples of borrowing from the *Elementa Astronomica* of Alfraganus occurs in the *Paradiso*, where Dante likens twenty-four spirits to as many brilliant stars.[96] He makes up the number by taking fifteen specially bright stars from different parts of the sky, and adding to them the stars of Ursa Major and two stars of Ursa Minor. The seven chief stars of Ursa Major are well known, but Beta and Gamma Ursæ Minoris are not conspicuous nor specially familiar; and why, if he takes fifteen unnamed, should he name any of the twenty-four? Turning to chapter 19 of Alfraganus' book, we find that he follows Ptolemy in enumerating fifteen first-magnitude stars in different parts of the sky, and then he gives as examples of the second magnitude (*i.e.* next in brightness) Benet Naax and Alfarcatein, Arab constellations which correspond with the tail of Ursa Major, and Beta and Gamma of Ursa Minor.

There is no reason to suppose that Dante knew Arabic, but there were several versions of Alfraganus in Latin, and it is even possible to determine with some certainty which of them he used. Speaking of the movements of the planet Venus, he says that they may be found, summarized from the best demonstrations of the astrologers, in the *Book of the Collection of the Stars*:—

"Li quali, secondochè nel *Libro dell' Aggregazione delle Stelle* epilogato si trova, dalla migliore dimostrazione degli astrologi ...." (*Conv*. II. vi. 133-136).

Although the usual name for the book of Alfraganus in Latin was *Elementa Astronomica*, the version for which mediæval students were indebted to the indefatigable translator Gerard of Cremona bears in the MSS. the title *Alfragani liber de aggregationibus Scientiæ Stellarum et de principiis cælestium motuum*. It is this book, therefore, that Dante means, though he has translated the title somewhat inaccurately.[97]

There were several other books known to Dante from which he must have gleaned information about Ptolemy's system, but Alfraganus we may regard as his standard reference book. Many versions of Alfraganus are still extant, and it is pleasant to feel that we can hold in our own hands Dante's text-book on astronomy. The best known is the printed edition of Golius, published in Amsterdam in 1669: it is in both Latin and Arabic, and opening in the middle we may turn the pages backwards to see the beautiful Arabic letters of Alfraganus' own language, which is written from right to left, or forwards to look at the Latin in which he became known to Dante.

Next in importance to Alfraganus among Dante's authorities on astronomy was Aristotle. We have seen how overwhelming his authority became in natural science in the second half of the thirteenth century, and how for some time scholars failed to distinguish between the system of spheres devised by Eudoxus, which was the scheme upheld by Aristotle, and the epicycles and eccentrics of Ptolemy. Among all Aristotle's admirers none was more devoted than Dante: for him Aristotle is not only "Il Filosofo," the Philosopher *par excellence*, as he was generally called in that age; he is "quello glorioso filosofo al quale la Natura più aperse li suoi segreti," his intellect was "quasi divino," and his words are of "somma e altissima autoritade."[98] Throughout Dante's own writings his references to Aristotle are so frequent that Dr. Moore observes: "The amount and variety of Dante's knowledge of the contents of the various works of Aristotle is nothing less than astonishing."[99]

He does not, however, consider Aristotle infallible as regards the details of astronomy, for in these the philosopher was only following the observers and mathematicians;[100] but in all matters which depend upon first principles, in which Dante would include the form of the earth, Aristotle's authority may not be called in question.[101] In cosmical physics, such as the doctrine of the four elements, and in meteorology, Dante follows "il mio Maestro"[102] implicitly, and it is also largely from the *De*

Cælo that he gained his knowledge of early Greek speculations regarding the universe.

Dante knew little more of Greek than of Arabic. His occasional use of Greek words is enough to prove this—for example when he discusses the Pythagorean theory of "Antictona," using the accusative as if it were a nominative.[103] In several passages he implies that the language was not known to him, and he distinctly states that he used translations of Aristotle's works. Here again we are able to identify the versions he used. In *Conv.* II. xv. 59-73 he complains that it is impossible to know what Aristotle believed about the Milky Way, because in the "old translation" he is made to say one thing, and in the "new translation" quite another. From his quotations of these two conflicting opinions, we are able to deduce that the "old translation" was that which Michael Scot made from the Arabic, and the "new translation" that of Aquinas, which was made direct from the Greek.

He had access, therefore, to more than one Latin translation of Aristotle; and beside this, his quotations seem to have been often taken from commentators and compilers of his own times, especially Aquinas and Albertus Magnus. This was frequently the case with the *Meteorologica* of Aristotle, for the *De Meteoris* of Albertus Magnus seems to have been much used by Dante in place of Aristotle's own work. Dante quotes it by name as *Meteora* in *Conv.* II. xiv. 169, and also in *Conv.* IV. xxiii. 125-126, where he mentions "Alberto" as the author.

These two, then, the Arab Alfraganus and the Greek Aristotle, were Dante's chief authorities in astronomy, both in Latin translations.

Besides these, there is evidence in his writings that he was familiar with several classical authors who treated of astronomy. He quotes frequently from many of Cicero's works, and we find echoes, in two passages, of the *Dream of Scipio*,[104] which he is almost certain to have read, since it was a favourite book in his time. There are several references to Seneca, but the only time that an astronomical phenomenon recorded by him is mentioned[105] Dante is evidently quoting at second-hand from Albertus Magnus. With all the works of Virgil, and with the *Metamorphoses* of Ovid (from which he took his astronomical myths)[106] he was evidently very familiar; and Lucan is one of his authorities for the position of the earth's equator. In this case the name of the author and the number of the book are precisely stated.[107]

Dante had a great reverence for Plato, whom he calls "uomo eccellentissimo,"[108] and of whom he repeats the mediæval legend that he was a prince who gave up all for the sake of acquiring wisdom.[109] Of his writings he only knew the *Timaeus*, probably the Latin translation and commentary of Chalcidius, which was widely known. Or he may have been acquainted with Aquinas' commentary, which has since been lost. He also knew something of Plato's teachings through Aristotle and Cicero, Albertus Magnus, and Aquinas, and perhaps St. Augustine. When he feels obliged to dissent from the great philosopher's doctrine that the souls of men come from the stars, he does so with reluctance and great gentleness.[110]

Among other Greek philosophers who speculated on astronomy, Dante mentions Thales[111] and Pythagoras. The date when the latter flourished he takes (he tells us) from Livy, and his theory that the Universe is governed by the principle of number from Aristotle's first book of Metaphysics. The astronomical theories of his school are also doubtless taken from Aristotle. Dante tells the story of Pythagoras that he was the first man to be called "philosopher," because when asked if he considered himself a wise man, he replied, "No, but only a lover of wisdom."[112] Dionysius the Academician and Socrates are referred to for their opinions on the influence of the stars upon human souls;[113] and Anaxagoras and Democritus for their galactic theories,[114] which Dante obtained from Albertus Magnus.

Except for Alfraganus, Dante refers but seldom to Arab astronomers. He does not seem to have known Albategnius. The *De Substantia Orbis* of Averroës is quoted in the *Quæstio de Aqua et Terra*,[115] and it is possible that from this book Dante derived a theory about the moon which he expounds in the *Convivio*. He quotes Alpetragius on circular existence (dependent on the spheres),[116] Avicenna on the Galaxy, and again with Algazel on the influence of the spheres,[117] and Albumassar on meteors.[118] The quotation from the latter, however, is second-hand from Albertus Magnus, and is a mistake, for the passage is not to be found in that astronomer's works. He was born in Turkestan, in 805.

We may here remind our readers that the three Latin poets, Virgil, Ovid, and Lucan, were among the world's five greatest poets met by Dante in Limbo; that next to Aristotle among the philosophers stood Socrates and Plato, and near these were Democritus, Anaxagoras, and Thales, Cicero and Seneca, Ptolemy, Avicenna and Averroës.[119]

Among Christian writers, Dante may have gathered some information about ancient Greek speculations from St. Augustine and Peter Lombard; and Orosius, the Spanish friend of Augustine, was his chief authority for that geographical system which, in connection with astronomy, plays an important part in the time-indications of the *Divina Commedia*. Orosius had a great reputation among geographers and map-makers. He had travelled much between Spain, Africa, and Palestine, and he devotes a chapter to describing the different parts of the earth in his history *Adversus Paganos,* which he wrote to prove that Christianity had not injured but benefitted the countries in which it had been accepted. This fifth century geography seems to have been largely based on the first-century geography of Strabo, who quotes Eratosthenes, Posidonius, and Hipparchus among his authorities. Orosius states that the land is entirely surrounded by Ocean, and is divided into three continents—Asia, with the mouths of Ganges in the middle of its eastern coast, Europe stretching vaguely very far to the north-east, and Africa a narrow and long strip from east to west between the Mediterranean and the southern ocean. This chapter is quoted in *De Mon.* II. iii. 87-90 and *Qu.* xix. 43; and Orosius is almost certainly "quell' avocato dei tempi cristiani,"[120] seen by Dante among the learned doctors in the Heaven of the Sun.

The speaker who points him out in these words also mentions his own name and that of the spirit nearest to him:—

"Questi che m'è a destro più vicino Frate e maestro fummi, ed esso Alberto È di Cologna, ed io Thomas d'Aquino."[121]

These two, so near in heaven, had been close companions on earth. The youthful Thomas, son of Count d'Aquino in southern Italy, after six years' study at the University of Naples joined the Dominican Order, and went to Cologne to learn from Albert, who was also of noble family but born in Suabia on the Danube. Together they went to Paris, together returned to Cologne, but after six more years there their paths separated: Albert rose to be Bishop of Ratisbon; Aquinas, after lecturing in Paris, Rome, and Bologna, became a professor at Naples, and died in Italy in 1274. His master survived him by six years, dying at Cologne at the advanced age of eighty-seven in 1280.[122]

These two are in the foremost rank among authors of his own time who influenced Dante. He quotes both writers and several of their books by name,[123] and though he never mentions it he was very familiar with the *Summa Theologica* of Aquinas. To the works of these famous authors he was frequently indebted for astronomical facts, theories, and history, as already noted. Besides this, his reverence for Aristotle, his belief in the essential harmony between religion and science, and his whole attitude towards knowledge, are greatly due to the influence of St. Thomas and his "brother and master," Albert of Cologne.

There are also three of Dante's own fellow-countrymen and contemporaries whose books he is most likely to have read, although he does not mention them:—Brunetto Latini the Florentine, Cecco d' Ascoli, professor at Bologna, and Ristoro the monk of Arezzo.

Brunetto Latini is only mentioned once in Dante's works, besides the passage in the *Inferno* already referred to. In the *De Vulgari Eloquentia* Brunetus Florentinus is mentioned with admiration as a distinguished man of letters, but blamed with other Tuscans for writing in his own local dialect.[124] Brunetto, who was born in Florence about 1210, was sent on an embassy in 1260 to Alfonso X. of Castile, the learned king under whose guidance the famous astronomical Tables had been drawn up. But on his way back to Florence he was met by the news that the Florentine Guelphs had been defeated at the disastrous battle of Montaperti,[125] and expelled from Florence; and as he belonged to this party he took refuge in France, first at Montpellier and afterwards in Paris. When the Guelphs had regained the ascendancy through their victory at Benevento in 1266 (where Manfred lost his life),[126] Brunetto returned to Florence, and his name subsequently appears in no less than thirty-five public documents as having been consulted by the government of his native city on various important matters. For the most part, moreover, it is recorded that his advice was followed. He died at a venerable old age in Florence in 1294, and was buried in the church of Santa Maria Maggiore.

It was while exiled in France that Brunetto wrote the Italian poem *Il Tesoretto*, in which he represents himself as a pilgrim on an allegorical journey; and from this Dante perhaps derived some suggestions for his Vision. The poem is incomplete, and breaks off at a tantalizing point, for Brunetto has just met Ptolemy on Mount Olympus, and has put a question to him: Ptolemy "rispose in questa guisa"[127]—and here the poem ends!

In France also he wrote *Li Louvres dou Trésor*, a great compendium of learning, in French prose. The first part treats of the Creation and Biblical history, and of the natural sciences; the second of ethics, rhetoric, and politics. In the section on astronomy which is included in the first part, Brunetto gives a very brief account of Ptolemy's system, and the sizes and distances of the heavenly bodies as estimated by the Arabs: it seems to be based chiefly on Alfraganus. But beside the periods of the planets he adds in each case their astrological properties, as for instance:—

"Saturnus, qui est le souverain sor tous, est cruex et felons et
de froide nature ... Jupiter ... ast dous et piteus et plains de
biens.... Mars et chaus et bataillereus, et mauvais, et est apelez
Diex de batailles."[128]

When speaking of the two principal movements of the skies, he quotes a curious old idea which had been mentioned by Isidore of Seville, that they are in contrary directions because the tremendous speed of the diurnal motion would shake the whole universe to pieces if it were not that the seven planets go as it were to meet it, and soften its vehemence:—

"Li firmamenz court de orient en occident entre jor et nuit
une fois, si roidement et si fort que sa pesanteur et so grandor la
feroient tout tressaillir, se ne fussent les VII. planetes qui vont
aussi comme a l'encontre dou firmament, et atemprent son cours
selonc son erre."[129]

The *Acerba* of Cecco d'Ascoli was another encyclopædic work, but in
Italian verse. It was very much the fashion in those days to undertake a
compilation of all kinds of knowledge, and this was doubtless very useful
when few books could be owned by any single reader. Cecco had a
considerable knowledge of natural science, and astrology was of great
importance in his eyes. Though Dante never mentions him or his work, he
can hardly have been ignorant of it, especially as he himself is mentioned in
it.

Nor does he mention Ristoro of Arezzo, yet we find many ideas and
expressions in Dante's writings which are also in the *Composizione del Mondo*.
Even if he did not borrow from it, the book is worth study by anyone
interested in the popular astronomy of the thirteenth century. We learn
from the author that he wrote his book in Arezzo, finishing it in the year
1282, that is about thirty years before Dante's *Convivio* appeared. The
opening words run:—

"Incominiciasi il libro della Composizione del Mondo colle
sue cagioni: composto da Ristoro di Arezzo in quella nobilissima
città ... etc."[130]

and the closing words are:—

"Compiuto e questo libro sotto li anni di Cristo nel mille
dugento ottantadue. Ridolfo imperadore aletto. Martino quarto
papa residente. Amen."[131]

We learn further that he was a monk and a native of Arezzo in an
interesting passage where he describes a total eclipse of the sun seen by
himself from his monastery: the sky was very clear, and it must have been a
specially dark eclipse, for besides Mercury he saw many stars; totality lasted
for as long as a man could easily walk 250 steps; there was a feeling of chill,
and birds and wild animals were so frightened that they allowed themselves
to be caught. Fra Ristoro made a calculation and found that on that day sun
and moon were in the same position in the sky. It was perhaps this event
which induced him to study astronomy.

The extent and depth of Ristoro's reading evidently bore no
comparison to Dante's, but he was able to sit writing in his quiet cell, day
after day, and had such books as he wanted at hand. The monastery seems
to have been furnished with astronomical tables, a celestial globe, and the

books of Alfraganus and other Arab writers. All these he had studied, and he was something of an observer too. Alfraganus must have been always at his elbow when writing, for he turns to it constantly, quoting (unlike Dante) the chapter referred to:—

"Alfragano pose nell' ottavo capitolo;" "è testimonio l'Alfragano nelli venti e due capitoli del suo libro;"[132]

one chapter is avowedly taken whole from Alfraganus, and where the opinions of "il grande Tolommeo"[133] are quoted they are evidently copied from the same source. Other Arab writers are quoted once or twice, for instance, Albumassar,

"il quale fu altissimo maestro d'astrologia,"[134]

and several lines are copied from a Latin translation of Algazel,[135] giving some astrological jargon about the twelve zodiacal signs.

We have already quoted from Ristoro's preface, showing the high opinion he held of astronomy. But it is disappointing to find that the greatest part of his book is devoted to the "cagioni," that is, to purely fanciful "reasons," for all the facts and fallacies concerning nature which he has here brought together.

He gravely argues about the constellation figures as if they were real pictures of animals and things pricked out by nature in stars on the vault of heaven, and not a human convention. He notes their paucity in the southern hemisphere, and that nearly all have their heads towards the north, and from this he draws the conclusion that the northern part of the sky is the nobler, for we can see it is the upper side, just as we know the top side of a book by the position of its letters.

It is for this reason, he thinks, that only the northern hemisphere of the earth contains land and is inhabited. Some, especially the great Averroës, had indeed held that there is inhabited land south of the equator, because the sun goes there. But this is because his movements of recession and approach are necessary to produce the seasons, and hence the growth of plant life, in the north. The constellations are upside down for the south, so they can have no effect there, therefore, there are no animals there (since every race of animals is under the protection of a constellation); therefore no plants, since they exist for animals, and therefore, Ristoro concludes his chain of argument triumphantly, there are no men, and no lands, for land without life would be useless.

Ristoro reproduces Alfraganus' description of the sun's movements as seen in different latitudes, quotes him as saying that the equatoral regions are inhabited, and Avicenna that the temperature there is equable because days and nights are always equal. The city Arym which is exactly on the equator has a perfect climate, and as it has two summers and two winters there are two harvests; moreover, all the stars in the sky are visible at the equator; therefore, the best astronomers, and the wisest and richest men, ought to live there, says Ristoro. When writing thus, he seems to have had before him one of the "climate maps" which were common in Europe in the eleventh and following centuries, having been introduced by Arab cartographers, who adopted the idea from Greek maps. For he goes on to point out that the first "climate" is much the longest, and that the others diminish gradually towards the pole, all the "terra scoperta"[136] being thus contained in one quarter of the earth, and having the shape of the moon when we see it half full.[137]

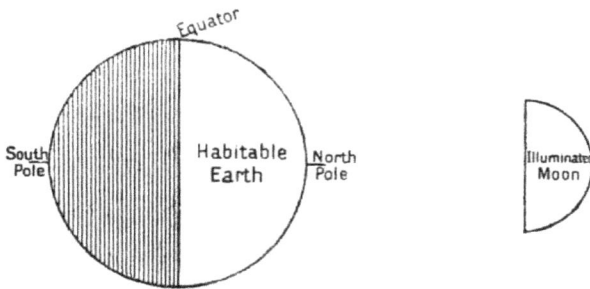

Fig. 38. The half-moon shape of the habitable earth. (Ristoro).

There is nothing but sea south of the equator, and the habitable earth lies wholly in one half of the northern hemisphere, the other half being sea. The habitable earth, therefore, is the same shape as the half-moon, *i.e.* a quarter sphere.

The accompanying climate map shows all these features, the famous city Arym—occupying nearly half the space, although it is supposed to be merely on the equator—the seven climates, and the region beyond the seventh, near the north pole, marked as uninhabitable through the cold. The half-moon shape of Earth's habitable quarter is obvious. The map betrays Arab influence in the orientation, the south being at the top; and Arym, or Aren, was an Arab myth, perhaps derived from the Mount Meru of the Hindus, which is said in the Vedas to be "in the middle of the earth."

According to Ristoro, the positions of the constellation figures do not only explain why one hemisphere of earth has land and the other none; they also explain the movements of the skies. For the figures of the zodiac face west (those that have faces), and this is why the diurnal movement of the whole heaven is towards the west, so that the constellations may move straightforward, in the natural way. On the other hand, the planets move in the opposite direction because it would not be seemly for them to pass each figure of the zodiac arriving first at the back and moving on to the head. This is the true reason of the contrary direction of the two principal celestial movements. Ristoro does not approve of the solution quoted by Brunetto Latini. He adds that the width of the zodiac was designed to contain the figures of the animals!

After this, we are not surprised to hear that the red colour of the planet Mars is the cause of its martial nature (not the reason for which such a nature is attributed to it). The markings on the moon Ristoro explains by saying that some body must exist in the World which is neither polished and shining all over, like the stars, nor altogether rugged and dark, as the earth was supposed to be, but partaking of both natures; and this body must be the moon, because of her intermediate position between the earth and the nearest of the other planets (Mercury).

[To face p. 248.

CLIMATE MAP, OF ABOUT A.D. 1110.

Reproduced by permission from Beazley's "Dawn of Modern Geography."

Of the influence of the heavens on the earth Ristoro has very much to say, from general statements such as that the spheres impress their influence on things of Earth, just as a seal impresses wax, down to details about the births of horses. If the moon is strong in Aries, and Aries is powerful, then, because the moon signifies white among the colours and Aries signifies the head, the horse will be born with a white mark on its head, and if the influence is very strong it will be a beautifully-shaped mark. Again and again he tells us about the various properties and "virtues" of the seven planets, but he does not omit to mention that the powers which move the planets and cause them to influence Earth are spirits, which we call Angels, and the philosophers call Intelligences. These have their several dwelling-places in the spheres, and the nobler their nature the higher is the sphere which they inhabit.

From these quotations the reader may judge of the general trend of the eight books comprised in *La Composizione del Mondo colle sue Cagioni*: we shall have occasion to refer to it again, by way of comparison and contrast, when dealing with Dante's works in detail.

It is rather curious that there is no trace in Dante's writings of acquaintance with the great Roger Bacon, or with Sacrobosco, especially as Brunetto Latini knew the former personally, and Cecco d'Ascoli wrote a commentary on the latter, but Dante may have known them, and also the works of Isidore and Bede, who are both mentioned in *Par.*, x. 131, as well as other books which are now completely lost. Those we have described above would, however, be enough to supply him with all the astronomical data, except one, which we find in his writings. It is now time to examine these, and see what use he made in literature of the knowledge he possessed.

IV.

ASTRONOMY IN DANTE'S WRITINGS.

If we except the seven *Penitential Psalms* and the *Profession of Faith*, which are only paraphrases, and are very doubtfully ascribed to Dante, no work of his can be mentioned which does not contain some reference to the heavenly bodies.

When writing the *Vita Nuova*, that story of his love for Beatrice which he calls the work of his boyhood,[138] he was haunted by the sublime idea of the rolling spheres. Beatrice at her first appearance as a little girl of eight, and Beatrice when she died, suggested thoughts of the nine heavens;[139] almost the first words of the book allude to the circling spheres of the sun and of the stars:[140] and the last sonnet soars beyond them all to the Empyrean.[141] Already he seems to have read Alfraganus, for he takes a suggestion which occurs in the first chapter of the *Elementa Astronomica et Chronologica*. Alfraganus there describes the different methods of reckoning days, months, and years, in use among different nations, and gives a special list of the Syrian months with their corresponding Roman months; and Dante, by using the Arabian system of days and the Syrian months, is able to prove to himself that the day and the month when Beatrice died were both at the sacred number of nine.[142] According to our method of reckoning, the time was sunset on the eighth of June, but this was the first hour of the ninth day according to Arabian usage, and our sixth month corresponds with the ninth of the Syrians.

Dante had also read at this time Aristotle's doctrines about the spheres and their Movers, for he quotes from the *Metaphysics*,[143] and this is the work in which they are found.

Yet a familiarity with Alfraganus and Aristotle so early as this is hardly consistent with the difficulty he says he found in understanding Latin books of philosophy some time after the death of Beatrice.[144] Perhaps the latter part of the *Vita Nuova* was written a good deal later; or it may be not altogether frivolous to suggest that in his youth Dante had read the first chapters of Alfraganus, which are easy, but did not master the rest of the book till later.

Out of his fifty-three short poems (including all in Moore's Oxford Edition, except those which form part of the *Vita Nuova* or *Convivio*), twelve, or nearly a quarter, contain some reference to sun, moon, or stars, planets, or spheres; and from these alone we could form some idea of Dante's acquaintance with astronomy. He speaks of the sun as measuring time,[145] and giving light to the stars[146] (according to the general belief); the moon[147] and each of the planets[148] is mentioned, some of the constellations also,[149] especially with reference to the seasons; the theory of the several heavens and their First Mover,[150] and the supposed influences of the stars and planets are alluded to.[151] These poems were written at different times, and some belong to a much later period than the *Vita Nuova.*

It is when we come to examine the *Convivio*, the work of his manhood,[152] that we find the clearest evidence of Dante's careful study of astronomy. This book was written with the professed intention (as one of its aims) of sharing with others the learning he had been happy enough to acquire—not indeed as one of the guests at the Table of Wisdom, but as one sitting at their feet and gathering up the crumbs.[153] The thread which connects his discourse is a collection of his own Odes, on each of which he had intended to write a treatise or commentary, but the book never advanced beyond the fourth treatise. These Odes were nearly all love-poems, but the poet explains that they are a figurative expression of his devotion to Philosophy, and the whole book is a glorification of the pursuit of knowledge. In the opening sentences, with the reference to Aristotle, the manner of reasoning, the assumptions of our innate desire for absolute knowledge and the bliss brought by its attainment, Dante is the spokesman of his period.

> "Siccome dice il Filosofo nel principio della *Prima Filosofia*,
> Tutti gli uomini naturalmente desideranno di sapere: La ragione
> di che puote essere, che ciascuna cosa, da providenza di propria
> natura impinta, è inclinabile alla sua perfezione. Onde, acciochè
> la scienza è l'ultima perfezione della nostra anima, nella quale sta
> la nostra ultima felicità, tutti naturalmente al suo desiderio siamo
> soggetti."[154]

And from a study of the *Convivio* we may learn much of the authors most esteemed, the methods of study pursued, and the results obtained, by thirteenth-century scholars seeking to gain the ultimate perfection of the soul.

Astronomy is frequently introduced, especially in the second and third treatises. The Ode which forms the text of the second treatise is that which is quoted by Charles of Hungary in *Par.* viii. 37—"Voi che intendendo il terzo ciel movete,"[155] and it gives Dante the occasion to speak first of the heaven of Venus, explaining Ptolemy's system of epicycles, and later of all the heavens in their order, and the celestial bodies contained in them. The ode of the third treatise contains the line "Non vede il Sol, che tutto il mondo gira ..."[156] and upon this he hangs a complete little essay describing the movements of the sun, and how they appear from different parts of the earth, as well as a short dissertation on the question whether it is Sun or Earth which actually moves.

The Latin works, *De Monarchia*, *De Vulgari Eloquentia*, the *Eclogues*, and the *Letters*, do not give much scope for astronomical references; yet there are many similes drawn especially from the sun and moon, and the thought of the spheres as instruments of God's Will in bringing about events on earth is constantly recurring. In the curious work *Quæstio de Aqua et Terra*, Dante argues learnedly about the spheres, the orbit of the moon, and her effect on the tides, and the influences of the stars. The book purports to have been written at Verona in 1320 as the outcome of a discussion in which Dante had taken part as to the respective heights of land and ocean, a problem which had been dealt with by Ristoro d' Arezzo some forty years earlier. Its authenticity has been questioned, partly under the mistaken idea that facts such as gravity, the spherical form of Earth, and the connection of tides with the moon were not known until much later; but as a matter of fact the discussion is quite in the manner of Dante's day, and no facts or theories are put forward which he could not have learned from books with which we believe he was acquainted.[157] The internal evidence of his authorship is strong, and belief in it seems to be gaining ground among experts.[158]

Finally, the *Divine Comedy*, the work of Dante's maturity, which took him so many years to write, and for which he had studied ever since he closed the *Vita Nuova* with the resolve to write more worthily one day of Beatrice,[159] focuses in one unique and finished work the thoughts and ideals, the knowledge and fancies, of the poet and his age. The subject of the poem, taken in its literal sense, is cosmical, for it describes a journey in which the author penetrates to the centre of the universe, and passes from planet to planet until he reaches the outermost sphere. Moreover, the journey is assumed to take a definite space of time, and the passing of the hours by day and by night is indicated by the successive positions of sun or stars, and the phases and movements of the moon. The *Divine Comedy* contains more than twice as many allusions to the heavenly bodies as all the other works of Dante put together.

But we must discriminate in drawing conclusions from these astronomical references so freely distributed in all Dante's writings. If he wrote the *Quæstio*, it is his only professedly scientific work, written for scientific men, and astronomy is only brought in incidentally. The *Convivio*, in which astronomical facts and theories are set forth in much greater detail and fulness than elsewhere, was a popular work meant for those who had no inclination, or else no opportunity, for prolonged serious studies. We expect, therefore, that technical details will be avoided. Still more will they be omitted in the *Divine Comedy* and other poetical works, where difficult problems and pedantical accuracy would be most unsuitable. We can *infer* the knowledge which lay in Dante's mind, behind his popular use of it in literature; and we can often find an explanation in the prose of the *Convivio* for a slight allusion in the poetical works; but we must not deal with any as if they were text-books, and set forth precisely and completely all that Dante knew of his favourite science. It is the poet's artistic use of the astronomy of his day which merits our admiration quite as much as the scholar's proficiency.

This being premised, we may now proceed to quote from his works, showing how he has dealt with first the facts, and secondly the theories, of astronomy. Finally we shall be able to form a clear mental picture of the universe as it was believed to exist by Dante, and not only by his contemporaries but by his successors for many generations.

V.

OBSERVATIONAL ASTRONOMY.

"The heavens declare the glory of God; and the firmament sheweth His handy-work. Day unto day uttereth speech, and night unto night sheweth knowledge. There is no speech nor language where their voice is not heard. Their line is gone out through all the earth, and their words to the end of the world. In them hath he set a tabernacle for the sun; which is as a bridegroom coming out of his chamber, and rejoiceth as a strong man to run a race."

1. MOVEMENTS OF THE MOON.

Of all the seven sciences of the Trivium and Quadrivium, Astronomy, Dante thought, was the noblest, and for two reasons. Aristotle had said that a science is noble in proportion to the nobility of its subject, and the certainty of its conclusions; and in both of these Astronomy excels. Its subject is the Movement of the Heavens, and its certainty is perfect. "E nobile e alta per nobile e alto suggetto, ch'è del movimento del cielo: è alta e nobile per la sua certezza, la quale è senza ogni difetto."[160] If astronomers are sometimes mistaken, the fault lies in them, as Ptolemy said, and not in the science.

This quotation from Ptolemy is from his book on "judicial astronomy,"[161] and it may be that Dante was partly thinking of astrological predictions, as he almost certainly was in the passage that follows, where he adds that astronomy takes a long time to learn, not only because of its great range, but because experience is necessary to form a correct judgment. Nevertheless, it is probable that by its flawless certainty he meant that unchanging laws govern the celestial phenomena, so that they may always be predicted without error when the laws are known.

His definition of astronomy was accurate, for it was not until the middle of the nineteenth century that physical astronomy began to take its place beside the old "astronomy of position," and in Dante's day practically the only subject open to research concerning the heavenly bodies was their movements. Of these, and especially the "prime motion" of the diurnal revolution, he was vividly conscious. His favourite name for the heavens is "wheels"—"le ruote magne," "eterne ruote," "stellate ruote";[162] and he often refers to them as a standard of motion. Thus when with Beatrice he was rapt from the summit of Mount Purgatory, to express the swiftness of their flight he says it was almost as rapid as the movement of the sky— "veloci, quasi come il ciel vedete."[163] Constantine moving the capital of the

Roman Empire eastwards is described as turning the Roman Eagle contrary to the course of the sky—"contra il corso del ciel,"[164] and the sky itself is described as a sphere which is ceaselessly at play like a lively child:—

... "la spera, Che sempre a guisa di fanciullo scherza."[165]

To Dante, as to the Greeks, it is not the unusual or startling that appeals, but the unfailing harmony of the regular celestial movements. Comets he mentions only twice, and shooting stars twice, eclipses seven times; but there are scores of allusions to the rhythmical progression of Sun and stars, Moon and planets. The skies were familiar to him in all their daily aspects—sunset and dawn and radiant noons; cloudy nights "sotto pover cielo,"[166] brilliant starlit nights, nights of clear moonlight. He had seen the stars fade one by one, till at last even the brightest vanished in the glow of dawn:—

"E come vien la chiarissima ancella Del sol più oltre, così il ciel sì chiude Di vista in vista, infino alla più bella."[167]

he had watched for their first appearance in the evening twilight:—

"E sì come al salir di prima sera Comincian per lo ciel nuove parvenze, Sì che la vista pare e non par vera...."[168]

Numberless other passages and expressions will occur to every Dante reader, proving how keenly he felt the beauty of the skies.

But they prove more than this. Other authors have felt and have described poetically the beauties of the skies, but they often remember imperfectly what they saw, or draw upon their imagination without any knowledge of the celestial movements, and so fall into absurd mistakes. The moon especially is a stumbling-block, and it is quite a rare thing for a modern novelist to introduce one without making it do something impossible. A new moon will rise at midnight, or a waning moon at sunset; she has even been known to rise and then to set in the dark hours of one short midsummer night;[169] and a well known author sees her in two phases at the same moment: "the full moon rose, yellow *and gibbous*!"[170]

Dante's moon does indeed give us a little trouble once or twice, but he never makes flagrant mistakes of this kind. His consistency and truth of description prove his knowledge of astronomy, and also imply intelligent thoughtful watching of the celestial movements; for it is notorious that book-knowledge, if unassisted by acquaintance at first-hand with the facts of a subject like astronomy, will not save a writer from glaring inaccuracies.

His allusions to the motions of the moon, both diurnal and monthly, show plainly that he understood them well. His new moon appears in the evening, his waning moon rises late at night and sets in morning sunshine,[171] and his full moon comes on the meridian at midnight, for when he wishes to describe her at her brightest, as a comparison with a very brilliant light, he places her in a clear sky at midnight in her mid-month.[172] In the middle of a lunar month the moon is full, and being exactly opposite the sun will reach the meridian at midnight: this therefore is the time when she gives all the light she possibly can. In the *Quæstio*[173] Dante mentions that the moon does not move in the celestial equator, but sometimes north of it, and sometimes an equal amount south; in the *Convivio* he indicates the length of her period, and briefly describes her phases and their cause—"Ora luce da un lato, e ora luce dall' altro, secondo che 'l sole la vede."[174]

Her position in the zodiac is mentioned several times. In one passage he indicates a very brief space of time in the following curious way:—

"Quando ambo e due i figli di Latona, Coperti del Montone e della Libra, Fanno dell' orizzonte insieme zona, Quant' è dal punto che il zenit inlibra. Infin che l'uno e l'altro da quel cinto, Cambiando l'emisperio, si dilibra, Tanto, col volto di riso dipinto, Si tacque Beatrice."[175]

That is to say, when Sun and Moon (the children of Latona) are both on the horizon, but one being in Aries and the other in Libra they are opposite one another, the zenith for a moment holds them, as it were, in balance; but the next moment one will drop below the horizon while the other rises above it, thus changing from the visible to the invisible hemisphere and *vice versa*, as each frees itself ("si dilibra") of the common horizon that girdled them. The pause of Beatrice was as brief as the time during which sun and moon would thus hang in the balance.

Yet, though Dante shows that he was familiar with the movements and appearances of the moon, his allusions to her are cold and comparatively rare. Unlike modern poets, moonlight does not seem to have had any great fascination for him. In all his works there are only fifty-one references to the moon, and far the greater number of these are remarks about the measurement of time, or else her phases, her markings, her share in causing eclipses etc. In all his short poems (including those of the *Vita Nuova* and *Convivio*) there are but two references, and those as dry as possible. "Più lune"[176] is once used, meaning several months, and in a sonnet which describes the influences of each heaven on his lady, all he can find to say of the moon's is "E 'l primo ciel di sè già non l'è duro."[177]

The chill of moonlight is spoken of as with a shudder on the hill of Purgatory;[178] and its usefulness is recognized rather grudgingly when Virgil remarks that yesternight the moon was full, and adds that Dante must remember it, since it did him no harm in the depths of the Forest.[179] It is true there is some beauty in the description of the aurora which preceded moonrise[180] on the first night in Purgatory, but Dante gives it an ugly name, and the waning moon when seen on the following night, four days after full, is oddly compared to a bucket in shape. Dante did not admire this gibbous form, for he instances the outline of the moon when not quite full as an ignoble curve, contrasting it with the beauty of a perfect circle.[181]

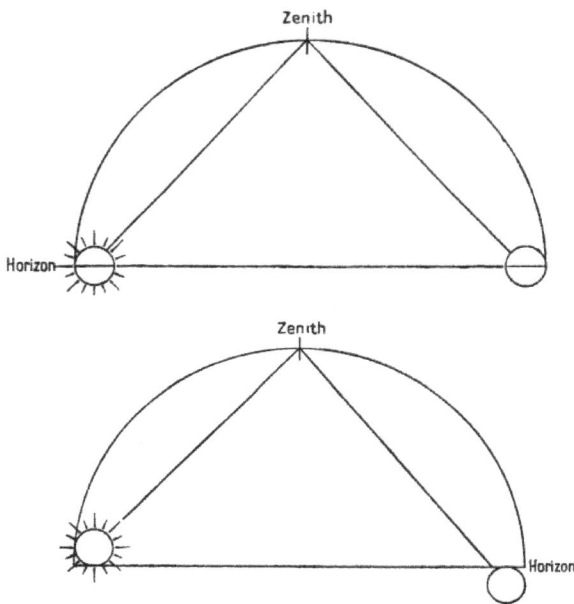

Fig. 39. Sun, moon, and zenith.

To illustrate *Par.* xxix. 1-6. (*See p. 265*).

In the first figure the sun and moon are balanced, as it were, being at equal distances from the zenith: in the second, a few minutes later, by the rising of the sun and the setting of the moon, the balance is disturbed, and each changes its hemisphere.

Hardly ever does he find a beautiful epithet for her: she is ruler of Hell,[182] Cain and the Thorns[183] (in reference to the legend which sees this figure in her dark markings); or she is simply Delia,[184] Trivia,[185] the daughter of Latona,[186] or with a little warmth the sister of the sun,[187] and with the sun one of the two Eyes of Heaven.[188] One exception to this harsh treatment is in the beautiful description of reaching the heaven of the moon,[189] for there she is the first star, the eternal pearl; and the sudden burst of eloquence in the midst of a grave argument in *De Monarchia* startles us doubly when we find that Justice is compared to the moon—not, as one might expect, resplendent in a dark sky, but Phœbe gazing at her brother opposite, in the purple of the morning calm.[190] The passage from *Par.* xxix. quoted above helps us to understand why the moon in this position should symbolize Justice, for here again is suggested the perfect balancing of the two great orbs, as the sun rises and the full moon sets.

Add to these passages two similes drawn not indeed from the moon itself but from the halo surrounding her,[191] and two descriptions of brilliant moonlight nights,[192] and we have all that Dante has written in praise of the moon. It is true that throughout the *Inferno* and *Purgatorio* the sun typifies Divine Grace, and the moon is his opposite, but this cannot altogether explain the difference in their treatment, even in the *Divine Comedy*. A deeper reason seems to be Dante's true southerner's love of the sun; and he doubtless had the astronomer's feeling that moonlight always means loss of starlight, which he dearly loved (see *Purg.* xviii. 76, 77). Perhaps too the astrological views of the moon influenced him, consciously or not. According to Ristoro, the moon represents the poorest and lowest classes of people, who are servants and messengers to the rest; and as all the planets were supposed to be either masculine or feminine, he ungallantly adds—"E questa Luna, a cagione di sua viltà, potemo dire per ragione ch'ella sia femmina."[193] Her only house, too, which is Cancer, is also poor and base. The moon, moreover, is nearer Earth than any other planet, and shows by her dark markings that she is less pure than the rest.

2. MOVEMENTS OF THE SUN.

Whatever the reason, Dante's fifty-one references to the moon are in strong contrast to his allusions to the sun. These are four times as many, and they include warm expressions of admiration and delight, and beautiful epithets and similes. The sun is Titan with his chariot and horses,[194] he is a Mirror, a Car of Light, the Lantern of the World, the Prince of the Stars, the Father of all mortal life.[195] His beautiful form is admired, "la perfetta sua bella figura,"[196] and his intense brilliance often noted.[197] He is the guide that leads all aright,[198] to whom Virgil appeals:—

"O dolce Lume, a cui fidanza i' entro Per lo nuovo cammin, tu ne conduci."[199]

Under his rays the rose expands, the air is gladdened, mists are dispelled, snow melts, and all things are quickened into life.[200] Unlike the moon, the sun is the bringer of warmth and comfort, chasing away cold and darkness,[201] sunrise is the hour of renewed hope and confidence,[202] it strengthens our limbs, paralysed by the cold of night,[203] it is eagerly awaited through the darkness by the little bird on her nest;[204] and delicate flowers, bowed and closed during the cold night, rise upright on their stems and open as soon as the sun shines once more upon them.[205] The exquisite sunsets and sunrises of the *Purgatorio* are favourite passages with all readers.

Love is compared with the summer sun,[206] the generosity of a noble nature resembles "the great planet."[207] The Emperor Henry VII., from whom Dante expected the regeneration of Italy and the whole Christian world, is likened to the rising sun;[208] the birth of St. Francis is described as "nacque al mondo un sole,"[209] and his birthplace ought rightly to be called not Assisi but Orient.[210] Virgil is addressed as "O Sol che sani ogni vista turbata;"[211] Beatrice is "Il sol degli occhi miei."[212] Finally the sun is the best symbol of God,[213] and so it is used many times in the *Divine Comedy*: "Il Sol degli angeli,"[214] "Il Sol che raggia tutto nostro stuolo;"[215] a spirit turns to "il Sol che la riempie,"[216] and when Virgil laments that he is shut out of heaven it is in these words: "Ho i' perduto Di veder l' alto Sol che tu disiri."[217]

But we may not dwell on those thoughts of the sun as the typical giver of light and life; we must ask what Dante says about his movements, since this was the chief subject-matter of mediæval astronomy.

His daily course, "il cammin del Sole,"[218] is often spoken of, and is indicated both by his position in the sky, and the length and direction of shadows. The second *Eclogue* begins with a description of a breathlessly hot midday, when objects, which are usually shorter than their own shadows, now surpassed them in length.

"Resque refulgentes, solitae superarier umbris, Vincebant umbras." *Ecl.* ii. 5, 6.

At this time the sun appears most brilliant, and seems to move most slowly, since he is scarcely changing his position with regard to the horizon.

"E più corrusco, e con più lenti passi, Teneva il sole il cerchio di merigge."[219]

The greatest number of references to the sun's daily journey occur in the *Purgatory*, and we shall quote them in a later chapter.

His yearly journey through the zodiac is also very often referred to. When relating his first meeting with Beatrice, Dante names his age, not by simply saying that he was nearly nine years old, but by counting the number of the times that the "heaven of light," that is, of the sun, had returned to the same point since his birth:—

> "Nove fiate già, appresso al mio nascimento, era tornato lo cielo della luce quasi ad un medesimo punto, quanto alla sua proprio girazione."[220]

The sun's "own revolution" is of course its apparent yearly movement, as distinct from the diurnal movement which it shares with all the heavenly bodies.

The same thought is expressed in a sonnet:

"Io sono stato con Amore insieme Dalla circolazion del Sol mia nona."[221]

We know further that Dante was born in either May or June, because he says, when he finds himself among the stars of Gemini, that the sun was rising and setting in this sign at the time of his birth.[222]

"O gloriose stelle Con voi nasceva e s'ascondeva vosco Quegli ch'è padre d'ogni mortal vita, Quand' io senti' da prima l' aer tosco."[223]

Boccaccio tells us that a Mayday feast, given by the father of Beatrice, was the occasion of Dante's first meeting with her, and also that not long before his death in Ravenna he told a friend that he had completed his fifty-sixth year in the preceding May, so we may conclude that May was the month of his birth.

Dante seems to have taken some trouble to find out the exact period of the sun's revolution. His text-book, the *Elementa* of Alfraganus, only gives it as 365¼ days nearly:—"Sol ... orbem confecit diebus 365 et propè ¼." Ristoro d'Arezzo and Brunetto Latini were content to repeat this rough estimate, and call the year 365 days 6 hours, but Dante wished to be more exact, and somehow contrived to obtain a value which was much nearer the modern estimate of 365 days, 5 hours, 48 minutes, 46 seconds. We know this from two passages. In the *Convivio* he gives the half period as less than 182 days 15 hours; and in the *Paradiso* Beatrice says that January will in time cease to be a winter month, because of a neglected fraction of time.

"Prima che gennaio tutto si sverni Per la centesma ch' è laggiù negletta."[224]

January was in fact receding from the winter, and steadily though slowly advancing towards the spring, because the Julian year (as we saw, p. 171) was 11 minutes 14 seconds longer than the true solar tropical year.

When Julius Cæsar reformed the Calendar, March 25 was made (as of old) to coincide with the spring equinox, but by Dante's time this date was 13 days late, and the true equinox fell on March 12. Consequently though the day called January 1 still came 84 days before March 25, it was only 71 days before the spring equinox.

If Beatrice meant to speak with scientific precision, and was correct in her hundredth part of a day per annum, that is, if the difference between the astronomical and the Julian year was one day in a hundred years, it would be 7100 years before the first of January came to coincide with the spring equinox, and January became a spring month. But the error quoted is really rather too large, for a hundredth part of a day is 14 minutes 24 seconds, and it was in fact only 11 minutes 14 seconds.

It would be interesting to know where Dante found this value, which is practically the only astronomical datum not given in Alfraganus. Ptolemy, following Hipparchus, had given the tropical year as $1/300$ part of a day less than $365\frac{1}{4}$; Albategnius gave a more accurate figure, and one which agrees well with the "centesma," for his tropical year differs from the Julian by a little less than $1/106$ of a day; but he also improved the value of precession, and Dante does not seem to have known this, so he was probably not acquainted with this astronomer's work. The Alfonsine Tables gave a value which was very close indeed to the correct one, making the tropical year only 30 seconds longer than the modern value, and the difference between the Julian and true year $1/134$ of a day. (The correct value is $1/128$). Either of these figures may have been known to Dante indirectly: perhaps he had obtained the Alfonsine value in conversation with Brunetto Latini, who had visited Alfonso's court. His fraction was evidently only approximate, but it is interesting to know that he had found out more than popular books could tell him about the period of the sun's revolution, and that he had been sufficiently struck by the small difference between the calendar and astronomical years to mention it in a picturesque way in his poem. It is possible that he had heard of Roger Bacon's appeal to the Pope in 1267 to correct the calendar because the date of the spring equinox was of importance in connection with the observation of Easter.

It is hardly necessary to remind my readers that this was done by another Pope three hundred years later, before January had made much more advance towards becoming a spring month. The hundredth part of a day, of which Beatrice complained, is now subtracted from the year by omitting to make century years leap years; but as the error was really not quite so great as she said, an exception is made for century years divisible by 400. Thus 1600 was a leap year, but not 1700, 1800, or 1900.

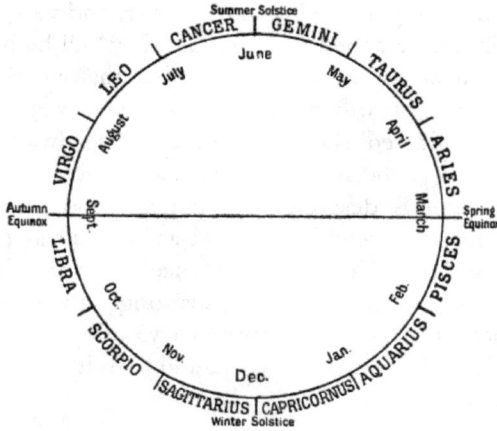

Fig. 40. The Zodiac and the Months.

Dante seems to have connected any season of the year with the sun's path in the zodiac as readily as with the name of a month. The sun's yearly course through the signs is shown diagrammatically in figure 40, and a glance at this will make clear a number of allusions without any explanation.

Starting from the first point of Aries at the spring equinox, which in Dante's time fell in the middle of March, and taking one-twelfth of the year to traverse each sign, it is evident that during the end of March and beginning of April the sun was in Aries; he entered Cancer and reached the summer solstice in June, the autumnal equinox in Libra in September, and the winter solstice when he entered Capricornus in December. The six signs above the horizontal line are northern, the six below are southern.

The following are a few typical examples.

When Dante wishes to express the transition of his feelings from despondency to courage, at seeing the stern and troubled face of his guide melt suddenly to a smile, he compares himself to a poor shepherd who wakes early and sees the fields covered, as he thinks, with snow, but in a short time finds that it is only hoar-frost which has put on the semblance of her white sister, and seeing the face of the world changed he gladly leads out his sheep to pasture. The time at which this may happen is

"Quella parte del giovinetto anno, Che il sole i crin sotto l'Aquario tempra, E già le notti al mezzodì sen vanno."[225]

That part of the youthful year when the sun cools his locks beneath Aquarius is clearly the end of January or beginning of February, but the last line may bear one of two meanings. If we translate "mezzodì" as "half the day" it means that the nights are growing shorter, and very soon (in March) when the sun will have reached the equinox, they will be just twelve hours long. If we take "mezzodì" as meaning "the south," we must interpret the passage in the light of others in the *Divine Comedy* where Night is personified, and considered as circling in the zodiac always opposite to the sun. For instance in *Purg.* ii. 1-6 (a passage which will be more fully explained later) Night is described as circling opposite to the sun, "La notte, che opposita a lui cerchia,"[226] and as the sun at that time was in Aries, Night was in Libra, "le Bilance." Henceforth for six months the sun will be in the northern signs, and the days longer than the nights, but as soon as Night begins to assert her supremacy, and the nights begin to be longer than the days, the sun enters Libra and then it is not visible all night. This is the meaning of

"le Bilance, Che le caggion di man quando soverchia."[227]

When Night becomes dominant, the Scales fall from her hand, for the sun is in that sign.

Applying this idea to the passage quoted above, it will mean "The nights are going southward," for the sun in Aquarius is not far from the point where he crosses the equator to the north, therefore Night circling opposite to him is nearing the point where she goes south. Either interpretation is faithful to fact, but perhaps the circumstance that the plural noun is used favours the first.

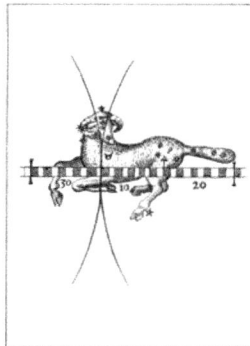

[To face p. 279.

Fig. 41. The Ram on the Ecliptic.
From the "Cosmi Historia" of Robert Flud, a.d. 1612.

(*Reproduced from Brown's Phainomena of Aratos.*)

Conrad Malaspina prophesied to Dante that his favourable opinion of the Malaspina family would be justified "before the sun had come to rest again seven times in the bed which the Ram covers and bestrides with his four feet."

"Ed egli: Or va, che il Sol non si ricorca Sette volte nel letto che il Montone Con tutti e quattro i pie copre ed inforca Che cotesta cortese opinione ..."[228]

The sun was at the time in the Ram, and Malaspina means that less than seven years would elapse, before his prophecy was fulfilled. The aptness of the description may be seen by the figure of Aries as portrayed in an old seventeenth century book in which the Ram is uncomfortably trying to sit on the ecliptic.

Another instance is in the beautiful midwinter Ode, "Io son venuto al punto della rota."[229] Each stanza pictures a feature of a severe winter—snow and rain, the absence of summer birds, bare trees and dead flowers, the favourite walk become a torrent; but the first describes the position of stars and planets. When the sun sets, the Twins appear on the eastern horizon, hence we know at once that the sun is in Sagittarius and the time is November or December. To confirm his melancholy, the poet adds that the Star of Love (Venus) is hidden from us by the sun's rays which now shine athwart her, and that the planet which strengthens the cold (Saturn, the "frigida stella" of Virgil) displays himself in that arc of the sky in which all the seven planets cast the shortest shadow;[230] that is, either he is on the meridian at sunset, or he is in the most northerly part of the zodiac, in Gemini or Cancer. A planet in this place, as for instance the sun at midsummer, is visible longer, describes a longer arc, and casts shorter shadows, than in any other part of the zodiac. And as this is the part which rises over the eastern horizon as the sun sets, Saturn would also rise at that time and remain visible all night.

The opening lines of the Ode run thus:—

"Io son venuto al punto della rota, Che l'orizzonte, quando il Sol si corca, Ci parturisce il geminato cielo; E la stella d'amor ci sta rimota Per lo raggio lucente, che la 'nforca Sì di traverso, che le si fa velo; E quel pianeta che conforta il gelo Si mostra tutto a noi per lo grand' arco, Nel qual ciascun de' sette fa poca ombra."[231]

It is astonishing that Giuliani should so completely miss the point of the "twinned sky," that he substitutes "ingemmato" for "geminato," and reads—

" ... la rota, Ch' all orizzonte, quando il Sol si corca, Ci parturisce l'ingemmato cielo."[232]

He understands the hour of evening to be meant, when the Wheel of Day and Night brings a jewelled, *i.e.* starry sky on the horizon as the sun sets. But it is not only on the horizon that stars appear at sunset; and the mention of Gemini to indicate the time of year is thoroughly characteristic of Dante. The Wheel is the revolving year which has carried the sun into Sagittarius.

Angelitti remarks that the whole description was literally true for December 1296, since Venus was then in conjunction with the sun, and Saturn in Cancer;[233] and as this poem is a complaint of the hardness of his lady, to whom the poet is nevertheless wholly devoted, it may well have been written in that time when the lady Philosophy refused to smile upon her lover.[234]

Further on in the same poem the effects of the sun in spring, when he is in the sign of Aries, are alluded to as "the virtue of Aries":—

"Passato hanno lor termine le fronde, Che trasse fuor la virtù d'Ariete, Per adornare il mondo, e morta è l' erba."[235]

Contrast with this the passage in the *Paradiso* where "nocturnal Aries" is used as a synonym for autumn.

"Questa primavera sempiterna Che notturno Ariete non dispoglia."[236]

For although Aries may be seen during some part of the night in the greater part of the year, it is most emphatically a nocturnal sign when it rises as the sun sets, and remains above the horizon until he rises; this happens when the sun is in the opposite sign of Libra, which he enters at the autumnal equinox.

A rather curious passage in the *Paradiso* is incomprehensible unless we think of the sun's path in the zodiac, and realize that Dante is comparing the brightness of the spirit of St. John in an indirect way with the brightness of the sun.

"Poscia tra esse un lume si schiarì, Sì che, se il Cancro avesse un tal cristallo, L'inverno avrebbe un mese d'un sol dì."[237]

If a light as dazzling as this spirit were to shine forth in Cancer, there would be perpetual day for a whole month in winter. For when the sun entered Capricornus, which he does in December, he would be exactly opposite, so that as one light set the other would rise, and there would be no darkness until the sun passed into another sign.

To many similar instances the diagram (p. 276) will be found to supply a key, and some we shall have occasion to notice presently in another connection.

3. THE STARS.

If our poet did not love moonlight, there is no doubt that starlight was very dear to him. Never are stars spoken of as cold, or placed in antithesis to the sun. Rather are they classed together, as in the pathetic letter, written when he heard of the possibility that the Florentines might receive him again if he would consent to return as a disgraced but pardoned criminal. If this is the only path to Florence, never will he re-enter the beloved city. "What then?" he cries. "Can he not see the mirrors of the sun and of the stars, wherever he may be? can he not meditate on precious truths under any sky?"[238]

He describes the Inferno as bereft of stars as well as of sun;[239] he hopes to escape those dark abysses to see the beautiful stars again,[240] and the terror of the sounds of weeping and crying is heightened indescribably by the simple words:—"risonavan per l'aer senza stelle."[241] When he does at length come forth to see "le cose belle che porta il ciel"[242] how eagerly he gazes, not only at Venus, but at the new stars in the south! how his "greedy eyes"[243] seek the same region as soon as dusk begins to fall! and how radiantly the stars look down upon him on that last night on the rocky slopes of the Mountain, near the summit![244]

This is all allegorical, no doubt, but it is because of Dante's feeling for the real "belle stelle"[245] that he uses them as symbols of truth and holiness, and concludes each Cantica of the *Divine Comedy* with their name:—

"E quindi uscimmo a riveder le stelle."[246] "Puro e disposto a salire alle stelle."[247] "L'Amor che move il sole e l'altre stelle."[248]

Here "stelle" has the wider meaning of the French "astres," for which we have no equivalent in English except the clumsy "heavenly bodies;" but most of Dante's hundred references are to stars in the more special sense, and often to particular stars or constellations. Over eighty occur in the *Divine Comedy.*

Stars are jewels, torches, flames, immortal nymphs adorning every region of the sky.[249] Heaven is made beautiful by their light, and the joy of the angels is expressed in their shining, as mortal joy shines forth in human eyes.[250] If the spirits in the fourth heaven are described as glowing suns,[251] the "splendours" that descend upon the golden ladder in the seventh, make Dante think that all the stars in the sky are gathered together there.[252]

The eyes of Beatrice shone brighter than stars;[253] light comes as from many stars in reading the sacred books;[254] faith gleams like a star in the sky;[255] and truth appears in a mind cleared of falsehood and error like the radiance of stars in a sky which has been wholly swept of cloud and mist by a north wind.

"Come rimane splendido e sereno L'emisperio dell' aer, quando soffia Borea da quella guancia ond' è più leno, Perchè si purga e risolve la roffia Che pria turbava, sì che il ciel ne ride Con le bellezze d'ogni sua paroffia: Così fec' io, poi che mi provvide La Donna mia del suo risponder chiaro, E come stella in cielo il ver si vide."[256]

Stars, as well as the sun, are used as symbols for the objects of Dante's deepest reverence; for the Blessed Virgin is called "la viva stella,"[257] and the Final Vision in which the redeemed find their ultimate bliss is a Trinal Light seen as a single star:—

"O trina Luce, che in unica stella Scintillando a lor vista sì gli appaga, Guarda quaggiù alla nostra procella."[258]

The diurnal motion of the stars is referred to many times. As Dante watched the mystical procession in the Earthly Paradise, and saw one group follow in the footsteps of another across the flowers and the grass, he thought of the stately procession of stars which we see here by night, star following star across the sky.[259] Their movement marks for him the passage of time, as we see in the first sonnet of the *Vita Nuova*:—

"Già eran quasi ch' atterzate l' ore Del tempo che ogni stella è piu lucente."[260]

and in the *Inferno*, where Virgil hastens Dante, saying that all the stars are sinking which were rising when he started on his long journey.[261]

The motion of those stars which neither rise nor set, but are always seen circling round the pole, is clearly described. In the heaven of the sun, spirits gather round Dante and Beatrice, and circle round them, "Come stelle vicine ai fermi poli;"[262] and the motion of the stars near the pole is said to be slow, like the part of a wheel which is near the axle.[263] We may compare these descriptions with the statement of Alfraganus:—"Eæ stellæ (*i.e.* the northern circumpolar) vertuntur omnes circa idem punctum. Et quei ex iis puncto huic est vicinior, minorem conficit circulum: motusque ejus appâret lentior."[264] Dante assumes that his readers understand this motion when he describes himself as seeing unknown stars near the South Pole. The Mount of Purgatory is supposed to be situated in the southern hemisphere, and at his first arrival, before dawn, he turns to the right, after looking at Venus in the east, therefore to the south, and sees near the pole four stars so brilliant that the whole sky seems to rejoice in their radiance, and he pities the "widowed" northern hemisphere because they are invisible there.[265] In the evening, as soon as it begins to grow dark, although he is engaged in an interesting discourse with a friend with whom he had exchanged warm greetings, he looks eagerly once more towards the pole (note the intensity expressed by the repeated "pure")—

"Gli occhi miei ghiotti andavan pure al cielo, Pur là dove le stelle son più tarde, Sì come rota più presso allo stelo."[266]

He now sees three bright stars which make this pole glow with light, and Virgil says that the stars he saw in the morning are low, while those now visible have risen higher to take their place.

"Ed egli a me: Le quattro chiare stelle Che vedevi staman son di là basse, E queste son salite ov' eran quelle."[267]

The diurnal movement of the starry heaven is also alluded to in the *Convivio*, and contrasted with that immensely slow movement which was discovered by Hipparchus, and is called by us Precession. The starry heaven, says Dante, displays one of its poles to us, and keeps the other hidden; and in like manner it displays only one movement to us, and keeps the other almost hidden. By the first, it revolves once in every day from east to west; the other is nearly insensible, being only one degree in a hundred years, and it is from west to east.[268] This is the value of precession as given by Alfraganus, following Ptolemy.

It was this stupendous and mysterious cycle which was used by Dante to measure the age of Beatrice, though for his own he used (as we saw) the ordinary measure of the sun's period. When he first saw "la gloriosa donna"[269] she had only been in this life so long that the starry heaven had moved towards the east one-twelfth part of a degree. Therefore, she was one-twelfth of a hundred years old, or 8 years 4 months, "so that it was near the beginning of her ninth year that she appeared to me, and I saw her nearly at the end of my ninth."[270]

Dante tells us how many stars had been counted by "the sages of Egypt," by whom he means Ptolemy and the other Alexandrians: he did not know that the star catalogue of the *Almagest* was originally made by Hipparchus of Rhodes.

> "Dico ch' il cielo stellato ci mostra molto stelle; chè, secondochè li savi d'Egitto hanno veduto, infino all'ultima stella che appare loro in meridie, mille ventidue corpora di stelle pongono, di cui io parlo."[271]

It will be noted that he carefully avoids saying that this is the number of all the stars visible from the whole earth. Had he, like Ristoro, run away with the idea that a blank space on the globe meant a blank in the sky, his night sky seen from Purgatory would have been strangely bare and dull! Yet both writers used the same text-book. Dante is here following Alfraganus closely, for he had written in his nineteenth chapter:—

> "Sciendum itaque sapientes inivisse mensuram stellarum fixarum omnium, quoad instrumentis observari eæ potuerunt, extremam usque meridiei partem, in tertio climate ipsis conspicuam.... Stellæ universae quarum agi mensura potuit sunt mille viginti duæ."[272]

Both in the *Quæstio* and the *Paradiso* Dante notes how stars differ, not only in their brightness or "magnitude," but also in the quality of their light, by which he probably means their colour:—

> "Videmus in eo [*sc.* cœlum stellatum] differentiam in magnitudine stellarum et in luce."[273]

"Lumi, li quali nel quale e nel quanto Notar si posson di diversi volti."[274]

It seems as if it were the beauty and the movement of the starry sky as a whole which appealed to Dante, rather than the distinguishing of special stars and constellations. Those which he mentions are almost all either in the zodiac, and used to denote the hour, the season, or the position of one of the seven planets; or else they are near one of the poles and illustrate the circumpolar motion.

All the zodiacal constellations are mentioned except Virgo and Sagittarius. Besides the ordinary names of "Ariete," "Libra," etc., Aries is called "il Montone;"[275] Gemini "il segno che segue il Tauro," "gli eterni Gemelli," and "il bel nido di Leda,"[276] in reference to the mother of Castor and Pollux, its two brightest stars; Libra is "le Bilance";[277] Scorpio "il freddo animale";[278] Capricornus is "il Capra del ciel";[279] and Pisces "la celeste Lasca."[280]

In one place Castor and Pollux are mentioned, but merely as a synonym for the whole constellation of Gemini.[281] "Il petto del Leone ardente"[282] is perhaps Cor Leonis, the Heart of the Lion, for this is the name given to Regulus by both Ptolemy and Alfraganus. The "Maggior Fortuna"[283] of *Purg.* xix. 4, is a group of stars belonging to the two constellations of Aquarius and Pisces, in which the geomancers, who told fortunes by means of certain points traced at random, thought they saw a special series of these points,-::..

It is generally agreed that the "gemme" of *Purg.* ix. 4, "poste in figura del freddo animale, Che con la coda percote la gente,"[284] are some stars of Scorpio which were shining on the eastern horizon just before the moon rose on the first night in the Island of Purgatory.

The zodiacal constellations in general are spoken of as "all the lights of his [the sun's] path:" "Tutti i lumi della sua strada." (*Par.* xxvi. 121, 122).

The pole star is described as the point of the axle round which the first sphere revolves.[285] The first sphere (or wheel, as it is called, with reference to its circling motion) is here the Primum Mobile, which was thought to cause the diurnal motion. This unique position gains for the pole star the name of "the star," as for instance in *Conv.* III. v. 84, 85, where Dante says that a man standing at the north pole would have "la stella" directly over his head. And in Paradise he hears a voice among the spirits which makes him turn in its direction as the needle turns to the star—

"Voce, che l'ago alla stella Parer mi fece in volgermi al suo dove."[286]

This property of the magnet was known in Dante's day, as well as its power of attracting iron, though only the latter had been known to Ptolemy and the classical world. Whether the discovery came from China, where it is said that some form of compass has been used since the second century A.D., or whether it had been discovered independently by Arab or Italian navigators, we do not know: but scholars and poets in the twelfth and early thirteenth centuries write of the "ugly brown stone" used by sailors to make an instrument that cannot lie. For a needle, rubbed by it and run through a straw, when floated on water turns so surely to "the star" that one need never doubt its guidance. Thus writes Guyot, the poet of Languedoc, about 1200:—

"Un art font qui mentir ne puet, Par la vertu de la manete, Une pierre laide et brunete, Ou le fers volentiers se joint, Ont: si esgardent le droit point, Puis c'une aiguile i ont touchie, Et en un festu l'ont couchie, En l'eve la metent sanz plus,

Puis se torne la pointe toute Contre l'estoile si sanz doute, Que ja nas hom n'en doutera."

And thus Ristoro:—

> "L'angola, che guidi li marinari, chè per la virtu del cielo è tratta e rivolta alla stella la quale è chiamata tramontana."[287]

Albertus Magnus speaks of it in the same way, as something well known to mariners.

It seems strange to us how Dante and his contemporaries failed to see the importance of this discovery; Brunetto Latini, when the ugly stone was shown to him on his visit to Roger Bacon at Oxford, even professing to regard it as a mere toy, of no practical use. Dante discusses the number of the stars known to Ptolemy, and describes the last mad voyage of Ulysses, who saw all the stars of the other pole, while ours sank low on the ocean floor;[288] but he does not seem to guess that the new toy would make possible even longer voyages than these, and that in time the blank in his celestial globe would be filled.

Yet even in his own life-time plucky little Genoa fitted out two galleys which ventured through the forbidden Straits, with intent to circumnavigate Africa and find a new route to India. And meanwhile, travelling by old overland or coasting routes, Italian missionary monks, and Italian traders, were visiting southern countries and describing southern skies. Friar Giovanni de Monte Corvino, who was in South India with Nicolo of Pistoia, writes home in 1291, telling of his disappointment that he had never been able to see "the other pole star" (l'altra tramontana), though he saw new stars moving round and evidently near to it, close to the southern horizon. A few years later Marco Polo the Venetian was dictating the story of his travels to a Pisan in Genoa: he had been further south than the missionaries, for in a certain island (probably Sumatra) he had seen the south pole "a spear's length" above the horizon; and in the land of Zinzi (Zanzibar?) he had seen a marvellous star as big as a sack (which was evidently the Greater Magellanic Cloud). This he drew a picture of with his own hand.

Dante's silence, and probable incredulity, regarding these experiences of his own countrymen and contemporaries is characteristic of his age; for scholars were too eager to explore the precious classical lore lately recovered from oblivion, to realize that they were on the threshold of a new era in knowledge, of which these men were pioneers.

It is sometimes thought, however, that Dante made use of contemporary observations of southern skies in his description of the stars he feigned himself to have seen when the Wain disappeared under his northern horizon.[289] Quite correctly, he does not place any single bright star to mark the south pole; his two constellations, one of four bright stars seen above the pole in the morning,[290] one of three which takes its place after sunset,[291] are these real or fictitious?

It is a fact that there are four bright stars in the form of a cross, lying between 56° and 63° south: are these Dante's "quattro chiare stelle," "quattro luci sante?"[292] They were not recognized as a separate constellation until the beginning of the sixteenth century, when Amerigo Vespucci described them in his letters about his southern voyages, and the Florentine Andrea Corsali wrote about the marvellous Cross which was so beautiful that in his opinion no other constellation in the sky was worthy to be compared with it. This, he believed, was the very cross of which Dante had spoken in a prophecy.

Even among Dante's most enthusiastic admirers, I suppose none will be found to-day to support this view; but many think that he must have heard of the Cross from travellers. True, these stars are visible during at least part of the year in all places south of 34° north, and therefore in North Africa, and they had been catalogued by Ptolemy as part of the Centaur, so that no astronomer could take them to be a newly-found constellation, but might not some unscientific traveller like Marco Polo have brought a vague report of their position?

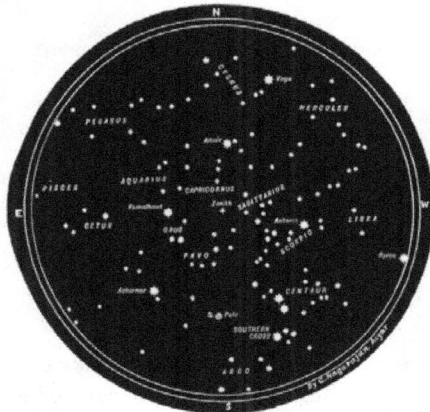

[To face p. 295.

MAP OF STARS VISIBLE BEFORE DAWN IN PURGATORY.

To this we might reply that Dante never says his four stars were in the form of a Cross; that there had to be four to represent the four Pagan virtues, the other constellation of three representing the three Christian virtues (compare the group of four handmaidens who sing, "Noi siam qui ninfe, e nel ciel siamo stelle,"[293] and are followed by a group of three);[294] and that this other constellation was certainly imaginary, since there is no group of three bright stars anywhere near the south pole.

But the fact is that conjectures and arguments are unnecessary, since Dante has expressly said that his four stars had never been seen before by anyone except the first people—"non viste mai fuor che alla prima gente," (*Purg.* i. 24)—that is, our first parents, in that Eden of his imagining which was in the southern hemisphere, and on the island where he was then standing in his vision.

As a matter of curiosity I have included a map showing what stars would have been really visible to Dante at the supposed latitude of Purgatory, when Pisces was on the eastern horizon, as described in *Purg.* i. It will be seen that the Southern Cross is low, and would have been hidden behind the Mountain of Purgatory at five o'clock in the morning.[295]

Among northern constellations, Ursa Major is frequently spoken of by Dante, under different names, but always with reference to its high northern latitude. As the Wain, "il Carro," it is said to be lying in the north-west when Pisces is on the eastern horizon, and to have disappeared from view in the southern hemisphere.[296] The wain-pole or shaft is pictured as sweeping round in the diurnal revolution, but always remaining above our horizon day and night, throughout its turning:

"Quel Carro, a cui il seno Basta del nostro cielo e notte e giorno, Sì ch' al volger del temo non vien meno."[297]

The same idea is expressed in the Midwinter Ode, where the constellation is called the Seven Frosty Stars, never lost to sight in Europe:—

"Fuggito è ogni augel che 'l caldo segue Dal paese d'Europa, che non perde Le sette stelle gelide unquemai."[298]

This is a reminiscence of some lines of Boëthius quoted by Dante in *De Monarchia*, in which the northern nations are described as under the sway of the "septem gelidi triones."[299] The Septem Triones, the Seven Ploughing Oxen, was one of the Latin names for Ursa Major, whence comes "septentrional" for North. This name also is used by Dante in *Purg.* xxx. 1, where the Seven Candlesticks of the mystic procession seen in the Garden of Eden are likened to these seven stars, and are named the Septentrion of the First Heaven (the Empyrean). This divine Septentrion was guiding the Procession, as the starry Septentrion of a lower heaven guides the mariner into port:

"faceva lì ciascuno accorto Di suo dover, come il più basso face Qual timon gira per venire a porto."[300]

Like the seven stars, also, the heavenly Septentrion is said figuratively to know neither setting nor rising, but unlike them it knows no cloud except of sin.[301] Because it is spoken of as guiding mariners, some commentators have taken the above to refer rather to Septentrio Minor (Ursa Minor), which also has seven chief stars, and is a better guide because nearer the Pole, as Thales taught; but the comparatively faint stars of the Little Bear would not be so apt a comparison with the celestial lights.

By the name of the Bears, both Ursa Major and Minor are referred to as guides at sea in *Par.* ii. 9. In the strange new seas on which Dante warns his readers he is about to enter, Minerva will blow a favouring wind, Apollo will steer the barque, and the nine Muses will guide his course by the Bears. They are also spoken of together in *Purg.* iv. 65, to indicate the northern part of the sky.

A fifth name for Ursa Major is derived from the fable (known to Dante probably from Ovid's *Metamorphoses*) of the nymph Helice, who was turned into a she-bear by Juno, and was hunted by her own son, Orcas. Jupiter transformed them into Ursa Major and Boötes.

"Se i barbari, venendo di tal plaga Che ciascun giorno d'Elice si copra, Rotante col suo figlio ond' ell' è vaga, Vedendo Rome e l' ardua sua opra, Stupefaciensi...."[302]

Here we have the same idea as of the Wain wheeling round but never setting, with the addition of a neighbouring constellation describing a circle in the same time.

The Barbarians who lived in a region always dominated by Ursa Major, and came to marvel at the mighty buildings of ancient Rome, are probably the races of northern Europe in general; and this reminds us again of the lines of Boëthius quoted above, for it is among the peoples ruled by Rome that he mentions "quos premunt septem gelidi triones."[303] If, however, Dante meant a country where the seven stars pass exactly overhead, the barbarians must have inhabited Scotland, or southern Scandinavia, or central Russia. If he means that Boötes also remained always above the horizon, they must have come from within the Arctic Circle, but this is not likely.

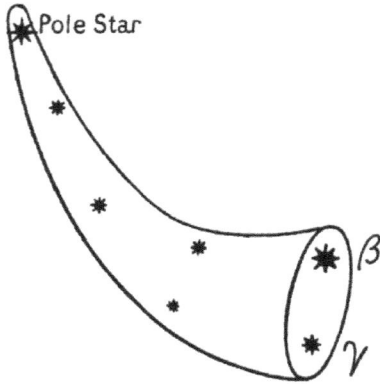

Fig. 42. Ursa Minor as a Horn.
Par. xiii. 10.

In one of the passages just quoted other constellations and stars are mentioned together with Ursa Major. In *Par.* xii. Dante has compared the two circles of spirits which surround Beatrice and himself to a double rainbow, and to two garlands of immortal roses. In the next canto he finds a new simile: in imagination he takes some of the brightest and most familiar stars from our sky, and makes of them two new constellations in the form of two crowns. The stars are these: fifteen from different parts of the sky, which are so brilliant that they shine through air dense enough to quench lesser orbs; the Wain, which never sets in our sky; and the mouth of the horn whose tip is the axis on which the Primum Mobile revolves. That is to say, as we have before remarked, Dante takes the fifteen first-magnitude stars and the stars instanced by Alfraganus as of second magnitude. Ursa Minor is aptly compared to a horn, the wide mouth of which is formed by its two bright stars Beta and Gamma,[304] while the narrow end is Alpha, the Pole Star. These twenty-four bright stars we must then imagine to group themselves into two constellations like that into which Ariadne, the daughter of Minos, was changed when she died; this is Corona Borealis or the Northern Crown, otherwise called Ariadne's Crown, which suggests a circle, though it is not a perfect one. Dante uses the word "segni" for any constellations, a custom we find in Ristoro also, although the modern usage is to restrict "signs" to mean only divisions of the zodiac.

The star-like spirits, thus grouped into a surpassingly brilliant double constellation, begin to sing their ineffable heavenly song, and to circle round the centre where Dante stands, in a marvellous dance, whose swiftness, when compared with anything known on earth, is as the movement of that swift heaven which carries with it all the rest, when compared with the flowing of the sluggish river Chiana.[305]

This passage helps us to understand expressions which often strike one as very strange, describing the circling movements of the spirits in Paradise. For the likeness to stars is clear throughout. Their brightness is an expression of their happiness,[306] and increases in each succeeding heaven. Only in the lowest can Dante see the forms and features of Piccarda and the other blessed spirits; in the next they clothe themselves in light, and although at first the radiant eyes of Justinian are seen, the first words addressed to him cause him to shine more brilliantly than before, and as the sun conceals himself by his own light, so the spirit conceals himself by the splendour which grows with his joy.[307] In the third heaven the spirits shine in the star of Venus like sparks seen in flame;[308] in the fourth they are called suns, and surpass the sun in brilliancy;[309] in the fifth they are compared with the Galaxy[310] and with shooting stars,[311] in the seventh and eighth with stars[312] and with spheres of light, turning on fixed poles and flaming like comets.[313]

In like manner the swiftness of their motion increases in proportion to the clearness with which each beholds eternal truth.[314] The almost incredible speed with which Dante himself soars from sphere to sphere, and with which the spirits move—whether in coming towards him, impelled by Divine charity, in returning to the heaven of heavens, or in circling with one another in mystic dance—is frequently dwelt upon, and illustrated by many striking similes. The dance is always a circling or wheeling movement ("il giro," "la rota," *Par.* viii. 20 and 26; ix. 65). And besides this, individual spirits turn with a rapid motion, remaining on the same spot. Descending the celestial Ladder in the heaven of Saturn, step by step, they revolve:—

"Vid' io più fiamelle Di grado in grado scendere e girarsi, Ed ogni giro le facea più belle."[315]

Now these two things, their light and their movements, were all the facts concerning the stars which mediæval astronomers could observe; and the movements were thought to be in circles and inconceivably rapid. Motion in circles, and especially the movement of rotation on an axis, remaining in the same place, had been said by Aristotle to be the most noble form of motion, and the fittest for celestial bodies. Therefore, when Dante represents the redeemed and the angels as expressing their bliss by radiance and marvellously rapid motion in circles, he makes them resemble the heavenly spheres and stars among which they manifest themselves to him. The motions are directly compared with those of stars in the passage already quoted:—

"Poi, sì cantando, quegli ardenti soli Si fur girati intorno a noi tre volte, Come stelle vicine ai fermi poli."[316]

Remembering this significance of circular motion we shall feel a little less strange the similes of a mill and a top applied to the spirits of Paradise.[317]

A third way in which the spirits express their joy is by the sweetness of their song,[318] and here also they resemble the heavenly bodies, which make sweet music as they circle. This doctrine of Pythagoras was very popular throughout the Middle Ages.

4. THE GALAXY.

The Galaxy or Milky Way is twice referred to in the *Divine Comedy*. In *Inf.* xvii. 106-108, it is in connection with the myth of Phaeton, who being allowed by his father Jove to drive the chariot of the sun for one day, lost control of his chargers, and going all astray, burned the sky in a part which still shows signs of this catastrophe, *i.e.* the Milky Way.

"Quando Fetòn abbandonò li freni, Per che il ciel, come pare ancor, si cosse."[319]

This of course is only a poetical myth. But in the *Paradiso*, in three lines the appearance, colour, and approximate position of the Galaxy are described, and allusion is made to the difference of opinions between learned men with regard to the puzzling question of its nature. Dante is in the heaven of Mars. He sees the soldier saints, star-like and fiery red, some larger and some less, thronging thick in two great bands, which, crossing one another in the depths of the planet, form a marvellous Crucifix. The crowded lights make him think of the Galaxy, with its multitudinous points of light, differing in brightness, although that is white, and lies in a great circle between the celestial poles.

"Come, distinta da minori e maggi Lumi, biancheggia tra i poli del mondo Galassia sì che fa dubbiar ben saggi, Sì costellati facean nel profondo Marte quei rai il venerabil segno Che fan giunture di quadranti in tondo."[320]

If we now turn to the fifteenth chapter of Treatise II. of the *Convivio*, we shall find the Phaeton myth and the various theories about the Milky Way set forth in due order, and shall learn which of these Dante considered to be true.

He is comparing the two sciences of Metaphysics and Physics with the Heaven of the Fixed Stars, and takes the opportunity to tell us some of the most important facts known about it. Here he calls the Galaxy a circle:— "La Galassia, cioè quello bianco cerchio che il vulgo chiama la Via de Santo Jacopo."[321] This popular name perhaps arose through a confusion of Galaxy and Galicia, where was a famous shrine of St. James, and hence came the belief in Italy that the *Galassia* was a sign by night to guide pilgrims on their way to this shrine at Compostella in *Galizia*.

Dante proceeds as follows with his comparison:—

"E per la Galassia, ha questo cielo grande similitudine colla Metafisica. Perchè è da sapere che di quella Galassia li filosofi hanno avuto diverse opinioni. Chè li Pittagorici dissero che 'l sole alcuna fiata errò nella sua via, e, passando per altre parti non convenienti al suo fervore, arse il luogo, per lo quale passò; e rimasevi quell' apparenza dell' arsura. E credo che si mossero dalla favola di Fetonte, la quale narra Ovidio nel principio del secondo di *Metamorfoseos*. Altri dissero (siccome fu Anassagora e Democrito) che ciò era lume del sole ripercosso in quella parte. E queste opinioni con ragioni dimostrative riprovarono. Quello che Aristotile si dicesse di ciò, non si può bene sapere, perchè la sua sentenza non si trova cotale nell' una traslazione come nell' altra. E credo che fosse l'errore de' traslatori; che nella Nuova par dicere che ciò sia uno raunamento di vapori sotto le stelle di quelle parte, che sempre traggono quelli; e questa non pare avere ragione vera. Nella Vecchia dice, che la Galassia non è altro che multitudine di stelle fisse in quella parte, tanto picciole che distinguere di quaggiù non le potemo, ma di loro apparisce quello albore il quale noi chiamiamo Galassia. [E puote essere che il cielo in quella parte è più spesso, e però ritiene e ripresenta quello lume] e questa opinione pare avere, con Aristotile, Avicenna e Tolommeo. Onde conciossiacosachè la Galassia sia uno effetto di quelle stelle le quali non potemo vedere, se non per lo effetto loro intendiamo quelle cose, e la Metafisica tratta delle prime sustanze, le quali noi non potemo simigliantemente intendere se non per li loro effetti; manifesto è che 'l cielo stellato ha grande similitudine colla Metafisica."[322]

This characteristic passage is interesting for several reasons. It illustrates the mediæval fondness for allegory, the reluctance to believe that Aristotle could ever be wrong, and the interest in ancient Greek speculations, in all of which Dante expresses the feelings of his age: and it brings vividly before us the methods and the difficulties of mediæval authors. For Dante is repeating from Albertus Magnus all these ancient speculations: the officious copyist who altered Aristotle's words, the fact that Aristotle could only be read in Latin, and the variety of translations, were all everyday obstacles to study; and Dante's own book has suffered from one of them in this very passage. The words placed between brackets are evidently an interpolation (as Mr. Wicksteed observes in the "Temple" *Convivio*): without them the passage goes on smoothly and logically to its end. On the assumption that the Galaxy is composed of a multitude of small stars, but not that the sky is denser in that part, the parallel with *Metaphysics* holds good.

Dante therefore held the true opinion about the Milky Way, but he was mistaken in thinking that he had the support of Aristotle. It was the "New Translation" (made by Aquinas from the Greek) which gave his opinion correctly; in the "Old" (made by Michael Scot from the Arabic) his statement had been apparently changed without comment into one which seems to Dante—and also to us—to represent the truth about the Galaxy.

5. THE PLANETS.

All the planets which were known before the days of the telescope are mentioned by Dante several times. Each has its peculiar beauty, but "la lucentissima stella di Venere"[323] is brighter and more beautiful than any other. He speaks of "la chiarezza del suo aspetto, ch 'è soavissima a vedere più che altra stella."[324]

No reader of the *Purgatorio* can ever forget that matchless morning on which Dante, escaping at last from the murky gloom of the *Inferno*, sees the sky once more, pure blue even to its furthest limits, and the planet of love shining in the east.

"Dolce color d'oriental zaffiro, Che s'accoglieva nel sereno aspetto Dell' aer, puro infino al primo giro, Agli occhi miei ricominciò diletto, Tosto ch' i' uscii fuor dell' aura morta, Che m'avea contristati gli occhi e il petto, Lo bel pianeta che ad amar conforta Faceva rider tutto l'oriente, Velando i Pesci ch'erano in sua scorta."[325]

Strange to say, the beginning of the above beautiful passage which conveys so direct and vivid an impression of blue sky and pure air, presents some difficulties when one comes to translate it word by word, especially as the texts offer more than one reading. By some commentators the "primo giro"[326] has been understood to mean the sphere of the moon,[327] or even the highest sphere (the Primum Mobile); but this is certainly inadmissible, as Dante supposed the atmosphere to reach only as far as the sphere of Fire, which came between it and the moon. The prime circle is here evidently the horizon, the fundamental circle of observers, the First Circle of the astrologers, from which they reckoned all the rest. It is here that the sky usually becomes pale and misty, however blue and clear it may be above. The colour was gathering and deepening,[328] especially in the east, where Dante was looking, as dawn began to appear in a perfectly cloudless sky; for during the darkness of the night it would have appeared almost black.

Dr. Moore thinks the authority better for a different reading, viz. "Dal mezzo"[329] instead of "Dell' aer."[330] In this case, we must take "mezzo puro" to mean mid-heaven, that is the zenith, "puro" being used as a noun, and meaning a cloudless sky (like "sereno" in *Purg.* v. 38); and "il sereno aspetto" must mean "the quiet scene."

Venus appears again in the *Purgatorio*, shining upon the Mountain from the east before sunrise; and this time she is called Cytherea, after the island near which she was fabled to have risen from the sea, and in which she received a special worship.

"Dell' oriente ... raggiò nel monte Citerea, Che di foco d'amor par sempre ardente."[331]

The position of Venus in her course through the zodiac is indicated in the previous passage by the expression "velando i Pesci."[332] The oscillating motion, by which she appears first on the morning and then on the evening side of the sun is referred to several times. In prose Dante speaks of her two apparitions, the morning and the evening: "la sua apparenza, or da mane, or da sera;"[333] in poetry, he describes her as wooing the sun, first following and then facing him:—

"la stella Che il sol vagheggia or da coppa or da ciglio."[334]

Mercury and Venus (called rather oddly after their mothers), are said both to move closely round the sun:—

... "vidi com' si move Circa e vicino a lui Maia e Dione."[335]

"Quel bel pianeta di Mercuro"[336] keeps closest, for he is more hidden than any other in the rays of the sun. Here again the fact is expressed in simple prose in the *Convivio*:—

"Mercurio ... più va velata de' raggi del sole che null' altra stella."[337]

which explains (if necessary) the more poetical language of the *Paradiso*:—

"La spera, Che si vela ai mortal con altrui raggi."[338]

In the same way we may compare the descriptions of Mars in the *Paradiso* with a statement in the *Convivio*:—

> "Questo fuoco,"[339] and "l'affocato riso della stella,"[340] and "Marte rosseggia,"[341] with

> "Esso Marte ... il suo calore è simile a quello del fuoco; ... esso appare affocato di colore."[342]

The famous occultation of Mars by the moon, seen by Aristotle, is quoted by Dante in *Conv.* II. iii. 59-65.

Jupiter is the "dolce stella"[343] and the "giovial facella."[344] He is silvery white, the whitest of all the stars:—"Intra tutte le stelle bianca si mostra, quasi argentata."[345]

Therefore when Dante saw the planet filled with brilliant spirits, he says: "Giove pareva argento lì d' oro distinto."[346] Comparisons between the colours of Jupiter and Mars are made in *Par.* xviii. 64-69, and xxvii. 13-15.

Finally Saturn is the slowest and the most distant of all the planets. These two characteristics are thus described in the *Convivio*:—

> "L'una si è la tardezza del suo movimento per li dodici segni; chè ventinove anni e più, secondo le scritture degli astrologi, vuole di tempo lo suo cerchio: l'altra si è che esso è alto sopra tutti gli altri pianeti."[347]

In the *Paradiso* Saturn is called a Mirror,[348] a Crystal,[349] a splendour;[350] his circling motion is alluded to, and his position at the time in the zodiac:—

"Cerchiando il mondo."[351] "Sotto il petto del Leone ardente."[352]

Some of the most beautiful similes in the *Divine Comedy* are drawn from the planets. The angel-pilot of Purgatory, when first seen far off over the sea, is likened to the planet Mars, glowing red through morning mists, low in the west above the ocean floor;[353] and the angel that welcomes the poet to the second circle on the Mountain is beautiful as the Morning Star.

"A noi venia la creatura bella, Bianco vestita, e nella faccia quale Par tremolando mattutina stella."[354]

Although any of the planets when rising just before the sun may be called a morning star, Venus is probably intended here, as also where St. Bernard, in his devotion to the Virgin, is compared with the Morning Star which takes its beauty from the sun:—

"Colui ch'abbelliva di Maria, Come del sole stella mattutina."[355]

All the seven planets (that is, including sun and moon), are occasionally mentioned together, "tutti e sette,"[356] and their movement in the ecliptic is referred to in *Par.* x, 14: "L'obbliquo cerchio che i pianeti porta."[357] In the 14th and 15th chapters of the second treatise of the *Convivio*, already so often quoted, Dante draws an elaborate comparison between the seven planets and the seven sciences of the Trivium and Quadrivium. The moon is Grammar, Mercury Dialectic, Venus Rhetoric, the Sun Arithmetic, Mars Music, Jupiter Geometry, and Saturn Astrology. The reasons given for the latter are the slow movement and the height (*i.e.* distance from Earth) of Saturn, which Dante compares with the length of the time taken in learning astronomy, and the loftiness of its subject.

In the same way the star sphere is said to resemble Physics and Metaphysics, the Primum Mobile Ethics, and the Empyrean Theology. The argument is very fanciful, but just what would appeal to readers of Dante's day, who loved to find allegories everywhere; and it gives him an opportunity of instructing them very simply in his beloved science.

He sometimes uses the planetary periods as divisions of time (as Plato said they should be used), for computing earthly events, and in him this does not seem affectation, as it would with almost anyone else. It seems quite natural that Cacciaguida, when speaking in the heaven of Mars, and answering a question regarding the date of his birth, should count the time not by solar but by Martian years, saying that from the beginning of the Christian Era to the day of his birth, Mars had returned to his Lion (the constellation of Leo) five hundred and eighty times.[358] The length of the Martian year according to Alfraganus is 1 Persian year, 10 months, and 22 days nearly. The Persian year (as he tells us in the first chapter of the *Elementa*) consists of 12 months of 30 days each plus 5 extra days, making 365 days exactly (not 365¼ days like the Roman), so the Martian year is 687 days (the modern estimate is 686·9). 687 days × 580 gives us 1091 a.d. as the date of Cacciaguida's birth. This is consistent with the date of his death, for he had just told Dante that he fell in the Crusade to which he followed the Emperor Conrad, and we know that this was fought in the year 1147.

Some commentators think that Dante did not intend the Martian year to be taken so precisely, but only as approximately two solar years, since he gave it thus in *Conv.* II. xv. 145. But there he specially states that he is quoting an approximate figure only ("uno anno quasi"[359] is half the period of Mars) whereas in the *Paradiso* it is used to fix a date. Since by taking 580 × 2 years we get 1160, an impossible date of birth for a man who died in 1147, the only way to support this theory that Dante was speaking loosely here is to adopt another reading of the passage, and substitute "tre" for "trenta."[360] Then 553 (instead of 580) × 2 years would give the date 1106; but the reading is not supported by good authority.

The period of Venus is used with equal appropriateness when Dante is relating how love for the Gentil Donna found a way to his heart, coming on the rays of Venus,[361] the star of love.

"La stella di Venere due fiate ere rivolta in quello suo cerchio che la fa parere serotina e mattutina, secondo i due diversi tempi, appresso lo transpassamento di quella Beatrice beata, che vive in cielo con gli angioli e in terra colla mia anima, quando quella Gentil Donna, di cui feci menzione nella fine della *Vita Nuova*, apparve primamente accompagnata d'Amore agli occhi miei, e prese alcuno luogo nella mia mente."[362]

Venus runs through her changes, appearing first as Evening and then as Morning star, in 584 days; and, as the reader will remember, this movement was ascribed by Ptolemy to her epicycle ("quello suo cerchio," "that circle of hers"). Dante would find 584 days given for it in his Alfraganus.[363] Twice 584 is 1168 days, or three years and a little more than 2 months. Therefore, since we know from *V. N.* xxx. that Beatrice died in June 1290, Dante wishes to say that he first saw the Lady of Pity in August 1293.

The period has been taken by some modern commentators to refer to the 225 days in which Venus revolves round the sun, but this period has no place in the Ptolemaic system. We cannot here discuss Dante's allegory of the Gentle Lady and Philosophy, but in this passage he has stated without ambiguity or uncertainty the date of her first appearance.[364]

6. ECLIPSES.

We find solar eclipses mentioned six times in Dante's works. As in all old records, there is no allusion to the features which chiefly strike modern astronomers—the pearly corona, and the blood-red prominences standing out like flames round the black disc. The terror of the sudden darkness, even when foreseen, seems to have absorbed all attention, and only the appearance of stars in the daytime and the frightened behaviour of animals was noted by Ristoro, and by others until quite modern times.

In his letter to the Cardinals of Italy, Dante speaks of the shameful and grievous removal of the Papacy from Rome to Avignon as an eclipse of the sun.[365] In his dream of the death of Beatrice, he sees the sun so darkened that the stars appear, to his great terror:—

"Poi mi parve vedere appoco appoco Turbar lo Sole ed apparir la stella."[366]

He also refers to the disputed idea that the darkness at the time of the Crucifixion was caused by an eclipse of the sun,[367] and to the figurative eclipse which then took place in heaven.[368]

If he had ever seen one in reality, it would have made more impression upon him, and we should surely have found it referred to several times and vividly described. The only suggestion of a personal experience is where he temporarily loses the power of sight by gazing too fixedly upon the spirit of St. John; and he compares this to a man who looks so intently at the sun, just before an eclipse, that he becomes incapable of seeing it at the critical moment.[369] This incident may have been told him by an observer of the total eclipse of June 1239, which was visible in Italy, and seems to be the one described by Ristoro.

Lastly, he mentions the discovery that the moon comes between us and the sun, which we have learned from eclipses, as an example of the proper method of reasoning: for we must first ask whether a thing is, and afterwards why it is. Thus man's wonder leads him to knowledge.[370]

The eclipsed moon is only once mentioned, viz. in *De Mon.* III. iv. 140-142, where she is described as not wholly dark. With his usual indifference to her Dante fails to note the often extraordinary beauty of her colouring when the eclipse is total.

7. COMETS AND METEORS.

The connection between these two classes of heavenly bodies seems to be real, though its exact nature is still a mystery. In old days little or no distinction was drawn between them, so that it is sometimes difficult to know whether a writer is describing a large meteor, or a comet, or a shower of falling stars.

Dante only mentions comets once, very briefly, when he describes spirits in Paradise as "fiammando forte a guisa di comete."[371] He says nothing of the magnificent comet described by Ristoro, which rose at three o'clock in the morning, huge as a mountain, and with great rays like a mane; but travelling always towards the south grew gradually fainter, and disappeared after 60 days. Yet it was seen in Tuscany shortly before Dante's birth, and it was thought to herald the death of the Pope (Urban IV. in 1264) and the terrible war in which first Manfred and afterwards Conradin were slain. These events are familiar to every reader of the *Divine Comedy*,[372] but there is never any mention of the comet.

Nor does he describe in any detail the fiery cross which was seen in the sky above Florence "in the beginning of her destruction," which does not mean, as one might think, the legendary sack of the city by Totila, but the treacherous entry of Charles of Valois, "Totila secundus,"[373] in November 1301. This portent was of course considered as of ill omen for the Republic, and although Dante was still absent on his embassy to Rome, and cannot have seen it himself, it is rather remarkable that he says so little about it. It is described by an eye-witness, "Io, che chiaramente la viddi"[374] as a red cross which appeared in the evening over the Priors' Palace; the vertical arm seemed to be 20 ells long, the transverse one a little shorter, and it remained visible for as long as it would take a horse to gallop two furlongs. Villani the historian describes it as like a comet with an immense trail, as of smoke, behind it. Dante merely mentions it, together with the ball of fire seen by Seneca at the death of Augustus, which, judging by Seneca's own words, was also a large meteor, for he describes it as flashing into sight and moving very rapidly, disappearing as it moved.[375]

Dante calls meteors "kindled vapours,"[376] in accordance with the doctrines of Aristotle, but the popular idea of them is shown by his observation that they might be taken for stars in motion, only that no star vanishes from its place, and they are only seen for a short time.[377] Ristoro d'Arezzo had already alluded to this fallacy:—

> "Alquanti non savi credono ch'elle sieno
> stelle, che caggino del cielo e vengano
> meno."[378]

He gives the same reason as Dante for disbelieving this, and Seneca had expressed it more forcibly, saying that if true there would be no more stars in the sky, for some meteors are seen every night.

Albertus Magnus, who seems to have had a genius for mistaking his authorities, quotes Alfraganus as stating that shooting stars are most frequent in evening twilight, though nothing of the kind is said in the *Elementa*. Dante perhaps accepted the statement, though it is by no means necessarily implied in his "di prima notte:"[379] moreover there is some authority for the alternative reading of "mezza notte."[380] Wherever the idea originated, its sole foundation must have been the circumstance that more persons are awake in the early than in the late hours of the night: the fact is that more meteorites are captured by the morning side of the earth, which is turned in the direction towards which she is moving in her orbit.

Shooting stars were necessarily familiar to a lover of starry skies, and Dante has given us an exquisite description of the suddenness with which they startle quiet eyes, while their brilliancy and swiftness of flight provides him with two beautiful similes.

In the first he compares them, and also flashes of summer lightning, with messenger spirits in Purgatory hastening to bring their friends to him:—

"Vapori accesi non vid' io sì tosto Di prima notte mai fender sereno, Nè, sol calando, nuvole d'agosto, Che color non tornasser suso in meno, E giunti là, con gli altri a noi dier volta."[381]

The second illustrates the brightness and beauty, as well as the movement, of the star-like spirits in Mars:—

"Quale per li seren tranquilli e puri Discorre ad ora ad or subito foco, Movendo gli occhi che stavan sicuri, E pare stella che tramuti loco, Se non che dalla parte ond' ei s' accende Nulla sen perde, ed esso dura poco ..."[382]

8. THE SUN'S PATH IN THE SKY SEEN FROM DIFFERENT PARTS OF THE EARTH.

We have now seen how familiar Dante was with the aspects of the skies above his head. His writings show also that he had pictured clearly to himself what they must be in other parts of the world—in regions far east or west, in the southern hemisphere, at the equator, and at either pole. Science could not tell him all that he would have liked to know about the stars, so that although he can speak of regions where Ursa Major passes overhead, and where it sinks out of sight below the horizon, he has to fall back upon his imagination and invent new constellations for the southern hemisphere.

But astronomy could tell him, and his habit of accurate thought, as well as his imagination, helped him to grasp how the sun would appear at any latitude on the earth, and what would be the results on the length of day and night, and on the seasons. A very interesting little disquisition on this subject is found in the *Convivio*, in illustration of some lines in one of his Odes, which speak of the sun circling the whole world. In order to understand this completely, Dante says, we must know exactly in what way the sun circles the world. The chief point to bear in mind while reading his description is that the sun, besides the simple daily motion in a circle, has a constant slow motion north or south, and therefore his path in the sky, as we see it day after day, is really spiral.

The passage is much too long to quote in full, but if the reader will follow me, taking his *Convivio* and opening it at chapter v. of Treatise III. and beginning at line 66, I will give a résumé which will form a running commentary on the text, and an explanation of any points which may not at once be clear.

We see the sky, says Dante, continually revolving round Earth as centre; and it has two fixed poles of revolution—the northern, which is visible to nearly all the land not covered by sea, and the southern, which is hidden from nearly all of it. And the circle which is equi-distant from these two poles [the celestial equator] is that part of the sky in which we see the sun when he is in Aries and in Libra.

Now if a stone could be dropped from this pole of ours [the northern], it would fall far away yonder in the ocean just where, if a man were standing, he would always have the Pole Star exactly overhead. ["La stella"[383] is the Pole Star, see p. 291]. And I believe (Dante says) that from Rome to the north pole, in a direct line, is a distance of about 2,700 miles. To fix our ideas, let us imagine that at this spot there is a city called *Maria*, and let us imagine another city, called *Lucia*, at the spot exactly opposite, where a stone would fall if dropped from the other pole. And I believe that from Rome southwards to this second place would be a distance of about 7500 miles. Thus the distance between the two cities, in whatever direction the measuring cord be stretched, would be 10,200 miles, that is, half the circumference of the globe,[384] and the inhabitants of *Maria* would have their feet opposite the feet of those of *Lucia*. [If any two spots on a sphere are exactly antipodal, but in no other case, the distance may be measured in an infinity of directions, and always come out the same.]

Lastly, let us imagine a circle on the globe, which will be at equal distances everywhere from *Maria* and *Lucia* [the equator]. According to the opinions of the astrologers, if I understand them aright, and according to what is said by Albertus Magnus and by Lucan, this circle would divide the dry land from the ocean there in the south, approximately along the extremity of the First Climate, where amongst others live the Garamantes, who go almost always naked, to whom Cato went when fleeing from Cæsar.

"Approximately," because the southern extremity of the First Climate, according to Alfraganus, lay a little north of the equator, although land extended, and was sparsely inhabited, as far the equator (see p. 186).

Dante speaks again of these extra-climatal races in *De Mon.* I. xiv. 43-51, where he contrasts the Scythians who live beyond the seventh climate, and therefore endure extreme inequality of days and nights, and suffer almost intolerable cold, with the Garamantes who live under the equator, where days and nights are always equal, and the heat is so intense that they can scarcely bear any clothing. Since "the astrologers who determine the climates"[385] had fixed their northern limit at 50½°, nearly the whole of Britain also lay, like barbarous Scythia, in this scarcely habitable region of long nights and bitter cold!

When we have marked these three places on the globe, *i.e.* the two poles and the equator, it is easy (Dante goes on) to see how the sun circles. I say then, that the heaven of the sun turns from west to east, not directly against the diurnal movement—that is, the movement which produces day and night—but obliquely against it, and so that its middle circle, which is similarly between its two poles, and on which is the body of the sun [*i.e.* the ecliptic], cuts in two opposite points the circle of the two first poles [the equator], that is, at the beginning of Aries and the beginning of Libra; it diverges from that circle in two arcs, one north and the other south. The highest points of these arcs are equally distant from the first circle in either direction, being 23 degrees and a little more; and the one summit is at the beginning of Cancer, and the other at the beginning of Capricorn.

Therefore, when the sun is in the beginning of Aries, travelling in the mid-circle of the first poles [the equator], *Maria* will see this sun circling the world, low down on the ground, or on the sea, like a mill-stone only half of which is seen, and day after day he will be seen to rise, like the screw of a press, until he has performed about ninety revolutions, or a little more. When these revolutions are accomplished, he will be as high in the sky above *Maria* as he stands in the sky of the Garamantes at middle-tierce at the time of equal days and nights.[386]

[From *Conv.* IV. xxiii. and III. vi., where Dante explains the use of temporal hours, we learn that mid-tierce on the day of the equinox is an hour and a half after sunrise, or 7.30 A.M. Since the sun moves through 360 degrees divided by 24, that is 15 degrees, in an hour, and his motion is vertical in equatorial countries at the equinox, it would bring him in 1½ hours to 22½ degrees above the horizon. His greatest height above the horizon at the pole is only one degree more than this, for it is obviously equal to the greatest distance between the equator and the ecliptic, *i.e.* 23½ degrees, since at the pole the equator coincides with the horizon. This gives a good idea, therefore, of the appearance of the sun in the sky to the people at *Maria*, and is a striking illustration of the difference between the polar sun and the equatorial. For at the time mentioned, the sun has only a quarter the height which it will attain at noon, when it will pass through the zenith of the Garamantes; yet this is the highest position in which the inhabitants of *Maria* can ever see it].

If a man stood upright in *Maria*, and kept turning his face to the sun, he would see it move to his right [as we do].

After reaching his greatest height in the sky, the sun would begin to descend again, in the same spiral way like a screw turning, for another ninety revolutions or so, until once more he was circling down on the horizon, only half his body visible.

Then he would be lost to sight altogether, and would begin to be seen in *Lucia*, where he would rise and descend in just the same way as in *Maria*. But if a man faced the sun in *Lucia*, he would see it moving to his left [as one does in Australia where it goes north at noon].

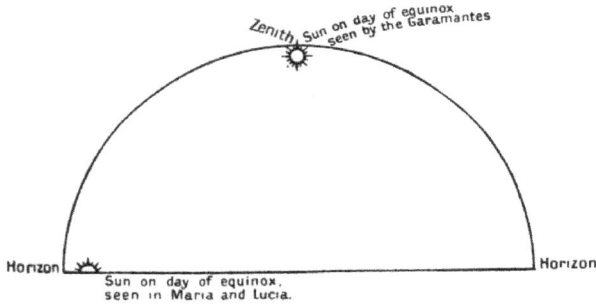

Fig. 43. The Sun at the Equinox, seen from the poles and the equator.
Conv. III. v.

(Maria, Lucia, imaginary cities at the North and South poles).

We see, then, that these places have one day in the year which is six months long, and one night of equal length, and when it is day with one, it is night with the other.

Now on the circle on the globe where the Garamantes live [the equator], the sun when in Aries goes exactly overhead, not circling horizontally like a mill-stone, but vertically like a wheel, and exactly half of this wheel is visible above the horizon.

After this, the sun is seen to depart and go towards *Maria* for about 91 days; then in about the same time it returns, and enters Libra; then it goes towards *Lucia* for about 91 days, and in the same period returns. And this place, which circles the entire globe, always has equal day and night, on whichever side of it the sun is; and twice in the year it has a very hot summer, and twice a mild winter.

The regions which are between the two imaginary cities and the mid-circle, see the sun in different ways, according to their distance from these places, but the details, our author says, he will leave to the ingenious reader,

> "Siccome omai, per quello che detto è,
> puote vedere chi ha nobile ingegno, al quale è
> bello un poco di fatica lasciare."[387]

He draws attention to the interesting fact, which follows from the above, that in the course of a year (when the heaven of the sun has made a complete revolution and returned to the same place), every part of this globe on which we live has received an equal amount of daylight. And the essay ends in an apostrophe which recalls the reproof of Virgil to those who will not look up to the skies:[388]

> "O ineffabile Sapienza che a così ordinasti, quanto è povera
> la nostra mente a Te comprendere! E voi, a cui utilità e diletto io
> scrivo, in quanta cecità vivete, non levando gli occhi suso a
> queste cose, tenendoli fissi nel fango della vostra stoltezza."[389]

It is interesting to compare this passage with the parallel passages in Alfraganus and Ristoro. The astronomer is concise and clear; the monk is diffuse and apt to repeat himself; the poet is quaintly picturesque. Both the Italians evidently draw the main facts from the Arab, and all use the illustration of a mill-stone to explain the horizontal motion of the heavens at the pole: here are the three corresponding passages:—

"Coelumque molæ trusatilis instar gyrum vertitur."[390]

"Lì si volgera il cielo attorno con tutte le sue stelle, in modo di macina."[391]

"Conviene che *Maria* veggia ... esso sole 'girare il mondo' ... come una mola."[392]

Dante alone uses the figure of a wheel to illustrate the vertical motion at the equator; he perhaps takes from Ristoro the idea of the spiral. For the latter describes the path of the sun as a "via tortuosa, la quale i savi chiamano spira;"[393] but he explains it as like a string wound round a stick, Dante like the screw of a press. Dante is dealing only with the sun, and does not enter into details about the visibility of the zodiacal signs like the other two authors; but on the other hand he goes south to the other pole, which apparently does not interest the others, perhaps because the southern hemisphere was thought to be uninhabited. But the poet has added life to his description by supposing both poles inhabited, placing there his imaginary, mysteriously-named cities of "Maria" and "Lucia." On the equator, where Ristoro has the mythical city of Arym, whose wise and prosperous citizens enjoy a perfect climate, Dante, relying more on the classics, lets his barbarous Garamantes run about naked most of the year, under the fierce equatorial sun.

Needless to say, his little homily in conclusion is all his own; it is more interesting to find that he alone mentions the opposite movements of the Sun, to right and left respectively in the north and south hemispheres,[394] and the fact that every part of the earth receives the same amount of sunlight in a year. This may have been suggested by someone else, but one would like to think he arrived at it independently, when thinking over the facts.

In the light of this treatise, which describes so truthfully the path of the sun, and its effect on the seasons in different latitudes, we may explain two rather puzzling passages in the *Divine Comedy*.

One is at the beginning of the tenth canto of the *Paradiso*, where Dante wishes his readers to realize the position of the sun when he entered it with Beatrice, and the great importance of this position. "Lift your eyes with me, reader," he says, "to that place in the lofty heavens where the one motion meets the other, and see how the oblique circle which carries the planets branches off from that point." The reader is now familiar with the "two chief motions," and knows that one is the diurnal, from east to west, the other the planetary periodical motions in the opposite direction; they meet at the equinoxes, where the ecliptic cuts the equator. It is the spring equinox which Dante is speaking of, for a little further on he says that the sun, situated in the place above-mentioned, was circling in those spirals which bring him to us earlier every day (lines 31-33). Now if the path of the planets were not thus oblique, he continues, much virtue in heaven would be lost, and almost every earthly power dead. Even if the obliquity were merely less or more, the universal order of both heaven and earth would suffer greatly. But why this should be so he leaves his readers to think out for themselves, having a greater matter in hand which demands all his attention.

It is easy to see that if the sun always moved in the equator, without departing from it either north or south, he would rise every day to the same height in the sky, and there would be no change of seasons anywhere on the earth; he would always be overhead at noon to Dante's Garamantes, living on the equator, and always just on the horizon, day and night, to the people of "Maria" and "Lucia" at the Poles. If the ecliptic made a greater angle with the equator the seasons would be more marked, and the sun would rise higher in summer in high latitudes. If the angle were less, the reverse would be the case. Moreover if sun and moon and all the planets followed the same track, they would be constantly eclipsing one another.

But if we would understand the full meaning of—

"Se la strada lor non fosse torta, Mota virtù nel ciel sarebbe in vano, E quasi ogni potenza quaggiù morta."[395]

we must turn to thirteenth-century Ristoro of Arezzo, who has set forth in detail the results which he believed would follow from a change in the ecliptic. If all the seven planets moved in the same narrow path, there would be much less variety in their "aspects," and therefore in their influences upon the earth; and in the frequent eclipses one would prevent the other from looking at the earth ("impedimenterebbe l'uno l'altro a guardare la terra"), and so hinder its action. And since it is the northward movement of the sun which causes our plants to blossom, and later to bear fruit, if there were no such motion we should see no renewal of life. He argues further that if the obliquity of the ecliptic were greater, countries in high latitudes would have too severe a winter to be inhabited; while if it were less, the summer there would not be warm enough for the ripening of harvests; therefore in whichever way it differed from the actual value, a smaller part of the earth would be habitable.

"Or ti riman, lettor, sopra il tuo banco, Dietro pensando a ciò!"[396]

How much of this was in Dante's mind it is hardly possible to say, but it is evident that the "via tortuosa" of the seven planets, and its effects on the earth, was a favourite subject for thought in his time.[397]

But the most famous passage relating to the sun's path and its aspect in strange latitudes is *Purg*. iv. 52-84.

Here Dante is able at once to understand the startling appearance of the sun travelling to his left instead of his right, as seen on the first morning in Purgatory, for he grasps the full meaning of Virgil's explanation, and caps it, in a way which seems astonishingly quick-witted to unastronomical readers!

The two poets had begun to climb the mountain just at sunrise, and sat down to rest on a ledge at about half-past nine, looking east. Dante at first looked down at the shore, whence he had so painfully climbed, then raised his eyes to the sun, and was amazed to find that its rays struck him on the left shoulder.

Virgil, seeing his astonishment, explains why the sun is going north instead of south as it climbs the sky. The sun is now in Aries, and therefore on the equator, as we know from several passages, but Virgil begins by saying that if it were in Gemini [a more northerly sign] and kept to its ancient path, Dante would see it travelling still further northwards in its daily course. "If Castor and Pollux [stars in Gemini] were in company with that Mirror [the sun] which carries his light up and down [*i.e.* goes north and south alternately], you would see the glowing zodiac [that part in which the sun is] revolving still closer to the Bears." Gemini is literally nearer to the constellations of the Bears than Aries, but this is not the sense of Virgil's statement: he means merely to indicate the north in general (see Ptolemy, p. 157). "To understand why the sun goes north," Virgil says, "you must know that this mountain and Zion [Jerusalem] are so situated on the earth that one is precisely antipodal to the other; and so, the sun's path being between them, when he is viewed in one direction from Jerusalem, he is viewed in the opposite direction from Purgatory. If you turn from the east to your right to see him from Jerusalem, you must turn to your left in Purgatory." Virgil speaks of the sun's path or the ecliptic as the path which Phaëton was not able to keep, and he describes the two antipodal places as having one and the same horizon but two entirely different hemispheres.

"Certainly," replies Dante, "I have never seen anything so clearly as I now perceive what at first puzzled me. For the middle circle of the celestial motion,[398] which is called the equator in a certain art [astronomy], and which always remains between the sun and winter, [the sun crosses it departing from us in either hemisphere at the autumnal equinox], here lies to the north of us, for the reason you have given [namely, that we are in the southern hemisphere], whilst the Hebrews [inhabitants of Jerusalem] see it towards the hot region [the south]."

Observe that the expression used of the equator "che sempre riman tra il sole e il verno,"[399] distinctly suggests the reversed seasons in the southern and northern hemispheres, so that Dante has no intention, as Dr. Moore suggests, of assuming by a poetic licence that it is spring in Purgatory as well as in Italy.[400] It is autumn in the southern hemisphere, but this need not distress us: the holy Mountain knows no inclement weather,[401] and in the Garden of Eden on its summit spring flowers and autumn fruits eternally flourish together.[402]

An allusion is made to a tropical summer in "la terra che perde ombra."[403] Tropical Africa is intended, where the sun at certain times passes directly overhead, and therefore casts no shadows, as the Alexandrian astronomers had observed (see p. 117).

Did Dante believe that no land exists anywhere in the southern hemisphere? There are some indications that he thought it possible to push the limits of the habitable earth a little beyond the equatorial boundary of Alfraganus. There would be no heresy in lengthening the southward extremities of Africa or Asia, because all that the Church forbade was to plant another land in the south, separated from us by a vast ocean or impassable torrid zone, and to people it with inhabitants, who could not therefore be descendants of Adam or have the Gospel preached to them. Nor would there be any inconsistency with the poet's own description of our hemisphere as that covered by "la gran secca,"[404] or with the voyage of Ulysses, on which he saw no land but the Mountain of Purgatory; for he sailed south-west, not due south, and the "mondo senza gente"[405] was "diretro al Sol."[406]

The indications are as follows:—

(1) The passage quoted above, where it is said that the equator "divides the uncovered land from the ocean"[407] is ambiguous in the original, and might mean "divides the land uncovered by Ocean."

(2) Shortly before, he has distinctly admitted that there is some land in the southern hemisphere, by saying that the north celestial pole is visible to *nearly all* the land not covered by sea, and the south is hidden from *nearly all* of it. The first is true in any case, since though the pole is theoretically visible as far as the equator, it is often hidden by mist or rising land on the northern horizon; but unless the second "nearly" has been added by a (somewhat natural) mistake of a copyist, it can only apply to lands in the southern hemisphere.[408]

(3) There is a stanza in *Canzone XV.* which suggests that Dante believed Ethiopia to be on the south side of the equator. Classical and mediæval cosmographers had very vague ideas about the extent and position of this country. Dante describes a wind raised in its deserts by the sun, which *now* is heating it (implying the time of summer?); but the same wind, crossing the sea to Europe, darkens all *this* hemisphere, bringing clouds which fall in snow.

"Levasi della rena d'Etiopia Un vento pellegrin, che l' aer turba, Per la spera del sol, ch' or la riscalda, E passa il mare, onde n'adduce copia Di nebbia tal, che s' altro non la sturba Questo emisperio chiude tutto, e salda; E poi si solve, e cade in bianca falda Di fredda neve, ed in noiosa pioggia."[409]

However, it may be said that "or la riscalda"[410] means merely that it is always hot in the tropics, even when we have winter, and that "questo emisperio"[411] cannot possibly mean the whole of the terrestrial northern hemisphere from equator to pole, but simply describes the appearance of the hemispherical sky,[412] completely covered with cloud.

These doubtful passages are not strong enough to set against the explicit statement in the *Quæstio* that the habitable earth extends from the equator to the Arctic circle and no further.[413] Letters and reports from missionaries and traders, none of them scientific, did not shake Dante's faith in the limits laid down by Ptolemy and Alfraganus. His geography was utterly wrong, but his astronomy was right; and therefore, although he had no idea of what he would see on our globe in untravelled latitudes, he knew exactly, and has described vividly, what he would see in the sunny sky.

9. CELESTIAL PHENOMENA AND TIME.

The foregoing quotations have shown how well Dante understood the changing appearances of the celestial phenomena as viewed from different latitudes: it now remains to be seen what were his conceptions of longitude.

A difference in longitude between two places makes no difference in the stars, nor in the apparent paths of the heavenly bodies, but they come to the meridian sooner in the more easterly place; and as it is the meridian passage of the sun which determines local time, this will vary in exact proportion to the distance east. When it is early morning in England, it is noon in Burmah, and evening in Fiji. As Alfraganus had observed (following Ptolemy), a celestial phenomenon such as a lunar eclipse, which is caused by an actual darkening of the moon's surface when immersed in Earth's shadow, is visible in every place where the moon is above the horizon, though the local hour differs.

It is otherwise with a total solar eclipse, for this is only visible from the very small portion of Earth on which the moon's shadow falls as she passes between us and the sun (see fig. 44); and as this shadow moves on, owing to her own motion and the diurnal revolution combined, the eclipse becomes visible successively from different parts of the earth. Dante realized this when he quoted the argument of some theologians that the miraculous three hours' darkness described in the Gospels could not have been caused by an eclipse of the sun, as Aquinas and others had suggested, because it was visible all over the earth at the same absolute time, that is, during the Crucifixion.[414] Some said, he tells us, that the moon went back and placed herself between us and the sun (for the moon at Passover time was always full, and therefore opposite the sun), so that the sun's light should not come down (to us on the earth); others, that the sun hid itself, for a corresponding eclipse was seen by Spaniards and Indians as well as Jews.[415] Spaniards and Indians were supposed to live at the extreme western and eastern limits of the habitable earth: they were 180° in longitude, or 12 hours in time, apart, and Jerusalem was midway between them. Therefore the sun would be visible to all, though only just rising in India and just setting in Spain;[416] but it could not have been an eclipse that darkened it to all at the same time.

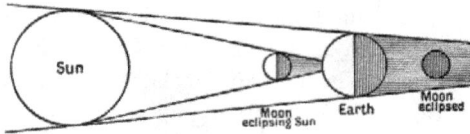

Fig. 44. Lunar and Solar Eclipses.

This passage shows that Dante was entirely orthodox and conservative in his geography, as far as longitude was concerned, and confirms us in the conclusion that he was so also with regard to latitude, and was not likely to accept contemporary evidence about lands south of the equator, or stars unknown to Ptolemy. In the *Divine Comedy* there are several passages which refer to the same system, as for instance where Night is described as covering the whole region from "the Shore" to Morocco,[417] or where the time is stated to be noon on the Ganges, sunrise at Jerusalem, and midnight in Spain.[418] And in the *Quæstio de Aqua et Terra* the system is thus clearly explained:—

> "As all these agree in believing [the naturalists, the astrologers, and the cosmographers], this habitable earth extends in longitude from Gades, which lies on the western boundaries of Hercules, as far as the mouths of the Ganges, as Orosius writes.[419] That longitude is such that at the equinox the sun is setting upon those who are at one of these boundaries, while it is rising upon those who are at the other, as astrologers have discovered by eclipses of the moon. Therefore the aforesaid boundaries must be 180 degrees distant in longitude, which is half the distance of the whole circumference."[420]

Gades is not the city of Cadiz, but two islands (the "Gades Insulæ" of Orosius), on which Hercules was said to have set up his Pillars, as a sign that no one should venture further; and they were thought to lie in the mouth of what we call the Straits of Gibraltar:[421]—

"quella foce stretta, Ov' Ercule segnò li suoi riguardi, Acciochè l'uom più oltre non si metta."[422]

Through these Ulysses sailed, never to return.[423]

It is difficult to believe that a lunar eclipse had ever been observed simultaneously, and the local time compared, on the Ganges and at Gibraltar, as the above quotation from the *Quæstio* declares: and if it had been, the astrologers, the cosmographers, and the naturalists would have found out that they were greatly mistaken, for the distance is little more than half what they thought. All that Ptolemy had said was that this was the right method for calculating longitudes, and he gave as an example an eclipse which had been seen at Arbela at the fifth hour and at Carthage at the second hour.[424] It was, in fact, an excellent method, and the only one before the day of chronographs and telegraphs; and it seems to have been assumed that Ptolemy's guess of 180° for the extent in longitude of the inhabited earth was based upon it.

Then this 180° was considered, in the Middle Ages, as divided into two equal parts, the western half containing Europe and Africa, the eastern containing Asia. On the western front of Asia lay Jerusalem, thus holding the central position in the inhabited earth which was thought to be her right, on the authority of certain Scripture texts, as we saw.[425] This gave the Mediterranean a length of 90°, which of course is vastly too great. Even Ptolemy had only estimated it at 62°, and the Alfonsine Tables made it 52°, while the true distance is only about 42°.

Latitude, however, was much more easily measured than longitude, for it was only necessary to take the height of the pole star above the horizon with an astrolabe; and the latitude of Jerusalem is really just about midway between the supposed limits of the habitable earth, the equator and arctic circle, for it is nearly 32° north.

Some daring spirits ventured to suggest that the dry land stretched much further round the globe than was commonly supposed, so that one might sail from Spain to Asia in quite a short time, and they sheltered themselves behind the great names of Aristotle, Seneca, Pliny, and Esdras.[426] Albertus Magnus quotes Aristotle to this effect, although in another part of the same book (his *De Cælo et Mundo*), he repeats the usual phrase about the habitable Earth being all contained within one of the northern quadrants of earth. Roger Bacon expresses himself more boldly, and it is possible that he was impressed by the long journey of the Franciscan friar Rubruquis, on his mission to Central Asia, from St. Louis of France to the Emperor of the Mongols. Besides relating the story of this journey, Bacon speaks of men who were known to live in tropical Taprobane (Ceylon), and if there, why not also south of Capricorn? There may be delightful countries there, since it is the better and nobler part of the world, as Aristotle and Averroës taught.

Moreover, Dante's own countryman, the learned Pietro d'Abano, maintained that there are habitable lands on, and even beyond the equator, and did not hesitate to quote conversations with Marco Polo as evidence, although this "most extensive traveller and most diligent enquirer" was accounted a romancer by most of his contemporaries. Had not Messer Marco told him that in islands south of the equator he had seen not only great rams with coarse stiff wool like the bristles of pigs, but human beings?[427]

Roger Bacon was banished from Oxford on a charge of heresy and witchcraft: Pietro d'Abano escaped the Inquisition by dying, in 1316. Men such as these belonged to the new age that was coming, the age that would throw away the old lesson-books, and begin to discover for itself—and they suffered in consequence. Dante belonged heart and soul to his own era, and clung passionately to its ideals and traditions. For him the Ganges still flowed into the all-surrounding ocean on the east coast of Asia, 90 degrees from Jerusalem, as Orosius had pictured it, and there is not the faintest echo in his writings of the voyages of his countrymen, Corvino and his bishops, or Polo and his companions,[428] between "Greater India" and "Zayton" (Amoy harbour), although they surely must have recognized, when they praised the great rivers of China, that these flowed into a sea many days' journey beyond that into which the Ganges rolled its waters. Nor is there, on the other hand, any hint of the monstrous races with which Dark-Age superstition had peopled the remote regions of the earth, although they were regularly represented on the maps of that time. It is true that the "portolani," or handy charts, of which the earliest known date from Dante's life-time, showed none of these things, but depicted the coast-lands of the Mediterranean and Black Seas with marvellous accuracy: but they were only for pilots and merchant captains, and had nothing to do with scholars. The maps Dante would see were hung on church or monastery walls, or illustrated learned books; and the mappe-monde of Cardinal Heinrich of Mainz may be taken as typical of them all. It much resembles our great Hereford Map of the World, and was probably copied from the same original design.

It is two hundred years older than Dante, but so stationary was geography throughout the Middle Ages that it might have been executed in his day, or, except for a few details, several centuries earlier.

Mediæval maps are as fascinating as a fairy tale, which indeed they much resemble. The shape is conventional, and not supposed to represent the true shape of the habitable earth; it is sometimes circular, sometimes quadrilateral, occasionally oval, as in the present instance. The east is here, as usual, at the top, and there we see the Paradise of Adam and Eve, and close beside it, to the north, the River Ganges, flowing into the great ocean which surrounds all the land. At the opposite or western extremity are the Pillars of Hercules, which two angels appear to be guarding. In the centre, but not so exactly or conspicuously central as in some other maps, we find "Hierusalem." Along the south is the long narrow continent of Africa, with the Atlas and Ethiopian mountains; and the Nile, rising far in the south-west, after mysterious submergences and reappearances, finally flows into the Mediterranean, past the pictured Pyramids, here strangely called (according to Dark-Age myth) the Barns of Joseph. In the north another Angel points to the country where dwell the Gog-Magogs, and not far off are the lands of the Hyperboreans, here described as untroubled by disease or discord, of the "most wicked" Gryphons, and of the Dog-headed folk, adjoining the Arctic Ocean.

Italy is a very large country, roughly triangular, and in the Mediterranean to the south lie Sicily, the whirlpool Charybdis, the rock Scylla, Sardinia, Corsica, and other islands. Rome is figured as a battlemented city, about halfway between Jerusalem and the Pillars of Hercules.

[To face p. 344.

MAP OF THE WORLD BY HEINRICH OF MAINZ,
ABOUT A.D. 1110.

*Reproduced by permission from Beazley's
"Dawn of Modern Geography."*

This is where Dante also places the imperial city, for in his system it is about 45°, or 3 hours, east of Gades and west of Jerusalem. He tells us in one place that it was vespers in Naples when the sun rose in Purgatory,[429] and in another that when it was vespers or three o'clock in the afternoon in Purgatory, it was midnight "here"[430] which may mean Florence or Italy in general; both these statements indicate a difference of nine hours or 135° between Italy and Purgatory. Purgatory in the *Divine Comedy* is exactly antipodal to Jerusalem, and therefore twelve hours distant in time;[431] and for the sake of his allegory Dante has so far departed from tradition as to place the Earthly Paradise here also.[432] These five places may therefore be diagramatically represented thus:—

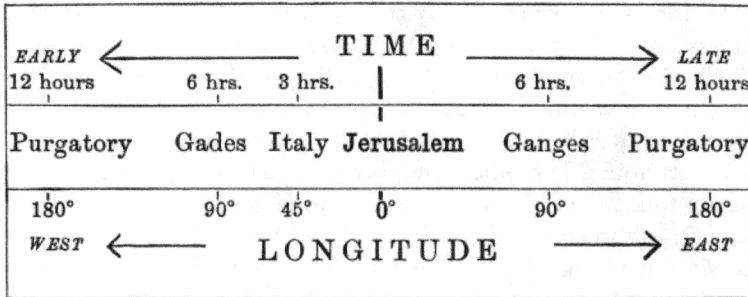

EARLY			TIME			LATE
12 hours ←	6 hrs.	3 hrs.	|		6 hrs.	→ 12 hours
Purgatory	Gades	Italy	Jerusalem		Ganges	Purgatory
180°	90°	45°	0°		90°	180°
WEST ←			LONGITUDE			→ EAST

With this key, Dante's fondness for describing the time at one place by mentioning one of the others need be no stumbling-block, as it is to many readers.

The fact is that though Dante had none of our opportunities for changing time-reckoning by long journeys, nor for communicating with friends at the other side of the world, who wake while we sleep, the marvellous daily revolution of the skies had a fascination for him, and he delighted to picture a glorious sunrise taking place in one country while in another it was high noon, or the middle of the night. Thus, the beautiful description of the stars fading in the growing light of dawn, which we quoted above, is preceded by a picture of the sun glowing on the meridian some 6000 miles away (rather more than a quarter the circumference of the globe), while with us Earth's shadow has sunk almost to a level plane.[433]

Again, Folco of Marseilles, when describing his birthplace, says that the greatest of all the seas (the Mediterranean) except the ocean which encircles all the habitable earth, stretches so far against the sun (i.e. from west to east), that the line of sky which is on its horizon at one end is on its meridian at the other.[434] If the sun were seen just rising on the eastern horizon at Gades, it would be on the meridian at Jerusalem and the time would there be noon.

Dante's hours are always reckoned from sunrise, and they are the "temporary" or Church hours. He defines both kinds in the *Convivio*. The same lines of his Ode which provided him with a text whereon to discourse of the sun's movements, give him occasion to treat of the division of each day into hours, thus:—

> "I say that the sun, in circling round the world, *ne'er beholdeth aught so noble as this lady*, from which it follows that she is, as the words affirm, the most noble of all the things on which the sun shines. And I say *in that hour*, etc. Wherefore we must know that the *hour* is understood in two senses by astrologers. One sense is employed when they assign twenty-four hours to the day and night, that is, twelve to the day and twelve to the night, however long or short the day may be. And these hours become short or long in the day and in the night, according as the day and night wax and wane. And the Church uses hours in this sense when she speaks of Prime, Tierce, Sext, and None; and these are called temporal hours. The other sense is employed when, of the twenty-four hours alloted to the day and night, the day has sometimes fifteen and the night nine, and sometimes the night has sixteen and the day eight assigned to it, according as the day and the night wax and wane; and these are called equal hours. And at the equinox these latter are always one and the same with those which are called temporal; for it must needs be so when the day and the night are of the same length."[435]

The choice of these two examples of unequal days and nights seems to be another instance of Dante's dependence upon Alfraganus, for in Chapter VIII. of the *Elementa Astronomica* we read that the middle of the seventh climate (in which the inequality is greatest) has a longest day of 16 hours, and the middle of the fifth, of 15 hours; and in chapter IX. we find that Magna Roma is situated in the fifth climate.

In the fourth treatise of the *Convivio*, Dante returns to the same subject, and enters into further detail about Church hours and divisions of the day.[436] Our lives, like the arch of heaven which is always above us, consist of a rise and a decline, with a culminating point, and this point he fixes as normally reached at a man's 35th year, as Ristoro had also done.[437] In the same way the day rises and declines, and culminates at the sixth hour; and it is the most noble hour of the day and the most virtuous (in the old Latin sense of virtue as strength, efficacy).

> "La sesta ora, cioè il mezzo del dì, è la più
> nobile di tutto il dì, e la più vertuosa."[438]

This means the whole of the sixth hour (as we should say, from 11 to 12 o'clock): the beginning of the seventh hour is the exact moment of noon, after which the day begins to decline. It was for this reason, Dante believes, that Christ chose to die during the sixth hour,[439] and during his 34th year, for then the day and his human life were both at their culmination, and had not begun to decline.

There was some difference in the methods of reckoning the hours for Church offices, but Dante tells in this same passage which of them he considered right, and he follows it consistently in his works. The Church day was divided into four parts, viz.:—

		At the equinox=
Tierce,	Sunrise to the end of the third hour	6 to 9 a.m.
Sext,	Fourth to end of sixth hour	9 a.m. to 12 noon
Nones,	Seventh to end of ninth	12 noon to 3 p.m.
Vespers,	Tenth hour to sunset	3 to 6 p.m.

Two of the Church offices were to be said at the beginning, and one at the end, of these periods, as follows:—

Tierce,	at the end	i.e. 9 a.m. at the equinox.
Nones,	at the beginning	i.e. 12 noon.
Vespers,	,, ,,	i.e. 3 p.m. at the equinox.

Thus sext is omitted altogether, and Dante says that the reason for the arrangement is to approximate in every case to that hour which is the noblest of the whole day, the sixth.

Mid-tierce was halfway through the period from sunrise to the end of the third hour, that is, 7.30 a.m. at the time of the equinox; and mid-nones and mid-vespers were counted in the same way.

As an instance of reckoning hours by the sun, we may quote the reply of Adam when Dante desired to know how long time he had spent in Eden:

"Dalla prim' ora a quella che seconda, Come il sol muta quadra, l'ora sesta."[440]

Just as the time of day may be told, if one is accustomed to watch the sun, by noting how much of the daily course has been run, so the time of day or night (according to her phase) may be told by the moon; only with her we must take into consideration the rapid and variable motion eastwards in the zodiac. Alfraganus says that her *mean* daily motion in longitude is 13° and nearly 11 minutes,[441] but that a small amount must be added or subtracted every day in order to find her true motion. As a clock, therefore, for use in daily life, she leaves much to be desired. This daily motion makes her fall constantly behind the sun, so that she loses time every day, and not even the same amount of time; for though she crosses the meridian on an average 50½ minutes later each day, the interval is sometimes only 38 minutes and sometimes as much as 66 minutes. And her times of setting and rising are even more variable.

This is easily understood if we remember how much the sun's time of setting varies throughout the year, according to the part of the zodiac in which he is travelling, and consider that what the sun does in a year the moon does in a month. And in her case the effect is sometimes exaggerated, sometimes diminished, for her path in the zodiac is inclined to the sun's path; moreover, her greatest departure from it to north and south takes place in different parts of the zodiac at different times. These facts were well known to Ptolemy and to mediæval astronomers, and everyone who watches the moon must have noticed how variable are the intervals between one moonrise or one moonset and the next.

The extent of variability also depends upon our latitude (just as with sunrises and sunsets); and in Florence the retardation in one day may sometimes be only twenty minutes, sometimes an hour and twenty minutes. The difference would be less in Dante's Purgatory, since this was in latitude 32° south, and the intervals between moonset and moonset become less variable in length as we approach the equator, just as the days become less unequal all the year round. Still, they would vary a good deal, so we must conclude that Dante only means to indicate the time quite roughly when he uses the moon as a clock. As a matter of fact, he seldom does so in the *Divine Comedy* without giving us another clue to the time as well.

These passages all belong to a most interesting series of time indications, which we may now proceed to examine.

VI.
DANTE'S JOURNEY THROUGH THE THREE REALMS:

INDICATIONS OF TIME AND DIRECTION BY MEANS OF THE SKIES.

1. TIME REFERENCES IN THE DIVINE COMEDY.

The plan of the universe through which Dante feigns himself to have journeyed is familiar to all readers of the *Divine Comedy*. Hell, as the Fathers taught, was a subterranean cavity, and Dante pictures it as an inverted cone, whose apex reaches the exact centre of Earth and therefore of the universe. It is situated vertically underneath Jerusalem, the centre of the inhabited Earth. He departs from the Fathers, however, in removing Purgatory from these dim regions, and placing it on an island in the midst of the ocean of the uninhabited hemisphere, exactly at the antipodes of Jerusalem. On this island rises a mountain whose immensely lofty summit reaches the upper regions of the atmosphere, and upon the summit is the Eden of our first parents. This original conception is an extraordinary gain, both from the artistic and the allegorical points of view; and it is in harmony with the idea of Aristotle, and of many mediæval writers, that the southern hemisphere was the "nobler" part of Earth. Here, then, man was permitted to dwell before the Fall, and hither come repentant souls, saved from Hell, but not yet pure enough to enter Heaven. Paradise consists of all the spheres of mediæval astronomy, and the poet rises from one to another until he finally reaches the all-embracing Empyrean, where his vision ends.

Nowhere does he describe this scheme in full, but it was evidently clear in his own mind, and by following him step by step in his journey it is easily reconstructed, and is represented in the accompanying diagram.

In the same way he never states how long a time he spent on this visionary journey, yet this also he had definitely determined, and in each Cantica he refers once to the period of time alloted to each realm.

In the last Circle but one of the Inferno, he is warned by Virgil that the time allowed is drawing to a close;[442] in the last but one of the Purgatorio Virgil urges him to make the best use of the time appointed;[443] and in the Paradiso, just before the final vision, St. Bernard tells him that the time of his trance is nearly over.[444]

In the first Canto, which is introductory to the whole *Commedia*, he mentions the time at which he assumes that his vision began. It was "nel mezzo del cammin di nostra vita,"[445] which may mean simply that he was middle-aged, or comparing it with the passage in the *Convivio* quoted earlier (see p. 347), we may suppose it to mean that he was exactly in his 35th year. The season was spring—"la dolce stagione,"[446] and when night was over the sun rose among those stars which were with him at the Creation, which was believed to have taken place at the vernal equinox some sixty-five centuries before.

"Il Sol montava su con quelle stelle Ch' eran con lui quando l'Amor Divino Mosse da prima quelle cose belle."[447]

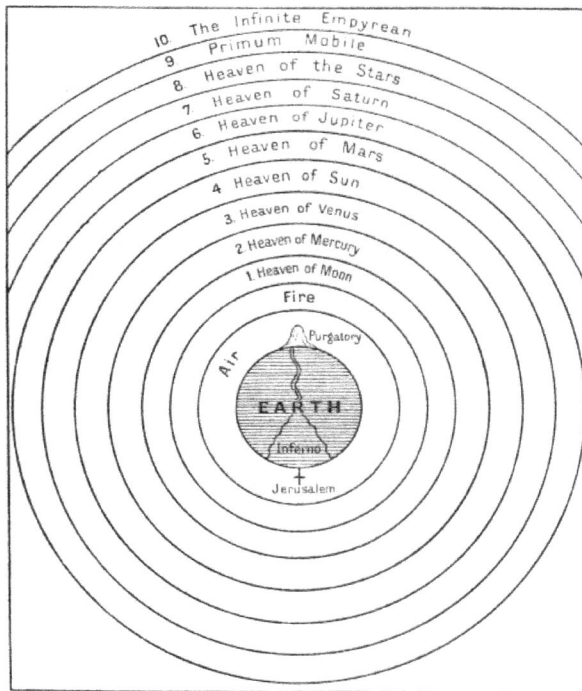

Fig. 45. The Universe of Dante.

The expression "quelle stelle,"[448] however, shows us that Dante does not intend to be pedantically accurate, which would indeed be inartistic in a poem. He speaks several times of the supposed slow motion of the star sphere,[449] and although he never mentions its effect in shifting the place of the equinox continually among the stars, we can hardly imagine that he failed to understand this. He knew quite well, therefore, that the stars which accompanied the sun in that first springtime of the world could not be the same stars which rose with him now. But astronomers had agreed to call the part of the sky which contains the vernal equinox by its old name, and it would be exceedingly tiresome if he were to distinguish, every time he mentioned them, between the zodiacal constellations and the zodiacal signs; there shall be no occasion, therefore, to stop and consider which he means, and no ambiguity: he assumes once for all that they always have been and always are identical, that the "first point of Aries" or vernal equinox is now and always at the beginning of the constellation Aries. To this assumption he is consistent throughout.

Another important point to notice is that, since it is the time of the equinox, there is no difference between "temporary" and "equal" hours.[450] Days and nights are of equal length; sunrise and sunset are nearly at 6 a.m. and 6 p.m., according to our modern method of reckoning.[451] In this way also, and without any sacrifice of astronomical accuracy, the poet has simplified his time references for his readers.

We must observe, however, that the day is as uncertain as the year; for, although it is certainly near the equinox, "quelle stelle" may mean exactly the first point of Aries or only the constellation Aries in general. From this Canto alone, therefore, we cannot be sure of the exact date.

From other allusions of various kinds, scattered throughout the *Divine Comedy*, we gather that Dante did intend a definite date to be understood, but unfortunately the data by which he means to fix it are themselves so ambiguous that commentators differ as to the day, the month, and the year. The question has so direct a bearing on Dante's astronomy that we cannot shirk it; but for the present, leaving it aside, let us deal merely with the hour, which is distinctly stated from the first, and indicated in a number of passages so clearly that readers with an elementary knowledge of the apparent movements of the heavenly bodies can find little difficulty in following Virgil and Dante through their journey, seeing exactly how many days it required, and frequently at what hour of day or night they were in a particular spot. If my readers are not now in possession of this elementary knowledge, this book has been written in vain.

I give below a list of all the hour-indications, astronomical and others, for the sake of clearness and completeness; but I omit for the present those that indicate only the date, and merely call the days the first, second, etc. These passages, though they are only to be understood in a popular sense, prove quite as clearly as the elaborate astronomical explanations in the

Convivio how deep an interest Dante always took in astronomy; and a study of them shows that he had carefully considered the positions of the heavenly bodies and their movements during the progress of the action in his poem.

TIME REFERENCES IN
THE DIVINE COMEDY

Night. In the Forest. Moon full

(see below, *Inf.* xx. 127-129, and *Purg.* xxiii. 118-121).

"La notte ch' i' passai."

Inf. i. 21.

"The night that I had spent."

1st day. Just before sunrise.

At the foot of the Hill.

> "Ma poi ch' io fui al piè d'un colle giunto,
>
> * * * * *
>
> Guardai in alto, e vidi le sue spalle
>
> Vestite già de' raggi del pianeta
>
> Che mena dritto altrui per ogni calle."
>
> *Inf.* i. 13-18.

> "But after I had reached a mountain's foot,
>
> * * * * *
>
> Upward I looked, and I beheld its shoulders
>
> Vested already with that planet's rays
>
> Which leadeth others right by every road."

Sunrise 6 a.m. [452]

Climbing the Hill.

> "Tempo era dal principio del mattino,
>
> E il sol montava su con quelle stelle
>
> Ch'eran con lui quando l' Amor divino

Mosse da prima quelle cose belle."

Inf. i. 37-40.

"The time was the beginning of the morning,

And up the sun was mounting with those stars

That with him were, what time the Love Divine

At first in motion set those beautous things."

Evening twilight.

At the Gate of Hell.

"Lo giorno se n'andava, e l' aer bruno...."

Inf. ii. 1.

"Day was departing, and the brown air...."

Midnight.

Crossing the fourth circle.

"Già ogni stella cade, che saliva

Quando mi mossi."

Inf. vii. 98, 99.

"Already sinks each star that was ascending

When I set out."

2nd day. About 2 hrs. before sunrise (4 a.m.)

On the edge of the cliff,

      preparing to descend to the 7th circle.

"I Pesci guizzan su per l' orizzonta,

E il Carro tutto sovra il Coro giace."

Inf. xi. 113, 114.

"Quivering are the Fishes on the horizon,

And the Wain wholly over Caurus lies."

(Dante refers to the previous day.)

In the 7th circle.

"Mi smarri' in una valle.

* * * * *

Pur ier mattina le volsi le spalle."

Inf. xv. 50-52.

"I lost me in a valley.

* * * * *

But yestermorn I turned my back upon it."

After sunrise, between 6 and 7 a.m.

Leaving the 4th Pit of Circle VIII.

(Virgil reminds Dante of the night before last, spent in the Forest.)

"Già tiene il confine

D' ambedue gli emisperi e tocca l' onda

Sotto Sibilia Caino e le spine;

E già iernotte fu la luna tonda;

Ben ten dee ricordar, chè non ti nocque

Alcuna volta per la selva fonda."

Inf. xx. 124-129.

"... Already holds the confines

Of both the hemispheres, and under Seville

Touches the ocean-wave, Cain and the thorns;

And yesternight the moon was round already;

Thou shouldst remember well: it did not harm thee

From time to time within the forest deep."

7 a.m.

On the bridge over the 5th Pit.

"Ier, più oltre cinq' ore che quest' otta,

Mille dugento con sessanta sei

Anni compiè che qui la via fu rotta."

Inf. xxi. 112-114.

"Yesterday, five hours later than this hour,

One thousand and two hundred sixty-six

Years were complete that here the way was broken."

About 1 p.m.

Leaving the 9th Pit.

"E già la Luna è sotto nostri piedi,

Lo tempo è poco omai che n'è concesso."

Inf. xxix. 10, 11.

"And now the moon is underneath our feet.

Henceforth the time allotted us is brief."

Nightfall.

At the bottom of the world.

"La notte risurge; ed oramai

È da partir, chè tutto avem veduto."

Inf. xxxiv. 68, 69.

"Night is reascending, and 'tis time

That we depart, for we have seen the whole."

3rd day. (no night intervening) 7.30 a.m.

In the new hemisphere

"Già il sole a mezza terza riede."

Inf. xxxiv. 96.

"And now the sun to middle-tierce returns."

4th day. About an hour before sunrise (5 a.m.).

On the island of Purgatory.

"E quindi uscimmo a riveder le stelle."

Inf. xxxiv. 139.

"Thence we came forth to see again the stars."

"Lo bel pianeta che ad amar conforta

Faceva tutto rider l' oriente,

Velando i Pesci ch' erano in sua scorta."

Purg. i. 19-21.

"The beauteous planet that to love incites

Was making all the orient to laugh,

Veiling the Fishes that were in her escort."

Dawn.

"Lo sol vi mostrerà, che surge omai."

Purg. i. 107.

"The sun, which now is rising, will direct you."

"L' alba vinceva l' ora mattutina."

Purg. i. 115.

"The dawn was vanquishing the matin hour."

Sun just below east horizon.

On the Shore of the Island.

"Già era il sole all' orizzonte giunto

Lo cui meridian cerchio coperchia

Jerusalem col suo più alto punto;

E la notte, che opposita a lui cerchia,

Uscia di Gange fuor con le bilance

Che le caggion di man quando soperchia;

Si che le bianche e le vermilie guance,

Là dove io era, della bella Aurora,

Per troppa etate divenivan rance."

Purg. ii. 1-9.

"Already had the sun the horizon reached

Whose circle of meridian covers o'er

Jerusalem with its most lofty point;

And night, that opposite to him revolves,

Was issuing forth from Ganges, with the scales

That fall from out her hand when she exceedeth;

So that the white and the vermilion cheeks

Of beautiful Aurora, where I was,

By too great age were changing into orange."

Sun risen.

"Da tutte parti saettava il giorno

Lo sol, ch'avea con le saette conte

Di mezzo il ciel cacciato il Capricorno."

Purg. ii. 55-57.

"On every side was darting forth the day

The sun, who had with his resplendent shafts

From the mid-heaven chased forth the Capricorn."

Soon after sunrise.

Walking westward, towards the Mountain of Purgatory.

"Lo sol che retro fiammeggiava roggio."

Purg. iii. 16.

"The sun that in our rear was flaming red."

Shortly after 6 a.m. (3 p.m. in Italy).

"Vespero è già colà dov' è sepolto

Lo corpo dentro al quale io facea ombra."

Purg. iii. 25, 26.

"'Tis vespers there already where is buried

The body within which I cast a shadow."

About 9.30 a.m.

At the foot of the Mountain.

"Ben cinquanta gradi salito era

Lo sole."

Purg. iv. 15, 16.

  "Fifty full degrees uprisen was

  The sun."

Midday.

On a ledge of the Mountain (Anti-purgatory).

"Vedi ch'è tocco

Meridian dal sole, e dalla riva.

Copre la notte già col piè Morrocco."

Purg. iv. 137-139.

  "See, the sun has touched

  Meridian, and from the shore the night

  Covers already with her foot Morocco."

Afternoon.

Climbing the eastern slope of the Mountain.

"Vedi omai che il poggio l' ombra getta.

* * * * *

Prima che sii lassù tornar vedrai

Colui che già si copre della costa,

Sì che i suoi raggi tu romper non fai."

Purg. vi. 51-57.

  "See, e'en now the hill a shadow casts.

  * * * * *

  Ere thou art up there, thou shalt see return

  Him, who now hides himself behind the hill,

  So that thou dost not interrupt his rays."

Evening.

"Vedi già come dichina il giorno."

Purg. vii. 43.

"See already how the day declines."

Sunset (6 p.m.).

In the Flowery Valley, on the Mountain-side.

"Era già l'ora che volge il disio

Ai naviganti e intenerisce il core,

Lo di ch'han detto ai dolci amici addio;

E che lo nuovo peregrin d' amore

Punge, se ode squilla di lontano

Che paia il giorno pianger che si more."

Purg. viii. 1-6.

"'Twas on the hour that turneth back desire

In those who sail the sea, and melts the heart,

The day they've said to their sweet friends farewell;

And the new pilgrim penetrates with love,

If he doth hear from far away a bell

That seemeth to deplore the dying day."

Evening twilight.

"Tempo era già che l' aer s'annerava."

Purg. viii. 49.

"Already now the air was growing dark."

(Dante refers to the morning of the same day.)

"Per entro il lochi tristi

Venni stamane."

Purg. viii. 58, 59.

"Through the dismal places

I came this morn."

Early night: about 7 p.m.

"Le quattro chiare stelle

Che vedevi staman, son di là basse,

E queste son salite ov' eran quelle."

Purg. viii. 91-93.

"The four resplendent stars

Thou sawest this morning are down yonder low,

And these have mounted up to where those were."

Shortly before moonrise, between 8 and 9 p.m.

"La concubina di Titone antico

Già s'imbiancava al balco d'oriente,

Fuor delle braccia del suo dolce amico;

Di gemme la sua fronte era lucente,

Poste in figura del freddo animale

Che con la coda percote la gente;

E la notte de' passi con che sale

Fatti avea due nel loco ov' eravamo,

E il terzo già chinava in giuso l' ale."

Purg. ix. 1-9.

"The concubine of old Tithonus now

Gleamed white upon the eastern balcony,

Forth from the arms of her sweet paramour;

With gems her forehead all relucent was,

Set in the shape of that cold animal

Which with its tail doth smite amain the people;

And of the steps with which she mounts the Night

Had taken two in that place where we were,

And now the third was bending down its wings."

5th day. Dawn.

At the Gate of Purgatory.

"Nell' ora che comincia i tristi lai

La rondinella presso alla mattina."

Purg. ix. 13, 14.

"Just at the hour when her sad lay begins

The little swallow, near unto the morning."

(See also ix. 52 "nell' alba che precede al giorno,"

and ix. 59 "come il di fu chiaro.")

(See also "At dawn, which doth precede the day."

and "As day grew bright").

More than 2 hours after sunrise (after 8 a.m.).

"Il sole er' alto già più che due ore."

Purg. ix. 44.

"And now the sun was more than two hours high."

About 9.30 a.m.

On the 1st Cornice of the Mountain.

"Pria lo scemo della luna

Rigiunse al letto suo per ricorcarsi,

Che noi fossimo fuori di quella cruna."

Purg. x. 14-16.

"Sooner had the moon's decreasing disc

Regained its bed to sink again to rest.

Than we were forth out of that needle's eye."

12 noon.

Ascent to the 2nd Cornice.

"Più era già per noi del monte volto,

E del cammin del sole assai più speso

Che non stimava l'animo non sciolto.

* * * * *

... Vedi che torna

Dal servigio del dì l'ancella sesta."

Purg. xii. 73-81.

"More of the mount by us was now encompassed,

And far more spent the circuit of the sun,

Than had the mind preoccupied imagined.

* * * * *

... Lo, returning is

From service of the day the sixth handmaiden."

3 p.m. (midnight in Italy).

Ascent to the 3rd Cornice.

"Quanto tra l'ultimar dell' ora terza

E il principio del dì par della spera

Che sempre a guisa di fanciullo scherza,

Tanto pareva già in ver la sera

Essere al sol del suo corso rimaso;

Vespero là, e qui mezza notte era.

E i raggi ne ferian per mezzo il naso,

Perchè per noi girato era si il monte,

Che già dritti andavamo in ver l'occaso."

Purg. xv. 1-9.

"As much as 'twixt the close of the third hour

And dawn of day appeareth of that sphere

Which aye in fashion of a child is playing,

So much, it now appeared, towards the night

Was of his course remaining to the sun;

There it was evening, and 'twas midnight here.

And the rays smote the middle of our faces,

Because by us the mount was so encircled,

That straight towards the west we now were going."

Evening.

On the 3rd Cornice.

"Noi andavam per lo vespero, attenti

Oltre, quanto potean gli occhi allungarsi

Contra i raggi serotini e lucenti."

Purg. xv. 139-141.

"We passed along athwart the twilight, peering

Forward, as far as ever eye could stretch

Against the sunbeams serotine and lucent."

Sunset at sea-level, 6 p.m.

"Io rividi

Lo sole in pria, che già nel corcare era."

Purg xvii. 8, 9.

"I saw again

The sun at first, that was already setting."

On the 3rd Cornice.

"I raggi, morti già, nei bassi lidi."

Purg. xvii. 12.

"Rays already dead on the low shores."

Ascent to the 4th Cornice.

"Procacciam di salir pria che s'abbui."

Purg. xvii. 62.

"Let us make haste to mount ere it grow dark."

Sunset on the Mountain, shortly after 6 p.m.

On the 4th Cornice.

"Già eran sopra noi tanto levati

Gli ultimi raggi che la notte segue,

Che le stelle apparivan da più lati."

Purg. xvii. 70-72.

"Already over us were so uplifted

The latest sunbeams which the night pursues,

That upon many sides the stars appeared."

Towards midnight.

"La luna, quasi a mezza notte tarda,

Facea le stelle a noi parer più rade,

Fatta com' un secchione che tutto arda;

E correa contra il ciel per quelle strade

Che il sole infiamma allor che quel da Roma

Tra i Sardi e i Corsi il vede quando cade."

Purg. xviii. 76-81.

"The moon, belated almost unto midnight, [453]

Now made the stars appear to us more rare,

Formed like a bucket that is all ablaze;

And counter to the heavens ran through those paths

Which the sun sets aflame when he of Rome

Sees it 'twixt Sardes and Corsicans go down."

6th day. Nearly 2 hours before sunrise, about 4 a.m.
On the 4th Cornice.

"Nell' ora che non può il calor diurno

Intepidar più il freddo della luna,

Vinto da terra, o talor da Saturno;

Quando i geomanti lor Maggior Fortuna

Veggiono in oriente, innanzi all' alba,

Surger per via che poco le sta bruno."

Purg. xix. 1-6.

"It was the hour when the diurnal heat

No more can warm the coldness of the moon,

Vanquished by Earth, or peradventure Saturn;

When geomancers their Fortuna Major

See in the orient, before the dawn,

Rise by a path that long remains not dim."

Early morning.

Ascent to 5th Cornice.

"Su mi levai, e tutti eran già, pieni

Dell' alto dì i giron del sacro monte,

Ed andavam col sol nuovo alle reni."

Purg. xix. 37-39.

"I rose, and full already of high day

Were all the circles of the sacred mountain,

And with the new sun at our back we went."

About 10 a.m.

On the 6th Cornice.

"E già le quattro ancelle eran del giorno

Rimase addietro, e la quinta era al temo,

Drizzando pure in su l' ardente corno."

Purg. xxii. 118-120.

"And four handmaidens of the day already

Were left behind, and at the pole the fifth

Was pointing upwards still its burning horn."

(Dante refers to "the other day" when he began the journey.)

"Di quella vita mi volse costui

Che mi va innanzi, l' altr' ier, quando tonda

Vi si mostrò la suora di colui—

E il sol mostrai."

Purg. xxiii. 118-121.

"Out of that life he turned me back who goes
In front of me, some days agone, when round
The sister of him yonder showed herself—
And to the sun I pointed."

Between 2 and 4 p.m.
Ascent to the 7th Cornice.
"Il sole avea lo cerchio di merigge
Lasciato al Tauro, e la notte allo Scorpio."
Purg. xxv. 2, 3.

"The sun bad his meridian circle
To Taurus left, and Night to Scorpio."

Evening.
On the 7th Cornice, west side of the Mountain.
"Feriami il sole sull' omero destro,
Che già, raggiando, tutto l' occidente
Mutava in bianco aspetto di cilestro."
Purg. xxvi. 4-6.

"On the right shoulder smote me now the sun,
That, raying out, already the whole west
Changed from its azure aspect into white."

Sunset at sea-level, 6 p.m.
About to cross the zone of fire.
"Sì come quando i primi raggi vibra
La dove il suo Fattore il sangue sparse,
Cadendo Ibero sotto l' alta Libra,
E l' onde in Gange da nona riarse,
Sì stava il sole; onde il giorno sen giva."
Purg. xxvii. 1-5.

"As when he vibrates forth his earliest rays

In regions where his Maker shed his blood,

The Ebro falling under lofty Libra,

And waters in the Ganges burnt with noon,

So stood the sun; hence was the day departing."

Sun about to set on the mountain heights.

At the foot of the last stairway.

"Lo sol sen va, soggiunse, e vien la sera."

Purg. xxvii. 61.

"The sun departs, it added, and night cometh."

Sun sets.

Climbing the last stairway.

"Io toglieva i raggi

Dinanzi a me del sol ch'era già basso.

* * * * *

Il sol corcar, per l' ombra che si spense

Sentimmo."

Purg. xxvii. 65-69.

"I cut off the rays

Before me of the sun, that now was low.

* * * * *

By the vanished shadow the sun's setting

Behind us we perceived."

Night.

On the last stairway.

"Vedev' io le stelle."

Purg. xxvii. 89.

"I beheld the stars."

7th day. About 4 a.m.

"Nell' ora, credo, che dell' oriente

Prima raggiò nel monte Citerea."

Purg. xxvii. 94, 95.

  "It was the hour, I think, when from the east

  First on the mountain Cytherea beamed."

Dawn.

"E già per gli splendori antelucani

* * * * *

Le tenebre fuggian da tutti i lati."

Purg. xxvii. 109-112.

  "And now before the antelucan splendours

  * * * * *

  The darkness fled away on every side."

Early morning.

In the Earthly Paradise.

"Vedi là il sol che in fronte ti riluce."

Purg. xxvii. 133.

  "Behold the sun that shines upon thy forehead."

"L' ôre prime."

  "The hours of prime."

Midday.

By the fountain of Eunoë.

"Teneva il sole il cerchio di merigge."

Purg. xxxiii. 104.

  "The sun was holding the meridian circle."

Ascent to Paradise.

"Fatto avea di là mane e di qua sera

Tal foce quasi, e tutto era là bianco

Quello emisperio, e l' altra parte nera."

Par. i. 43-45.

"Almost that passage had made morning there

And evening here, and there was wholly white

That hemisphere, and black the other part."

8th day.

A lapse of 6 hrs. since entering the 8th heaven.

(*Cf. Par.* xxii. 151-153).

Sunset in Jerusalem (sunrise in Purgatory).

In the eighth heaven among the stars of Gemini.

"Dall' ora ch'io avea guardato prima

Io vidi mosso me per tutto l'arco

Che fa del mezzo al fine il primo clima;

Si ch'io vedea di là da Gade il varco

Folle d'Ulisse. e di qua presso il lito

Nel qual si fece Europa dolce carco;

E più mi fora discoperto il sito

Di questa Aiuola, ma il sol procedea

Sotto i miei piedi un segno e più partito."

Par. xxvii. 79-87.

"Since the first time that I had downward looked

I saw that I had moved through the whole arc

   Which the first climate makes from midst to end;

So that I saw the mad track of Ulysses

Past Gades, and this side well nigh [454] the shore

Whereon became Europa a sweet burden;

And of this threshing-floor the site to me

Were more unveiled, but the sun was preceding

Under my feet, a sign and more removed."

2. THE INFERNO.

Using the above table as a guide, we may now see how Dante uses the movements of the heavenly bodies both for time and direction, and we shall discover how long his journey takes.

Dante has spent a night in a dark forest. The moon was then full; but we only learn this later, when Virgil reminds him of it, and he himself mentions it to Forese, when talking in the Purgatorio. At dawn he finds himself at the foot of a hill on which are already shining the first rays of the morning sun—the planet which leads all aright by every path. The sun rises: it is in Aries, consequently the moon, being full, must be in Libra, and must have set as the sun rose, i.e. at about 6 a.m. Virgil appears, rescues the poet from three terrible beasts, and announces that he has come to conduct him through Hell and Purgatory, and to lead him to Beatrice, who will guide him to Paradise.

On the evening of the same day, at twilight, the poets enter the Inferno. This great subterranean cavity, shaped like an inverted cone (see fig. 45), is divided into nine circles, in which the different classes of sinners are punished. In this blind world of eternal darkness no star shines, and the sun is never once mentioned as timegiver; when Virgil wishes to remind Dante that they must hasten, he refers to the position of moon or stars, and the former is called "La Donna che qui regge."[455]

After passing through four circles, Virgil says that every star is now sinking which was ascending when he started, indicating that it is now six hours since they entered the Inferno, or soon after midnight. He does not mean that all these stars are setting, but that they have crossed the meridian, and are at some point of their descent towards the west. Strictly speaking this does not apply to stars north of the equator, since they take more than six hours to reach the meridian, being above the horizon for more than twelve. Perhaps Virgil is thinking chiefly of the constellations of the zodiac.

The poets cross the fifth circle, enter the "basso Inferno"[456] or city of Dis, and on the edge of the precipice which borders the circle of the Heretics they sit while Virgil explains the system of the several circles. Then he expresses a wish to continue the journey, for the Fishes (of the zodiac) are quivering on the horizon, and the Wain lies to the north-west. As it is the time of the Vernal Equinox, and the sun is in Aries, Pisces is the zodiacal sign immediately preceding the sun, and begins to rise about four in the morning. The part of Ursa Major called the Wain is somewhat less than 180° distant from Pisces; hence when the latter appears on the eastern horizon the former will be in the west (and of course always in the north). The "Caurus" of the Latins was the wind from the north-west.

They descend a cliff over loose stones which roll under the unaccustomed weight of Dante's living feet. Virgil tells him that both here and elsewhere in Hell the ancient rocks were riven by an earthquake which immediately preceded the triumphant descent of Christ after His crucifixion. So mighty was the shock that he thought the universe was dissolving into chaos by the force of love, as Empedocles had said must happen periodically. At the base of the cliff they cross the river Phlegethon, then pass through the wood of the Harpies, and walk along the stone banks of a conduit which afterwards falls with thunderous roar down a steep precipice. On the burning sand through which the conduit runs, Dante sees Brunetto Latini, and tells him that it was only yesterday that he escaped from the Forest. The monster Geryon carries them down the precipice, and they reach the ten Pits of Malebolge, which like so many moats surround the central Depth.

When looking down upon the astrologers in the fourth Pit, Virgil again hastens Dante on, telling him that now Cain and the thorns—by which he means the moon, alluding to the legend[457]—is on the boundary of the two hemispheres, and already touching the waves below Seville. The two hemispheres are those of the habitable earth whose centre is Jerusalem, and the uninhabitable with Purgatory at its centre: the moon, therefore, is just setting as seen from Jerusalem, and in the Inferno the time is Jerusalem time. It is here that Virgil adds, "And yesternight the moon was full:" did we not know this, the fact that she was setting would give no clue to the time. When full, she set as the sun rose, that was at 6 a.m.; now, therefore, she is setting after sunrise, between six and seven in the morning;[458] and this is how Virgil tells Dante, without mentioning the sun, that a new day has begun.

The poets accordingly hasten on to the bridge above the next Pit, the fifth, and here the demons treacherously mislead them by assuring them that the next bridge directly on their way is broken down, but the one to the left is standing. The anniversary of the catastrophe is precisely determined by Malacoda, the chief demon of the band, as five hours later than the present hour on the preceding day. There can be no doubt that he refers to the same earthquake as Virgil described when descending the stony cliff, and all commentators are agreed to this. The time now, therefore, is five hours before midday, *i.e.* 7 a.m., since it was at midday that Dante believed Christ to have died.[459]

The scenes with the demons follow, the breathless escape into the next Pit, of the Hypocrites, and the horrors of the Pit of the Thieves.

Then for a short space, when the suffering and the murk of Hell is becoming almost intolerable, we are suddenly lifted into the upper air by the narrative of Ulysses—into the pure air and the vast spaces of the southern ocean, and we see once more the skies. We follow the adventurers to the regions beyond the sun, the world where no one lives; we see all the stars of the other pole, and the moon shines down upon us. Too soon comes the tragic end of the story, and the waters close over the ship; the shade of Ulysses passes on, and we are once more in the "buio d' inferno."[460]

At the next Pit, that of the Schismatics, as Dante stands ready to weep for pity on the bridge above it, Virgil reproaches him for delaying, for little remains now of the time permitted for seeing the Inferno, and already the moon is beneath their feet. This is about an hour after midday; for if the moon were full she would be just opposite the sun, hence under our feet when the sun is on the meridian, *i.e.* at noon; but this is one day after Full Moon, and therefore she does not reach the position till about an hour later.

The tenth and last Pit is traversed, and when the giant Antæus has lifted them down into the sorrowful pit towards which all the rest of Hell converges, the poets have nearly reached the centre of the earth. For they have been descending continually, finding a slight fall between each circle, and between the great divisions of different classes of sin, three sharp steep falls—the cliff descending to the River of blood and other circles of the violent, the precipice down which Geryon carried them to the Malebolge of Fraud, and the Dark Pit to the bottom of which Antæus conveyed them, where in the Frozen Lake treachery was punished. Last of all, at the very centre and bottom of the whole "dolorous kingdom" of Lucifer, they see its King, and Virgil says all has been seen, and it is time to depart, for Night is rising again.

Thus we see that the journey through the Inferno has taken exactly the time from one sunset to another, twenty-four hours.

Now Virgil seizes the shaggy side of the fiend, while the monstrous wings are raised, and with Dante clinging to him climbs down the side, as far as the hip; here with difficulty and labour the master turns round, so that his head is where his feet were, and climbs up, panting, so that Dante thinks they are returning. But when the ascent has been made, and both poets sit down to rest, Dante sees to his astonishment that the frozen lake is gone, the head and wings of Lucifer are gone, they are looking down upon him, and see only his legs which are directed upwards! To add to Dante's bewilderment, Virgil now says, "Rise, for the way is long and the road is bad, and the sun has already returned to middle-tierce," which, as we have seen, means at the time of the equinox 7.30 in the morning. Night was just beginning in the dreadful place they have so recently left, but now for the first time Virgil mentions the sun as giving them the time, and it is early morning.

The explanation is that the point which they passed with so great an effort is the exact centre of the world; they are now on its further side, and above them is the other hemisphere, where all is ocean save the island of Purgatory. Virgil explains how it was through the fall of Lucifer from Heaven that all the land except Purgatory is in one hemisphere, and that there is a cavity through which they may now climb to the opposite hemisphere. When Lucifer fell, all the land which then existed in the unknown hemisphere fled for fear of him, sinking under the sea, and then escaping round the globe settled in the inhabited hemisphere. It may be, he adds, that then the earth in the interior fled away from him upwards to form land in this hemisphere (Purgatory) and thus an empty space is left here. Up this, which is the same length as the tunnel of the Inferno, they climb, guided through the darkness by the sound of a little brook which has worn its winding way down the rock. On this route there is nothing to detain them, for they are alone in the darkness, but it presumably takes much longer to ascend than it did to descend, for climbing without a pause it takes about twenty-three hours (from the centre) to reach that round hole in the rocks through which at last they once more see the stars, just before the dawn of a new day.

3. PURGATORY.

Leaving Earth's centre at sunset by Jerusalem time, sunrise by Purgatory time, the poets emerge next morning by starlight when the morning star is making all the orient smile. Henceforth Dante does not need to be told how moon or stars are standing in the sky: he has but to look up, and see the sun and sometimes the moon by day, the stars and sometimes the moon by night. The Inferno was like one long dark night, Paradise will be one brilliant day; in Purgatory the sun rises four times and sets three times, before Dante reaches the summit, and thence rises to Paradise with Beatrice.

"Lo bel pianeta che ad amar conforta"[461] is of course Venus: if there were any doubt about this, the words used concerning the dawn of the last morning on the Mountain would solve it, for there "Cytherea" is spoken of as shining in the east on the mountain. She is in Pisces, the sign which rises about two hours before Aries, or 4 a.m. at this season, and as the stars are still clearly visible, it is about 5 a.m. The poets have evidently come up from the subterranean passage facing east, for they see Venus first, then Dante turns to his right (south) and sees the four brilliant stars near the south pole, then turning towards the north pole he sees that the Wain has disappeared beneath the horizon, and close beside him he finds Cato, on whose face the Four Stars are shining. Cato tells them that the sun, which is about to rise, will be their guide to reach the Mountain, and as they first go down to the shore, following his instructions, daylight begins to appear.

While they are still on the shore the sun reaches the point just below the horizon, for he is described as just touching the western horizon of the other hemisphere, whose zenith is above Jerusalem, and night is just rising on the eastern horizon of that hemisphere. Where the poets are the rose colour in the east is turning to orange, just before the sun rises. It rises over their horizon while they are watching a light far over the sea, which with almost incredible rapidity draws near, becomes visible as a vessel, touches land, the spirits disembark, and the celestial pilot "sen gì, come venne, veloce."[462] The sun has now driven Capricorn from the mid sky with his bright beams, that is, the stars have become invisible in bright daylight, and fig. 46 shows that when Aries rises over the eastern horizon Capricorn will be overhead. But the sun is still very low, so swift was the coming of the spirits, for Dante's shadow is not yet visible to them: it is by his breathing that they perceive he is a living man. It is only after the meeting with Casella, and the dispersal of the spirits by Cato, that the sun, still red with sunrise tints, makes Dante's shadow fall in front of him as the poets are walking westward towards the mountain. He is alarmed not to see Virgil's shadow on the ground beside it, and is told that the body within which Virgil once had cast a shadow lies buried in a country where it is now the time of vespers. That is, in Italy it is three o'clock in the afternoon: hence in Purgatory, where they now are, it is six in the morning.

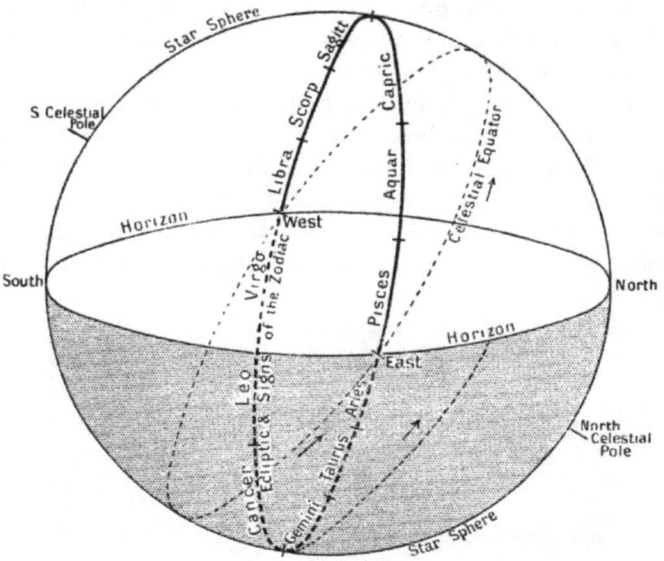

Fig. 46. The Signs of the Zodiac as seen at sunrise from the Mountain of Purgatory at the autumnal equinox there (the *vernal* equinox in the Northern Hemisphere). *Purg.* ii. 55.

At 6 a.m. Aries is just rising, Libra the opposite constellation is just setting, while Capricornus has just reached the meridian. The signs follow one another in their diurnal course over the sky in the direction shown by the arrows, circling parallel to the celestial equator.

At the foot of the mountain they are joined by a band of spirits, and walking very slowly with them, conversing with Manfred, so much time passes that to Dante's surprise, when they reach the place where the ascent is to be made (still on the eastern side), the sun has travelled fully fifty degrees since rising. As the sun passes over 360° in 24 hours, this indicates about 3½ hours after sunrise, or 9.30 a.m.[463] The climb to the first ledge is slow and very arduous, for as Virgil explains, the lower slopes are the most difficult, since the act of climbing becomes more and more easy as the ascent is made. Resting on a ledge, they look back and at the sun, and note how it has travelled north, so that it now strikes their left shoulders; and after a conversation with Belacqua Virgil begins to mount again, saying that it is noon. The other hemisphere consequently is all in darkness, from Ganges ("la riva"[464]) to Morocco.

Other spirits have been met and spoken with, and they are still climbing when evening comes, and Dante, thinking he is to reach the summit of the mountain and see Beatrice that day, entreats Virgil to hasten. "See," he says, "how the mountain now casts a shadow." For the sun has travelled round to the west, and the poets, still on its eastern side, are in deep shadow: this explains how, when they meet Sordello, he does not recognize that Dante is a living man, as all the spirits until now have done. But Virgil assures him that the way is longer than he knows: before they reach the summit, the sun, now hidden behind the mountain and not causing Dante to cast any shadow, will return again to the east.

Fig. 47. Northern slope of the Mountain of Purgatory, up which Dante climbed. Being in the southern hemisphere, this was the sunny side, and he followed the sun's course, from east through north to west.

Immediately after this, they meet with Sordello, who tells them that it is impossible to climb even one step on the mountain after sunset, and leads them to a flowery valley, in which to rest for the night. The spirits in the valley sing the evening hymn while the sun sets; and it begins to grow dark while Dante is talking with judge Nino, explaining whence he came that morning. As soon as the stars become visible, his eyes seek the southern pole, and fix themselves on three brilliant stars which have taken the place of that constellation of four which he saw some fourteen hours before. The latter have travelled to the south-west, and are hidden behind the mountain, and the new constellation is above the pole towards the east.

Night has risen two steps and part of a third in this place when Dante falls asleep, and the most natural meaning of this is that it is somewhat more than two hours after sunset, or after 8 o'clock, in Purgatory. The moon has not yet risen, since she is like a clock which loses on an average 50 minutes every day.

Since the first morning in the Forest, when she was exactly 12 hours behind the sun and set at six o'clock, three whole days have elapsed,[465] and if she has lost 3 times 50 minutes, or 2½ hours, she will rise 14½ hours after the sun rises, that is 2½ hours after sunset, or at about 8.30. As the retardation varies much, let us say simply that she rises to-night between eight and nine o'clock. She has also moved among the stars, about 13 degrees each day, and as 3 × 13 = 39, and there are 30° in a zodiacal sign, she has passed out of Libra and is in Scorpio. Libra begins to rise at 6 p.m., and Scorpio at a little before 8, therefore part of Scorpio is now above the horizon.[466] This, then, is the meaning of the first two stanzas of Canto ix.: the mistress of Tithonus is the aurora before moonrise (the solar aurora being his wedded wife, according to classical mythology), and this pale aurora is showing white in her balcony of the eastern sky, while on her forehead shine the stars of Scorpio, the cold creature with the stinging tail.

This curious passage has been otherwise interpreted by some commentators, and Scartazzini calls it "oscuro al superlativo,"[467] a riddle which still waits its Œdipus. According to him, "Titone" should be read "Titan," the sun-god, and his mistress is Tethys the ocean, which is now shining not under his rays ("fuor delle braccia del suo dolce amico"[468]) but under the rays of the rising moon. This seems far-fetched, and the reading doubtful. In any case, it would indicate the same hour and the same phenomenon (moonrise).

Others think that the solar aurora is intended, and explain the "freddo animale"[469] as Pisces, and the "steps" as the watches of the night. But why a terrible tail should be ascribed to a fish is not clear, nor is it easy to see how the third step with which Night ascends could be the hour of dawn. Night, which circles opposite to the sun, surely reaches the culminating point and begins to descend at midnight, just as midday is "il colmo del dì."[470] Moreover, the third watch would bring us at latest to a little before 3 a.m. To make the "steps" signs of the zodiac is no better; for if Libra, Scorpio, and Sagittarius have risen, it is midnight; what then can be shining in the east? and on whose forehead are the stars of Scorpio glittering?

On the other hand, all accords well if we accept the first interpretation—the light of the moon just about to rise whitening the eastern sky, some brilliant stars of Scorpio shining there, and Dante falling asleep at the third hour of the night, wearied after his sleepless nights, and the hardest day's climb that he will have. We need not grudge him his long sleep of nearly twelve hours, as those do who take this passage to indicate midnight or early morning.[471]

When Dante wakes the sun has already been up for more than two hours, but a dream which had come to him in the early dawn, at the time when the swallow begins to twitter, has had its counterpart in the truth, that the lady Lucia came at that time and carried him up the mountain side, leaving him outside the Gate of Purgatory when the day was bright.

When this Gate has been passed through, a narrow chasm between the rocks has to be climbed, and the passage is so difficult that it is at least an hour before the poets emerge on the open space of the First Cornice. For the waning moon has already set, and as she has lost another half hour or so during the night, the time of her setting this morning was about 9 a.m.

When Dante next notices the position of the sun, which he is too intent to do for some time, they have travelled some distance along the First Cornice, with its outer edge on their right, and the sixth handmaid is returning from the service of the day, that is, the sixth hour is ended: it is midday, and they have spent nearly three hours here, looking at the rock sculptures and conversing with the spirits.

The ascent to the Second Cornice is made with marvellously little fatigue, as Virgil had promised. There they at first find no one to direct them, so Virgil looks at the sun, and turns westward towards it, taking it as their guide, and they walk for about a mile before encountering any spirits in this circle.

The stairway to the next Cornice is reached at the time of vespers, that is, 3 p.m.; and Dante describes the position of the sun by saying that the portion of sky which it still had to traverse before evening was equal to the portion which is brought into view between the beginning of the day and the end of the third hour, that is, it was three hours before sunset; and he compares this ceaseless motion of the sky, which according to the Ptolemaic system caused the motion of the sun, to an ever-playing child. We need not particularly enquire whether he is thinking of the heaven of the sun, or the Primum Mobile: it may be either, but is most likely simply the diurnal movement in general, made strikingly evident by the sun. The sun's rays now strike their faces, for they have so far circled the mountain that they are walking towards the west; "Vespero là"[472] means in Purgatory, and "qui"[473] is Italy, where it is midnight now.

The sun is low when they reach the third Cornice, and when they emerge from the dense cloudy fog which here envelopes the spirits, Dante gradually sees the ball of the sun as when mists clear away, and he sees that it is just about to set. In fact the low shores at the foot of the mountain are already in shadow. Virgil urges that they make the ascent to the next Circle before darkness comes, and, as each ascent is easier than the last, they are able just to reach the summit before the sun is quite gone. Here Dante sees the stars begin to appear, while he is wondering at the strange law which robs him of power to proceed (as Sordello had warned them) after the sun has set.

They employ this time of forced inaction by talking of the sins which are punished in the several Circles, until all Dante's questions are answered, and he is beginning to fall asleep. It is at this moment that he first mentions the moon: she is glowing, shaped like a bucket in her present phase, and the stars are paled by her light. She is "belated almost till midnight," if the epithet "tarda" belongs to her; if it applies to "mezza notte," then it is almost late midnight when she is seen paling the stars. She ought to have risen about ten o'clock, but Dante does not speak of her rising, so this may mean simply that it was moonlight and starlight at the time of which he is about to speak, viz. nearly the late hour of midnight; or else that since he is now on the northern slope of the mountain, and the moon is in a southern sign, she did not become visible to him till towards midnight. In any case, the "quasi" shows that he is only giving us a rough idea of the time.

But in what sign was she? Her position in the zodiac is described in a peculiarly round-about fashion. Dante says that her course *against the sky*, that is, her monthly motion which is contrary to the diurnal, had taken her to that part of her track in which the sun is when from Rome he is seen to set between Sardinia and Corsica. The straits between these islands lie west and a little south of Rome, therefore the sun sets in that direction when he is going south after passing the autumnal equinox in Libra, and this is just what the moon is doing now. (The sun passes the same point again when coming north, before entering Aries at the spring equinox, but Dante has either forgotten this, or considers it obvious that the first position is meant here).

This is probably all that he means to convey, and if we ask for more exact information, it is almost impossible to tell what degree or even what sign in the zodiac he intends us to understand; for as the straits are nearly two hundred miles from Rome they are quite invisible thence, and we do not know what he thought was their precise direction. The mediæval commentator Benvenuto da Imola says that the position intended is in Scorpio; the moderns usually say Sagittarius, but without explaining how they draw this conclusion. Benvenuto's is probably simply the position which he thought the moon ought to have in order to agree with what Dante says about her, just as he states that Venus was in Pisces in March 1300, although this was not the case at all, as we shall presently see.

A calculation based on the difference in latitude between Rome and the Straits shows that they are only about 9° south of west, and that the sun, when setting in this direction is about 6° south of the equator: this is the case 15 days after the equinox, which in the beginning of the fourteenth century fell on September 14. Therefore the date at which the sun was seen from Rome setting in the direction between Sardinia and Corsica was the 29th of September, and its position in the zodiac was then the middle of the sign Libra. (Owing to precession, it would be in the *constellation* Virgo, but we have seen that Dante ignores precession in these references.)

But the moon ought now, on this second night in Purgatory, to have reached at least the last degrees of Scorpio, since on the previous night she rose when the first part was already above the horizon. Dante has therefore made a flagrant mistake with his moon, running her back into Libra again, and falsifying all his previous descriptions—*if* he knew as much about geography as we do now. But it is only necessary to glance at the map of Henry of Mainz to see that such a supposition would be grotesque. The portolan maps did indeed show the relative positions of Rome, Sardinia, and Corsica, with remarkable accuracy;[474] but as we have already said, it is not in the least likely that Dante had ever seen one of these sailing charts. It was on the classics, and especially on Orosius, that he relied for his geography,[475] and the suggestion of Dr. Moore seems most plausible, that he was here thinking of the Spaniard's description of Sardinia:—"Habet *ab oriente et borea* Tyrrenicum mare quod spectat ad portum urbis Romae."[476] This implies that Sardinia lay south-west of Rome, in a direction such as the sun might have had when setting in November, in Scorpio or Sagittarius.

Or possibly, when the poet was in Rome in 1301, he may have been watching a sunset in the second week in November, and may have been told, among other pieces of misinformation which are so readily given to visitors to great cities, that he was looking towards the straits which separate Sardinia from Corsica. We can imagine his replying, or thinking: "Then the sun sets behind them when he has nearly finished his course in Scorpio, and is about to yoke his steeds under another star!"[477]

It is while the moon is shining on this Cornice that the spirits who purge their sins on it appear, rushing past in haste. Then at last Dante sleeps, and dreams again in the cold moonlit dawn when Aquarius and Pisces are on the horizon (see p. 290).

The mountain is full of daylight when Virgil wakes him, and the newly-risen sun shines behind them as they continue their journey, for they had reached the north point of the circle, and are walking now due west. They mount almost immediately to the Fifth Cornice, where presently they are startled by a violent trembling of the mountain followed by a mighty shout of "Gloria in excelsis" from all the spirits. Statius joins them, and from him they learn that this was no earthquake, for since passing the gate of Purgatory they are in a region too lofty for such disturbances to be felt, or for rain or snow or dew to fall; the mountain trembles, and the shout of praise follows, when a soul rises, purified and ready to ascend to Paradise. With Statius they climb to the Sixth Cornice, and then, looking round, they see that the fifth handmaid of the sun's chariot is directing its upward course, that is, it is the fifth hour, or between 10 and 11 a.m. As usual, they walk on with the outer edge of the circle on their right. It is here that we are reminded, as once in the Inferno, that only a certain time is allowed, for while Dante is looking eagerly at the first mysterious tree Virgil gently urges him onward:—

"Lo più che padre mi dicea, Figliuole, Vienne oramai, chè il tempo che c' è imposto Più utilmente compartir si vuole."[478]

It is here also that Dante informs his old friend Forese Donati and other spirits, who are all astonished at the sight of his shadow, that he is a living man who was rescued by Virgil some days ago, when the sister of the sun was round (referring of course to the full moon in the forest), and he points to the sun. Now a curious question arises: Dante was in sunshine in the early morning; he is in sunshine now, and has been walking all morning in the same direction in which the sun was moving; how is it then that Statius did not notice his shadow, but had to be told that he was not a spirit?[479] If the Mountain had been a small hill, it would have been quite easy to walk so large a part of a circle round it, in a short time, that the poets would have been in shadow until the sun caught them up again when they lingered talking on the Sixth Cornice; but the Mountain was so large that a whole day's walking only brought them a quarter of the way round, *i.e.* from the east side to the north, and it was evidently a distance of many miles.[480] It is difficult to believe that Dante has forgotten, because he is so careful throughout to mention the surprise of the spirits except when his shadow could not be seen for one reason or another. It is true that in the latitude of Purgatory (32° south, as that of Jerusalem is 32° north), the sun rises rather steeply, which makes it easier to get into the shade in the early morning while circling a mountain.[481] On the other hand a simpler solution offers itself in the text. Statius approaches Dante and Virgil from behind, and they only stop and turn their faces to him for a few moments while Virgil returns his greeting, then walk on quickly while Statius makes his surprised reply: "What! if you are spirits...." If he has not yet quite overtaken them he cannot see Dante's shadow, which Dante himself is screening from view as it falls in front of him, the morning sun being still at his back, and Virgil explains at once that his companion is a living man.

In spite of their haste, the three poets remain about four hours in this sixth Cornice; and when they reach the ascent to the next it was indeed time to climb without any delay, for the sun had left the meridian to the sign which follows Aries, viz. Taurus, and Night, "che opposta a lui cerchia,"[482] had left the meridian of the opposite hemisphere to Scorpio. Aries at the time of the equinox begins to cross the meridian at midday, Taurus at two o'clock, and Gemini at four[483]: therefore it is now between two and four in the afternoon. The three poets hasten therefore (Statius still accompanying Virgil and Dante), like men who pause for nothing on their way, whatever the road be like; and a question Dante is longing to ask with regard to the form of purgation last seen is put and answered as they follow one another up the narrow stair; as soon as they reach the top, they turn to the right, as usual, without hesitation or delay. As Dante walks cautiously between the outer edge of this Cornice and the fire which bursts forth from the inner wall of rock, all the blue of the western sky is already changing into ash-white, and he has come so far round the Mountain that the evening rays strike his right shoulder. With one of the many touches which show Dante's close observation and his vivid realization of the scene he is describing, he says here that his shadow falling on the flame makes it appear a deeper red, and this is observed by the spirits. When they reach the spot where, on the further side of the flames, the chanting Angel guides them across to the last ascent, the sun is very near to its setting. Its burning noonday beams are shining on the waves of Ganges, its first morning beams quiver over Jerusalem, and Spain lying under Libra is wrapped in midnight gloom; hence it is beginning to set at sea-level in Purgatory. High on the Mountain it still shines, but when the poets have crossed the flames, the voice of the Angel persuades them to hasten their steps before the darkness comes. The stair mounts through the rock on the west side of the Mountain so that the low sun casts Dante's shadow in front of him as he climbs. And he has only climbed a few steps when they are all aware that the sun has completely set because the shadow disappears. Before the colour has all faded from the west, and the whole vast range of the horizon seen from this height has become one uniform tint, each of the three makes a bed of one step, and between the high and narrow enclosing rocky walls Dante sees the stars come out, much brighter and much larger than he has ever seen them before. No moonlight dims them, for moonrise will be late to-night, long after Dante has fallen asleep.

Once again, for the third and last time, he has a prophetic dream in the early dawn; and if we compare these three parallel passages, we find that each begins with the same words "Nell' ora."[484] The first time the ideas connected with the time are wholly sorrowful, for it is said to be the hour when the swallow begins to twitter, perchance in memory of her former woes—alluding to the tragic tale of Procne and Philomel. The second is a depressing time, for it is the cold hour when, the day's heat being all spent, the chilled earth cannot overcome the influence of the cold planets such as the moon or Saturn, but there is a thought of hope in the stars of the Greater Fortune seen rising in the east, in a sky which is now scarcely dark to them. The third is all joy, for it is the hour when Cytherea (the planet Venus), burning ever in the fire of love, shines first upon the mountain. Dante's sleep flees from him as the shades of darkness flee on all sides, chased by the brightness before sunrise; and when Virgil tells him that to-day he shall reach the goal of his desires, he speeds up the remaining steps of the stair, almost as if on wings.

The poets arrived on the shores of Purgatory, as we saw, facing east, and now that they have attained the summit of the mountain, coming up the western slope, they again face east, and the just risen sun shines in their faces. Dante walks towards it: a breeze fans his forehead and causes all the trees to bend in the direction of the shadow thrown at sunrise by the Mountain (*i.e.* westwards), and the murmur of this gentle wind in the branches forms a bass to the treble of the singing birds who are greeting the morning hours. For a few steps the stream of Lethe stays his steps in this direction, but he soon reaches a bend, and once more "a levante mi rendei."[485] The symbolical procession which accompanies Beatrice comes towards him from the east, and after he has crossed the river Lethe and joined the procession beside the Car in which Beatrice is throned, all turn to the sun, which by this time must have travelled some distance towards the north. After the vision of the disasters which befell the holy Car, and after Beatrice has spoken with Dante again, they reach the fountain whence Eunoe flows, and now the sun is on the meridian and it is noon, though (as Dante reminds us) different places have different meridians, so that it is not noon everywhere else. He is bathed in the holy waters, and thus, all having been accomplished for which he has visited these regions, he is "puro e disposto a salire alle stelle."[486]

Hell was entered at twilight, Purgatory at dawn, the ascent to Paradise is made at noon, "il colmo del dì."[487] The time spent in Purgatory is three days and the morning of a fourth, which is the seventh from the beginning of the vision.

4. PARADISE.

That the time of ascent to the spheres was noon is again stated in the opening Canto of the *Paradiso*, where Dante says that the sun had brought morning there (Purgatory) and evening here (the inhabited hemisphere), and that now the whole of that hemisphere was bright while the whole of this was dark.[488] The sun is therefore now due north. While Beatrice stood beside the waters of Eunoe, after walking towards the sun, it had moved to her left, so she now turns in that direction to gaze upon it.[489] She, like Virgil, takes as her guide that sun which Dante has said is the fittest symbol of God, and it is through the power of this heavenly light that he is raised from earth. Compare

"Amor che il ciel governi ... *col tuo lume* mi levasti."[490]

He is not conscious of any effort, or even movement, but is astonished by a great light and strange music, and learns that he is in the sphere of fire and listening to the harmony of the spheres. His soul, freed from all that bound it to earth, is soaring towards its natural goal, as fire mounts up and rivers run down.

The upward gaze to the sun or to the skies, and its power to raise them, is again described in *Par.* i. 142: Beatrice "rivolse inver lo cielo il viso,"[491] and in ii. 22, "Beatrice in suso ed io in lei guardava."[492] Thus they rise to the First Heaven, of the moon. In answer to an eager question of Dante's about the markings of the moon, Beatrice is led to discourse on the spheres and their movers. Spirits appear, and Dante learns that although all the redeemed dwell in the Empyrean, those of different degrees of blessedness will manifest themselves to him in each sphere. After explaining this and other things, Beatrice once more

"si rivolse tutta disiante A quelle parte ove il mondo è più vivo,"[493]

and swift as an arrow that strikes its mark before the cord has ceased to quiver, they ascend to the Second Heaven, of Mercury. To understand the above lines we must remember that the sun, as Dante told us in the introductory Canto,[494] is at the equinox, or near it, and therefore on the equator; also that the equator of a sphere, where motion is quickest, is its most living part.[495] Beatrice is therefore looking again at the sun. Some understand "quella parte"[496] to refer directly to the sun, "quegli ch' è padre d'ogni mortal vita,"[497] some to the equinoctial point;[498] others think that the east is meant; others the Empyrean, in which is the true life of all the Universe.

Equally swiftly and unconsciously, Dante ascends to the successive heavens of Venus, the sun, Mars, Jupiter, and Saturn, finding himself always in the planet which occupies each sphere. Mounting the mysterious golden ladder which is set up in Saturn, with a rapidity inconceivable on earth, the poet finds himself in the Heaven of Stars, in the constellation of Gemini. Now that he is so near the final blessedness his eyes are clear and keen, and that his heart may be filled with joy Beatrice bids him look down and see how much of the universe is already beneath his feet. There lie the seven spheres, and there is the little globe of Earth, so small that, like Scipio, he smiles at its paltry appearance; there is the moon, showing him the face turned away from Earth, which has no dark markings: there is the sun, at which for once he is able to gaze; and near it circle Mercury and Venus; the temperate Jupiter moves between his father Saturn and his son Mars; with a new power of vision Dante can grasp all their sizes and distances and the various speeds of their movements; and the poor little disc of the earth, small and round as a threshing floor,[499] about which we quarrel so fiercely, is all visible to him at a glance, its highest hills and its lowest river-valleys running down to the sea. He turns away his eyes to look into the beautiful eyes of his guide. She is standing, erect, intent, watching eagerly for that glorious Procession, whose coming is making a growing brightness in the sky.

"La donna mia si stava eretta Ed attenta, rivolta inver la plaga Sotto la quale il sol mostra men fretta."[500]

No other intimation is given as to the direction from which the Procession comes. The sun's daily motion appears slowest on the meridian, as Dante remarked in *Purg.* xxxiii. 103, 104, because he is scarcely changing his position with regard to the horizon; or in the south (where we in this hemisphere see him when on the meridian); he shows least haste to leave us in the long summer days, near the solstice. If the Procession descends from the heaven of heavens to show itself here symbolically to Dante, as we know that the spirits in other heavens did,[501] then Beatrice was gazing upwards, and the meridian may be meant, particularly that part of it which passes directly overhead. If the Procession came from the east, like the Procession in Eden,[502] the summer solstice in Cancer is meant, which lies east of Gemini. It is difficult to see any special meaning in its coming from the south, for although some commentators say that this suggests the consummate glory of the midday sun, the same idea is conveyed by the meridian. A constellation like Cancer perhaps corresponds better than a circle like the meridian with the expression "plaga,"[503] and the idea of the east makes more perfect the beautiful simile which the poet has just drawn between the little bird on the spray beside its nest waiting for the sun to rise, and Beatrice watching for the advent of Him who is likened to a sun among the starry company of His saints.[504]

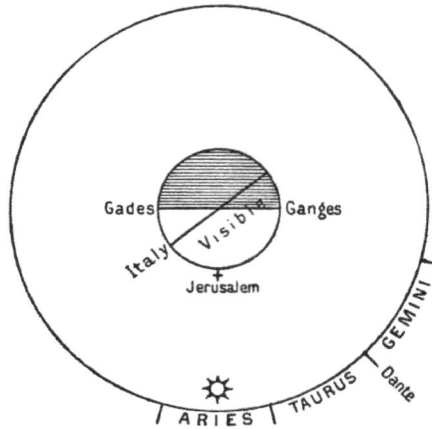

Fig. 48. Dante's first view of Earth from the Stars.

Par. xxii. (see p. 397).

When this glorious company has at length vanished, she bids him look down once more, this time that he may realise how he has been revolving with the spheres, the little earth lying immoveable at his feet. He looks down, and notes what parts of Earth are lighted by the sun, and what parts are wrapped in darkness, for now he is able to see what he had often imagined, the shadow of night sweeping westward over the globe. The Atlantic Ocean, where Ulysses sailed beyond Gades, is all in daylight, and so is the Mediterranean, but the sun is setting over Jerusalem, for he says that its position prevents him from seeing beyond the coast whence Europa was carried (*i.e.* Phœnicia or Palestine). It must therefore be midday at Gades, and the sun is on the meridian there; it is "a sign and more," or something between 30° and 60°, advanced to the west beyond the meridian over which Dante stands.

He does not define his own position more exactly than this, but we may conjecture that he was 45° (a sign and a half) east of the sun, since this would be the meridian of Rome, and there would be a peculiar fitness in such a position at such a moment, when he has just received a message of warning and prophecy from St. Peter to bear to this city.[505]

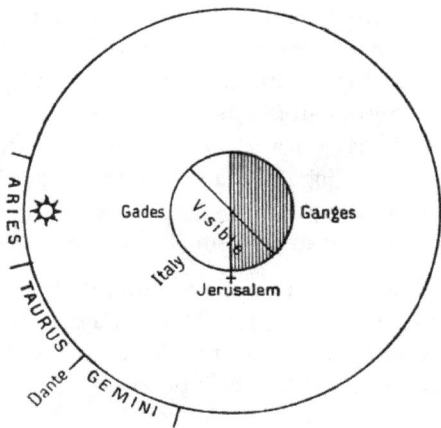

Fig. 49. The view six hours later.

Par. xxvii.

Fig. 49 shows these positions of Dante and the sun, and the consequent limits of his vision. Jerusalem is "near," in comparison with the much more distant sunlit space of ocean in the opposite direction. We place the sun in the middle of Aries, because we shall see presently (when discussing the date) that this is its most probable position, and Dante in the first degree of Gemini because this is 45° east of the sun, and also because the fact that he is above Earth's "first climate" suggests that he is in the beginning of this sign. For the northern limit of the first climate was $20\frac{1}{2}°$, according to Alfraganus, and the ecliptic in the sign of Gemini rises gradually from 20° north of the equator to $23\frac{1}{2}°$ at its end, where it adjoins Cancer.

Since the time when he looked down before, he sees that he has revolved over an arc equal to half the climate, that is 90°, for the climates only extended over the habitable earth, which was 180° in longitude. His former position, then, must have been 90° further east, halfway between Jerusalem and Ganges; and it was then noon at Jerusalem, while all the ocean east of Ganges lay in the black darkness of night. Fig. 48 shows this position.

In this interpretation of a difficult passage, the meaning given to the first lines seems a little forced; but if we take them in the more natural sense that Dante had moved from above the centre to the end of the climate, that is, from the meridian of Jerusalem to Gades, we are met by a difficulty in the following lines. For to say that he is standing over Gades, and in the same breath that the sun is more than 30° further to the west, means that it is setting more than 30° west of Jerusalem: he cannot therefore possibly see Jerusalem, nor is it "near" him. It does not help us to take "presso" in its alternative meaning, and say that he could see "nearly as far as" Jerusalem, for however little more than 30° "a sign and more" may mean, it brings the "terminator" (as astronomers call the dividing line between light and darkness on a planet) a long way from Jerusalem and much nearer Italy, according to Dante's reckoning of distances in the Mediterranean. This is shown on fig. 50, where the data are made as favourable as possible, by putting the sun at the very end of Aries.

On the whole, it is easier to accept a round-about expression in the first lines than an inaccurate one later. The former is quite characteristic of Dante, and we have already met some like it in the *Divina Commedia* in descriptions of the positions of heavenly bodies.[506]

Looking back to the earlier position, as described in Canto xxii., we find no light thrown on the question; for here he does not mention any particular place as visible, nor say where the terminator lay. His words are:—

"L'aiuola che ci fa tanto feroci, Volgendom' io con gli eterni Gemelli, Tutta m'apparve dai colli alle foci."[507]

This seems to express simply a bird's eye view of Earth, as it lay stretched out like a map below the observer.

It is, however, often taken to mean that the whole habitable earth was visible. This could never be unless the sun, as well as Dante, was over the meridian of Jerusalem, which is obviously impossible while one is in Aries and the other in Gemini; and as he refers back to the first passage from the second he cannot have forgotten this.[508]

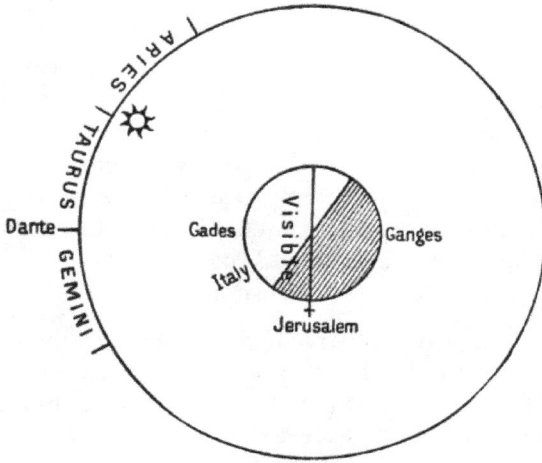

Fig. 50. An impossible interpretation of *Par.* xxvii.

A book which aims at explaining Dante's *Paradiso* to young people[509] suggests an original way of reconciling the two passages, for the author gravely asserts that the sun was at first below Dante's feet in Gemini, but that by the time he looked down again it had moved a good distance onward, and was then in the sign of Taurus! According to this, a whole month must have elapsed, and time had gone backwards, from May to April!

A suggestion much more worthy of attention is that of Professor Rizzacasa d'Orsogna, who thinks the key is to be found in the line, "Volgendom' io con gli eterni Gemelli." This, he suggests, is in anticipation of the second passage, for although Dante could not see the whole habitable earth at any one time, his change of position, as he revolved with the Twins, enabled him to do so in the course of time which elapsed between his two views. A comparison of figures 48 and 49 will make this clear.

One more suggestion may perhaps be made, though it is rather a hazardous one. "Dai colli alle foci"[510] may mean from the Apennine Hills to the mouths of Ganges. This would be decisive for the correctness of the explanation illustrated by our diagram. (See fig. 48).

In any case, the points which Dante specially wishes us to understand are clear and simple: at the first glance he was struck by the insignificance of the Earth; at the second, he became conscious that he was moving with the spheres, having passed over 90° of arc, or a quarter of the diurnal revolution, and he saw that it was now sunset at Jerusalem.

He next ascends to the Primum Mobile, though he cannot tell us to which part of it Beatrice chose to lead him, since it is perfectly uniform throughout. Beatrice explains that this most swift and living sphere, although it contains no star by which its motion is made visible, causes the (diurnal) motion of all the lower spheres; and thus it is like a flower-pot in which are hidden the roots of Time, while in the others we see the leaves. Here he sees the nine hierarchies of angels who direct the nine moving spheres; and finally he soars to the Empyrean, where Beatrice returns to her seat among the blessed spirits who form the Celestial Rose, and St. Bernard encourages him to lift his eyes to "L'Amor che move il sole e l' altre stelle."[511] This is the end and consummation of the Vision.

The reader will have observed that although there are several indications of direction in Dante's journey through Paradise, all referring to the heavenly bodies, there are no time references after he has left Earth and entered the regions of eternity, except the passage discussed above, which indicates a period of six hours spent in the eighth heaven, since 90° of the diurnal revolution correspond with six hours of time. But it tells us also that at the close of this period the sun was setting over Jerusalem: consequently another morning was dawning in Purgatory. Eighteen hours have passed, therefore, since Dante left the Earthly Paradise at noon; and he rises to the Primum Mobile and thence to the Empyrean on the eighth day of his Vision.

The whole Vision, therefore, occupied an octave, or eight days.

5. POSITIONS OF SUN, MOON, AND PLANETS AMONG THE STARS DURING DANTE'S VISION.

The allusions to sun, moon, and stars in the *Divine Comedy*, when thus brought together, are seen to follow one another according to a regular scheme, and they form a very good guide by which to time Dante on his journey. There is a little inaccuracy in one passage about the stars, and the moon is ambiguous once, but quite explicable. Taking the allusions in a simple popular sense, as he seems to intend us to do, assuming that the signs and the constellations of the zodiac are identical, and that the moon, which at first is opposite the sun, traverses about 13° in the zodiac daily, and loses about an hour of time more or less, there is very little difficulty in following him. It is quite enough to know that the sun is somewhere in Aries throughout, and that the moon is therefore in Libra at the beginning.

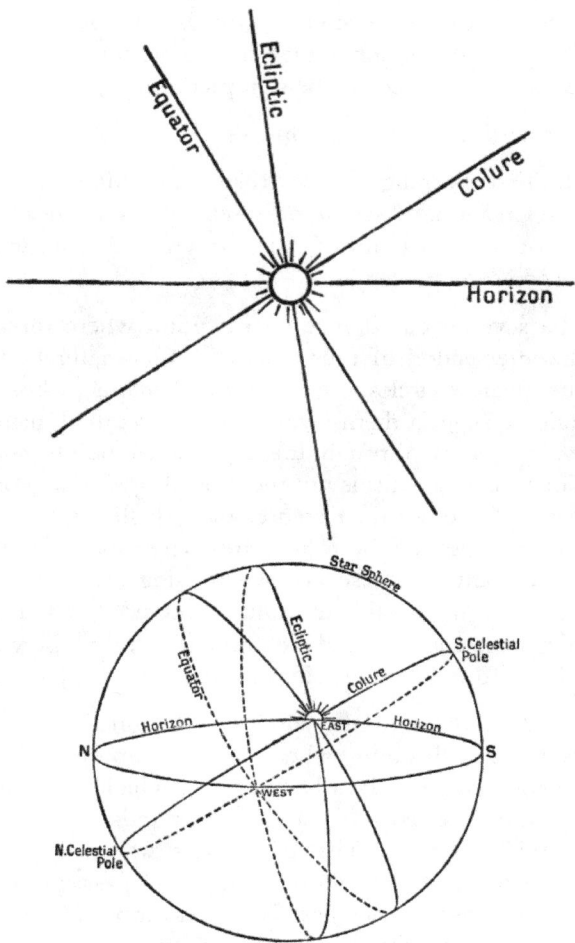

Fig. 51. The rising sun at the Spring Equinox.

But if we wish to know a little more accurately the positions of sun and moon among the stars, we must carefully compare the passages bearing on this question. Commentators differ so much that while some think the sun's annual motion is ignored, and it is assumed to be at the equinox all the time, others maintain that it was at the equinox on the first day and moved on through Aries at the rate of about one degree a day (as it should do, since there are 365 days in a year and 360° in the circumference of the sphere);[512] while others again say that it was merely somewhere in Aries from the beginning.

What are the actual indications of its position? No one questions that it was in Aries throughout the Vision, for this is often indicated, sometimes directly, as in *Inf.* i. 38-40, sometimes by mentioning Taurus as the sign following the sun,[513] or Pisces as the sign preceding.[514]

There are four detailed descriptions:—

(1) On the first morning, the sun rose among those stars which were with him at the Creation.[515] As we have seen, this may mean merely some point in Aries, or the first point of Aries, at which the sun was thought to have been placed when created.

(2) On the seventh day it rose near a point where three crosses are formed by the intersection of four circles.[516] One naturally thinks of the horizon as one of these circles; the equator and ecliptic, whose intersection define the equinox, suggest themselves as two more; the fourth is doubtless the colure which passes through the equinoctial points and the poles, although curiously enough this is not mentioned by Ptolemy or Alfraganus, who only speak of the solstitial colure. Fig. 51 shows how the horizon makes a cross with each of the other three when the equinox is upon it. The sun is at the point of intersection when rising with the equinox, and it was *near* there when it rose this morning—"quasi." It was therefore near, but not exactly at the equinox, and we know that it was the vernal, not the autumnal equinox, from the many allusions to Aries.

(3) When Dante rises to the sun (on the same day) he directs our attention once more to this point where "the one motion strikes the other," that is, where the ecliptic—the oblique circle which bears the planets—branches off from the equator, and he reminds us of the supreme importance of this divergence. The sun, he says, was in this place. He does not mean at the exact point of intersection, or this passage would be in flat contradiction to the last: "quella parte"[517] is no more precise than "quelle stelle,"[518] and means only this part of the sky, or the constellation of Aries. The words which follow not only specify that it was the spring equinox, but suggest that the sun had already traversed several of those spirals which bring him to us earlier each morning, *i.e.* it was several days after the equinox.[519]

(4) On the same day the sun was "one sign and more" removed from some point in Gemini.[520] This is very vague, but at least the sun cannot now have been at the first point of Aries, or it would have been two whole signs from even the first degree of Gemini.

It seems, then, that we cannot possibly agree that the sun was at the equinox throughout the poem. It is clear that it was past the equinox towards the end; but from these four passages we cannot be sure whether it was there on the first morning, or whether it moved through some undefined part of Aries during the Vision.

One consideration gives us some help, however. We know from *Inf.* xxi. 112-114 that the first day was either March 25, which was believed in the Middle Ages to be the true anniversary of the Crucifixion, or Good Friday, which was kept by the Church as such. March 25 is four days after the date which the Church still kept as the day of the Equinox (March 21), and 13 days after the true equinox, which at this time fell on March 12. Good Friday also almost inevitably must fall after the equinox, since Easter follows the full moon which follows the equinox. It is true that some of Dante's readers may still have regarded the old traditional date of March 25 as the true day of the equinox, and possibly it was to suit all shades of knowledge that he left the sun's position vague; but he himself certainly knew better, and probably had March 12 in his mind. Then those who held to March 25 might consider the sun as exactly at the equinox on the first day of the Vision, but all better-informed readers and Dante himself must have regarded it as somewhat past this point. If we take it that, *e.g.*, it was a week past the equinox, the sun was in about 7° of Aries, and this position agrees with the four crucial passages mentioned above. The stress laid on the equinox seems to demand that the sun should be at least as near it as this on the first day; and a week later it would have reached the middle of Aries, which is the position we assigned to it from a discussion of *Par.* xxvii. 86, 87.

This also agrees well with the course of the moon in the zodiac, which must depend upon the sun's. We have seen that she was opposite the sun on the first morning, that her light paled the first-risen stars of Scorpio while she was yet beneath the eastern horizon three days later, and that she was proceeding south on the following night. If the sun was in the 7th degree of Aries on the first morning, the moon must have been in the 7th degree of Libra; her mean daily motion of 13° would bring her to the 16th degree of Scorpio three days later and to the last degree on the day following; so even if we allow a deviation of a few degrees from the average, she was just where she should have been on each occasion. We have assumed for convenience that she became full and was opposite the sun just as she set on the first morning, but this may have happened at any time during the previous night (in the Forest): Dante has not specified the hour, and it is of no consequence, since the indications of time by the moon must necessarily be vague.

The time of sunrise will still differ little from 6 o'clock, a week after the equinox. In London it is a quarter of an hour earlier, but the difference is less in Italy, and still less in the latitude of Jerusalem.

Besides the sun and moon, two of the planets are mentioned as in a particular place among the stars.

Venus, before sunrise on the fourth morning, is seen among the stars of Pisces;[521] and she is also mentioned as a morning star before sunrise on the seventh day.[522]

Saturn is in Leo. "Il petto del Leone"[523] may mean the western part of the Lion, the part which first rises and contains the head and breast, or it may refer to the star Regulus, which is called by Ptolemy and Alfraganus "Cor Leonis."[524] "Sotto il petto"[525] may mean the position of the planet as seen projected in the sky or a map, that is, a little south of this part of the constellation; more probably, perhaps, Dante is thinking of the supposed position of Saturn's sphere *below* that of the stars.[526]

Mars is spoken of as having returned to his own Lion a number of times: "al suo Leon ... venne,"[527] but this does not necessarily imply that he was in that constellation at the time.

Mercury and Jupiter are not noted as seen anywhere in the sky, nor does Dante say in what constellation they were when he entered them with Beatrice.

There are a number of pitfalls into which Dante might have stumbled, with this disposition of the skies, and the change to the southern hemisphere; and that he has avoided all is as clear a proof of his familiarity with astronomical phenomena as the positive evidence of his truthful, consistent descriptions. For instance, he might very well have represented the cold planet Saturn as actually aiding the cold moon in driving away the heat of the day during the night, in *Purg.* xix. 1-3, instead of only saying "talor da Saturno;"[528] but if so, this would have been inconsistent with the position of Saturn in Leo, since Leo was below the horizon when dawn was drawing near and Pisces was rising. Again, we have already mentioned that he evidently did not forget that it was autumn in the southern hemisphere. He never suggests that it was spring in Purgatory, and in the passage quoted above, where he describes sunrise at the March equinox, saying that the sun then moves "con miglior corso e con migliore stella,"[529] this happiest effect on the world is restricted to "mortals," who as he thought all dwelt in the northern hemisphere, "surge ai mortali."[530] I do not mean that the expression was used for this reason, but it is an example of how instinctively he avoided mistakes into which any other author would almost certainly have fallen.

6. THE ASSUMED DATE OF THE VISION.

Even among Dante enthusiasts there must be many who are astonished at the warmth of the controversy concerning the assumed date of his Vision. To know whether he feigned himself to have entered the Inferno with Virgil in March or in April, in the year 1300 or in 1301, really does not help us to appreciate the music of his verse nor the depth of his thought; yet a wealth of learning and research has been poured out upon this problem, and some commentators seem ready to defend their opinions with their lives, if that would avail to convince their readers!

The date is indeed intimately connected with the narrative of the *Divine Comedy*, since there are constant allusions to current events. By choosing a year earlier than the time at which he was writing, Dante is able not only to tell the spirits what is passing on the earth they have left, but to endow them with the power of prophesying to him events which are to come.[531] The peculiar interest of the problem for us just now is that astronomy has been called in to solve it. We have seen that the positions and movements of the heavenly bodies are all consistent with one another in his poem, but it is interesting to ask whether they are merely a consistent fiction or whether they correspond with facts. Did Dante arrange his moon and planets according to his feeling for scenery and the convenience of his time references, or did he take the positions as they really were at some particular date? If so, the full moon, Venus a morning star in Pisces, and Saturn in Leo, all together near a spring equinox, ought to help to fix that date.

The year which is almost universally accepted, by the oldest as well as the most modern commentators, is 1300, chiefly on the authority of *Inf.* xxi. 112-114. Here Malacoda says that yesterday (the day of Dante's entering the Inferno) it was 1266 years since the earthquake which accompanied the Crucifixion (see p. 370). Adding to this the 34 years which Dante has told us he believed to have elapsed between the Nativity and the Crucifixion,[532] we get 1266 + 34 = 1300. And the day is either the Church Festival commemorating the Crucifixion, *i.e.* Good Friday, which in 1300 fell on April 8, or March 25, which was believed to be the actual anniversary.

If this is the right interpretation of the passage, we may say at once that Dante's astronomical data are imaginary, for the moon was not full on the night preceding either of these dates, but on March 5, and April 4: Saturn was indeed in Leo, but Venus was so near the Sun as to be invisible in March, and began to appear as an evening star in Taurus on about the 9th of April.

These positions were all carefully calculated by Prof. Angelitti, who is a professional astronomer as well as a Dante scholar; and he made the thrilling discovery that the positions agree wonderfully well for the year 1301! In this year the moon was full on the early morning of the 25th of March; Venus entered Pisces three days later, and was a morning star, near her greatest elongation west, and very brilliant; Saturn was not only still in Leo but was just three degrees west of Regulus; and Mars was also in Leo.

If the anniversary of the Crucifixion means Good Friday, the day would be March 31 in 1301, and the moon would be wrong, but March 25 has some claims to be considered the more probable date. A fixed calendar date is more appropriate to use for calculating an exact number of years than a changing festival, which might rather be regarded as an "ideal date."[533] Moreover, March 25 is suggested by the allusion to the Creation in the opening Canto; this day was also the traditional date for the spring equinox,[534] and it was one of the most important days in the year to Florentines, being their New Year's Day as well as Lady Day.

We know that the idea of writing his great work was in Dante's mind even when he wrote the last words of the *Vita Nuova*,[535] and we need only suppose that he was thinking of it when he looked at the sky on the evening before Lady Day in 1301, that he saw the full moon, and the two bright planets Mars and Saturn making a striking asterism with Regulus; also that in the early dawn about the same time he rejoiced in the sight of Venus veiling the Fishes in the east.

But is this attractive theory admissible? Can we take Dante's words elsewhere as referring to the year 1301?

Take first the one passage in which he explicitly states the day and year, the words of Malacoda quoted above. This is usually taken to indicate the year 1300, as shown above (p. 410), but it is easy to interpret it another way—thus: 1300 years had been completed since the Nativity on the day Dante entered the Inferno. Now the year 1 is not one year after the Nativity, but the year of the Nativity itself; hence the year 2 is one year after, the year 3 two years after, ... the year 1300 is 1299 years after, and the year 1301 is 1300 years after. The year 1300 was completed on March 25,[536] and the date then changed, according to Florentine usage, to 1301. (It had already been called 1301 for three months, according to Roman usage, so fortunately for us, there is no disagreement between the two calendars from this part of the year).

It must also be noted that in several old MSS., considered authoritative by some of the very early commentators, the passage reads:—"Mille dugent 'uno con sessanta sei."[537] Although it is impossible to believe that Dante wrote such a faulty line, this shows that some early copyist felt a difficulty either in accepting the date 1300 or in understanding how Dante arrived at it by Malacoda's computations. This passage, therefore, is inconclusive for 1300 or 1301.

Then there is the first line of the *Divina Commedia* "Nel mezzo del cammin di nostra vita,"[538] which is usually taken to mean that Dante was in his 35th year, since in *Conv.* IV. xxiv. 30, 31, and xxiii, 88-94, he says that in normal natures the "punto sommo," the "colmo,"[539] of the Arch of Life is reached at this age. But is it to be taken as indicating a precise year? If so, it points to one which has never been suggested for the Vision, so far as I know, for in another passage of the *Convivio* he says when speaking of his exile, that he had lived in Florence "fino al colmo della mia vita,"[540] and we know that the year of his exile was 1302. But if we look again at the discussion concerning the arch of human life, we find that the whole period of "gioventute,"[541] which is from the 25th to the 45th year "veramente è colmo della nostra vita"[542] (*Conv.* IV. xxiv. 22, 23). If, nevertheless, it is said that the "mezzo del cammin di nostra vita" means Dante's 35th year, we must know the date of his birth with certainty before we can deduce the year at which he had reached this age. And this is not quite easy. This very line is used indifferently to show that he was born in 1265 because the date of the Vision was 1300, or that the date was 1300 because he was born in 1265! No registers of births or baptisms were kept in those days. Most of the old biographers give 1265 for the date, but not all, and some of those who do betray ignorance or confusion in the matter: Boccaccio says the year 1265, when Urban the Fourth was Pope, yet Urban we know died in 1264; Lionardo Bruni says "1265, a little after the return to Florence of the Guelfs, who had been in exile by reason of the defeat of Montaperti," but this return was in September 1266! Dante himself says that his ancestors shared in this exile,[543] and as he also often refers to the fact that he was born in Florence,[544] it is difficult to see how that could have been in 1265.

Again, he says that there was a difference of less than 8 months between the age of Beatrice and himself (she was 8 years 4 months old when he was nearly nine),[545] that she died in June 1290,[546] and was then on the threshold of her "second age,"[547] that is in her 24th year, or 23 years old. This gives 1267 for the year of her birth, and he must have been born in the same or the preceeding year. Historical documents show that Dante spoke in the Council of the Hundred in 1295, and Brunetto Latini says no man could hold office in Florence under thirty years of age; but Brunetto wrote in exile in France and died in 1290, and in a translation from his original French into Italian the age has been altered to 25, so it is possible that the age had been lowered before 1295.[548] There remains the story told by Boccaccio, that a Ser Piero di Messer Giardino (a notary whose name occurs in legal documents of the time), heard from Dante himself shortly before his death that he had been 56 years old in the preceding May. The date of his death is unanimously given as 1321, so this story, if we may accept it as authentic, is one clear piece of evidence that Dante was born in 1265.

On the whole, we must say that it is doubtful whether the "mezzo del cammin" means Dante's 35th year, and if it does, we are not certain that he reached that age in 1300.

Another passage which is often quoted as clearly indicating the year of the Jubilee, 1300, for the assumed date of the Vision is Casella's explanation of his late arrival in Purgatory.[549] Dante is surprised to witness this arrival, knowing that it is some time since his friend's death; and Casella replies that the Angel of the Passage had refused many times to bring him, until three months ago, when every one who wished to come had been graciously received. This is believed to refer to the Bull of Pope Boniface published at Christmas 1299, that is just three months before March 1300. But this Bull gave no indulgence to the dead; it was in favour of those who performed certain spiritual exercises in the churches of Rome during the Jubilee year. On the other hand, there was a Bull published just a year later, which was for the benefit of departed souls: it extended the indulgences to those who had died on their way to Rome, or before the spiritual exercises were completed there. We do not know when Casella died, but if it was some time in 1300, with his purpose of gaining the indulgence unfulfilled, the Angel's refusal at first, and acceptance afterwards—after December 1300—and his arrival in Purgatory in March 1301, would be explained. The fact that his arrival was still three months after the Bull, and that all without exception were welcomed by the Angel, are difficulties which apply to both interpretations. But perhaps that just given is a little too clever. It is most natural to consider the Indulgence as the famous one of 1300, and to understand that in the Jubilee year the way of salvation was made easier for all souls, living or dead.[550]

If however these, and many other passages quoted in the controversy, are unconvincing, the same cannot be said of Cunizza's "Questo centesim' anno."[551] Prof. Angelitti's suggestion that she was reckoning by the Easter year, which was used in some parts of France and Italy, so that the date was still 1300 for her, though for most Italians it was 1301, is too ingenious; and it is hard to believe that she means "this century year" applying it to 1301. Though this might be correct, it is clear that 1300 was regarded as the century year, because the words "centesimo anno" occur in the Bull of Boniface VIII. with reference to the year 1300. This then speaks strongly and clearly against our astronomical theory of 1301.

The historical events alluded to in the *Divine Comedy* are so often of unknown date to us, that few of them help as much as might be expected.

Judge Nino's widow remarried on June 24, 1300, and as he speaks of her as having already put off her widow's weeds,[552] this might indicate that we are in 1301, but it is not conclusive, since they would probably be discarded three months before she was actually the wife of another. On the other hand, Can Grande della Scala, who is said in a fourteenth century chronicle to have been born on March 9, 1291, is described as only nine years old.[553] Dante might possibly have been mistaken by a year in the date of his birth, but he cannot have been ignorant of the death of his friend Guido Cavalcanti. The cry of anguish from Guido's father will be remembered by all readers of the *Inferno*, after Dante has spoken of his friend in the past tense:—

"Di subito drizzato gridò: Come, Dicesti *egli ebbe*? Non viv' egli ancora? Non fiere gli occhi suoi lo dolce lome?"[554]

And as Dante hesitates a little, Cavalcante sinks back into his tomb of punishment, and does not reappear. Before Dante leaves the place, feeling compunction, he charges Farinata to tell the father that his son is still among the living, and that the sinister hesitation was due to another cause than he thought. Now it is beyond dispute that Guido died in August 1300, for he was buried in the cemetery of St. Reparata in Florence on August 29, 1300, according to official records still extant;[555] and Dante had a share in banishing him amongst other exiles to the unhealthy place where he contracted the illness of which he died. The year 1300, in which Dante held the official position which obliged him to do this, was burned into his mind, as we see in his letter, mentioned by Lionardo Bruni, in which he says that all his misfortunes sprang from the time of his priorate.

How then can we believe that he here imagines himself speaking in the spring of 1301, unless we assume that Cavalcanti had really guessed the true cause of his hesitation, and that what Dante said afterwards was a kindly lie? Is this in character with his uncompromising, even ruthless straightforwardness? Unless therefore some really satisfactory way of escape can be found from this incident and the "centesim' anno" I fear we must reluctantly abandon the theory of 1301, championed so skilfully by Angelitti and others, and must agree with the majority that Dante means us to understand the year 1300.

The early commentators seem to have taken for granted that Dante was correct in his astronomical data, without taking any particular trouble to verify them. Jacopo della Lana in commenting *Par.* xxi. 14 says that in 1300 in the month of March Saturn was in Leo, and was in the 8th degree of that sign; and in order that we should know what was the disposition of the whole sky at that time he further informs us that Jupiter was in Aries in the 24th degree, and Mars in Pisces in the 11th degree, that the sun was near the beginning of Aries, and Venus in Pisces (these two are evidently taken simply from Dante). Mercury was in Virgo (with the sun in Aries! in this case Mercury would have been visible nearly the whole night), and the moon in   . Here there is said to be a blank in all the MSS. The moon was too much for Della Lana, whether the difficulty was in finding her position or in reconciling it with that given by Dante. The Ottimo and Benvenuto da Imola copy Della Lana with little change, both giving the positions of Saturn, Jupiter, and Mars in degree and sign, but the sun only as near the beginning of Aries, Venus in Pisces, and Mercury in Virgo. (It is astonishing that neither should have noticed this gross error.) Benvenuto boldly adds that the moon was in Libra, doubtless because Dante's description so required it at the beginning of his poem.

Nor do these early commentators seem to feel any difficulty about the date on which the Vision began. From *Inf.* i. 38-40 they generally conclude that it was the day of the equinox, which they say vaguely fell in the middle of March "mezzo Marzo," and from *Inf.* xxi. 112-114 all except Boccaccio (who interprets March 25) take it to be Good Friday.[556] No discrepancy seems to have been felt here: they merely remark that in 1300 Easter fell in March, and more than two hundred years passed before anyone pointed out that this was wrong.

At last a work appeared in 1554, called *"Del Sito, Forma, et Misure dello Inferno di Dante,"* in which the author, Giambullari, remarked that according to astronomical calculations the Paschal Moon was not full on the night between Thursday and Good Friday in 1300, but on the morning of April 5th. This excited a good deal of interest, and many theories have since been put forth to reconcile Dante's astronomical data with the facts of the year 1300.

Dr. Moore thinks the date of entering the Inferno was Good Friday, April 8, 1300, but that Dante uses instead of the real moon the ecclesiastical moon, which, as we know, is a conventional body revolving according to Meton's cycle, and sometimes appearing on tables as full two or three days before or after the real Full Moon. Dr. Moore finds that this ecclesiastical moon, was full on April 7, 1300.[557] But would Dante trouble about the ecclesiastical moon, and would his readers know anything about it? Its only use was for determining Easter, and this was done for the people by the Church. The real moon was extremely important for astrologers, for whose benefit Tables and Almanachs were chiefly drawn up; the real moon was useful to everyone as light-giver and time recorder. It is surely inconceivable that a Florentine who wished to arrange a midnight festival, or to start on a night-journey, would consult a calendar to find out on what date the *ecclesiastical* moon would be full! And the ecclesiastical moon does not alter the position of Venus.

According to another suggestion, Dante followed a tradition as to the position of all the "seven planets" at the Creation, and for this reason not only was his sun in Aries, but the moon full and opposite, and Venus in Pisces. Venus is said to have this position in a picture of the Creation in a Church at San Gemignano.[558] Compare also Milton's moon at her creation:—

" ... Less bright the moon, But opposite in levelled west, was set, His mirror, with full face borrowing her light From him."[559]

Or again, it is sometimes held that the day and all the celestial positions are ideal. Dante means us to understand that the first day of his Vision was Good Friday, but without troubling as to what day it fell on in 1300; the equinox and the full moon are naturally connected with Good Friday, since they determine Easter; Venus is connected with the idea of dawn, and when the Sun is in Aries the morning star may well be in Pisces.

Either of these views might be correct, and no one will dispute a poet's right to arrange his skies as he thinks fit, if he is consistent with them, as Dante was. But how did these positions, if they are purely imaginary, come to coincide exactly with the real positions of 1301?

If we examine the positions of 1301, and accept nothing that is doubtful in Dante's own words, what does the coincidence amount to? Saturn was near Regulus, but he was also in the forepart of Leo in 1300, and we are not sure that Dante means more than this; Mars was in Leo, but we are not sure that Dante wanted him there;[560] the moon was full on the morning of March 25, but we are not sure whether Dante means March 25 or Good Friday. Venus was a morning star in Pisces, and this till very lately was the one astronomical fact incontestably in favour of 1301 and against 1300.

But in 1908 there was printed in Florence, for the first time, a mediæval Almanach, which suggests so simple an explanation of this most striking coincidence that it seems to leave nothing more to be said for the 1301 hypothesis. This Almanach, which has been published under a title which perhaps takes things a little too much for granted—*Almanach Dantis Aligherii*—was compiled by a learned Jew, Jacob ben Machir ben Tibbon, who was born in Marseilles about 1236, and died at the beginning of the fourteenth century. Using the Toledan Tables, he constructed this so-called perpetual almanach, which gives the positions of sun, moon, and planets, at intervals of a few days for whole cycles of years. Each has a cycle of a different length, that of Venus for instance being 8 years, as on the Babylonian tablets of ephemerides; and at the end of each cycle nearly the same positions come over again. It was written in Hebrew, and translated into Latin by an unknown author, and must have been popular, since there were an immense number of Latin MSS. in the fourteenth century. In the original, all the cycles begin in 1301, but in the Latin they begin in 1300, *except the cycles of Venus and the sun, which begin in 1301.* Yet 1301 is not written over the first column for Venus, but merely 1, or nothing at all, or even in some copies 1300, and in both the Hebrew and Latin prefaces the beginning of all the cycles is stated to be 1300.[561]

If, therefore, Dante looked up the position of Venus in this almanac, he would very naturally take the 1301 positions for 1300, and he would find that Venus was in Aquarius throughout March, until the 27th, when she passed into the sign of Pisces, and therefore was a morning star during the whole month and also in April. And he would find that Saturn was in the 7th and 8th degrees of Leo in March and April 1300 (which was very nearly correct).[562]

It appears then that while certain expressions used in the *Divine Comedy* prove clearly that 1301 is not the year assumed as the date of the Vision, the astronomical argument in favour of it is not strong enough to be convincing, especially in the light of this recent discovery.

Granting that 1300 was the year, what was the day? Probably not March 25, for the early commentators, excepting only Boccaccio in one passage, are all in favour of Good Friday, and Dante might see in the Almanac that the moon was not full on that date. But neither was she full on the actual date of Good Friday, which fell on April 8 in 1300; moreover the sun was then so far advanced in Aries that seven days later, when Dante describes it as a little past the equinox, it would really have been in Taurus. The third and only possibility is to accept the theory that his moon and his day were ideal; that is, he simply assumed Good Friday and a full moon to occur together soon after the equinox, with which both are so intimately associated. The date of Easter is not given in the old almanac, and we have seen that the early commentators who knew that the equinox fell in "mezzo Marzo"[563] were content to accept Easter as falling in that month also. We must not forget the convenience of an equinox and a full moon for simplifying the time references also.

We conclude, then, that the year chosen by Dante was 1300 (as has been most generally believed), and the day Good Friday, which he assumes to have fallen about a week after the equinox, and at a time of full moon. Saturn in his poem, and also in fact, was in Leo; Venus was not actually a morning star, nor in Pisces, but he may have been led to assign this position to her through an error in a contemporary almanac.

I must confess to a great feeling of disappointment in coming to the conclusion that Prof. Angelitti's enticing theory must be rejected, but a careful study of Dante's own words, and Prof. Boffito's discovery of the "Almanach perpetuum ad annum 1300 inchoatum" seem to leave no other choice. Last April,[564] when helping to photograph Halley's Comet in beautiful early dawns, this problem was much in my mind. Venus once more had the Fishes in her train, and oddly enough Good Friday had fallen on March 25! I could not help wishing that these things had happened in 1300 as well as in 1910!

If in spite of all arguments to the contrary, any readers are convinced that Dante's celestial positions were taken from the real skies, it remains to them to imagine that he saw them in March of 1301 and wove them into the narrative of his poem, although for some reason he adopted another date.

The following tables (pp. 425-26) summarize the foregoing pages. It will be noted with what appropriate symbolism Dante enters Hell on the evening of Good Friday, spends that night and the whole of Saturday in the depths, and rises out of them on Easter Day before the dawn.

For although he reaches Earth's centre on Saturday evening by Jerusalem time, the change of hemisphere there makes his time suddenly go back twelve hours, so that it is again Saturday morning, and he spends the day and night of Saturday (Purgatory time) in climbing to Earth's surface.

DAYS OF THE VISION AND POSITIONS OF THE SUN, MOON, AND PLANETS.

Dante's Journey.	Jerusalem Time.	Purgatory Time.	Positions of Sun, Moon and Planets in Zodiac.	References.
	Night		Moon full in Libra, about 7th degree	Inf. xx. 127.
Forest	Day I. Good Friday		Sun in Aries, about 7th	Inf. i. 38-40.

Dante's Journey.	Jerusalem Time.	Purgatory Time.	Positions of Sun, Moon and Planets in Zodiac.	References.
	(1300)		degree	
	Night			
Inferno	Day II. Saturday			
	Night	Day III. Saturday		
Earth's	Day	in		
Interior	(Sunday)	Purgatory		
	Easter Day	Night		
		Day IV. (Sund.)	Venus in Pisces at dawn	*Purg.* i. 19-21.
		Easter Day in Purgatory		
		Night	Moon in the middle of Scorpio	*Purg.* ix. 1-6.
Purgatory		Day V. (Mond.)		
		Night	Moon at end of Scorpio or beginning of Sagittarius	*Purg.* 79-81.
	Day VI. (Teus.) Night			
Earthly Paradise,		Day VII.	Sun in the middle	*Par.* i. 37-44;

Dante's Journey.	Jerusalem Time.	Purgatory Time.	Positions of Sun, Moon and Planets in Zodiac.	References.
planetary &		(Wed.)	of Aries	x. 7-33;
stellar spheres			Saturn in Leo	xxvii. 86,87.
Primum Mobile		Day VIII.		*Par.* xxi. 13-15.
and Empyrean		(Thur).		

Summary.

Arguments for 1300 or 1301.
From Expressions and Incidents in the D.C.

Malacoda.	Inconclusive: method of reckoning uncertain.
Inf. xxi. 112-114.	
"Nel mezzo del cammin."	Inconclusive: a precise year possibly not indicated,
Inf. i. 1.	and date of Dante's birth not certainly known.
Casella.	Inconclusive: may refer to Bull of 1299,
Purg. ii. 98-99.	but possibly of 1300.
Can Grande.	For 1300: Can Grande 9 years old in March 1300.
Par. xvii. 80, 81.	
Cunizza.	For 1300: "questo centesim' anno" can only mean 1300.
Par. ix. 40.	
Nino's widow.	Inconclusive: might put off widow's weeds
Purg. viii. 74.	before re-marriage.

Guido Cavalcanti.	For 1300: Guido, Dante's intimate friend,
Inf. x. 111.	died in August 1300.

The Astronomical Data.

(True Equinox, March 12. Ecclesiastical, March 21. Traditional, March 25).

In D.C.	*On Good Friday, 1300.*	*On March 25, 1301.*
SUN near EQUI NOX.	Day assumed to fall soon after equinox (actually April 8.)	Date of traditional equinox (13 days after true equinox).
MOON full on first day.	Assumed full, since Easter associated with equinox and full moon. (Actually full March 5, and April 4.)	Full.
VENUS in Pisces on 4th day.	Supposed in Pisces from error in contemporary almanac, confounding 1301 with 1300.	Entered Pisces March 27.
SATURN "on the Lion's breast.	In forepart of Leo, 8th degree of the sign.	In Leo, 21st degree of the sign, lose to

In D.C.	On Good Friday, 1300.	On March 25, 1301.
,,		Regul us.
MARS possib ly intend ed to be in Leo.	Not near Leo.	In Leo.
Early commentators understand Good Friday, not March 25.		

VII

THEORETICAL AND SPECULATIVE ASTRONOMY.

"Thou shalt learn the burning nature of the pure clear sun, thou shalt hear the wandering nature of the round-eyed moon, and thou shalt learn of the surrounding heaven, whence it arose, and how Necessity guiding it compelled it to hold fast the bounds of the stars."

Parmenides.

1. THE SPHERES.

Dante's astronomical phenomena, though they are just the same that are familiar (or ought to be familiar) to us, have led us into some strange by-ways, because they are so intimately connected with his works of imagination. His astronomical theories, by which these phenomena were explained, will also take us into paths which are now seldom trodden by astronomers or others, because mediæval astronomy was mingled with magic, with metaphysics, and with religion, in a way which seems very strange to us.

The second treatise of the *Convivio* has for its text the Ode which is quoted in *Paradiso* viii. 37, "Voi che intendendo il terzo ciel movete."[565] When speaking of the literal meaning of this Ode, Dante takes the occasion to discourse on the system of astronomy which he had learned, and describes how all the apparent movements of the heavens are explained by postulating the existence of spheres and epicycles, entering with special detail into the system of Venus, the third heaven; and later on, when expounding the allegorical meaning of the Ode, he draws an elaborate comparison between the spheres and the different branches of science and philosophy, in which he incidentally gives his readers a good deal of information about the heavens.

From this interesting treatise, therefore, as well as from many scattered passages in all his works, we learn what Dante thought were the real movements of the heavenly bodies, their dimensions, their nature, and the forces which move them; and it forms a valuable commentary on the *Paradiso*, in reading which he supposes us to understand the elements of the Ptolemaic system, and the ideas of the philosophers and the Fathers of the Church concerning the organization of the Universe.

Of these ultimate truths concerning the heavens, Dante says, very little can be known, but the little which is within the reach of human reason, gives more delight (as Aristotle says) than the abundance and the certainty of other things easier to be understood.[566] This thought finds an echo in the *Paradiso*, where the poet reminds us of the beautiful order to be found throughout the universe, and invites us to contemplate the skill of the architect as shown in the movements of the planets.[567]

Dante knew that great men in ancient days had speculated on the possibility that it was Earth, and not the skies, that moved in the diurnal period, but the balance of authority was against them, and they had been confuted by the Master of those who KNOW,[568] that glorious philosopher to whom Nature most fully revealed her secrets,[569] Aristotle. In Treatise III. of the *Convivio* he gives a clear and concise account of the theory of Philolaus, and of what he understood to be the theory of Plato.[570] In the former he seems to have been struck with the deduction that, if Fire were at the centre of the Universe, the real motion of fire on Earth, which to us appears always to ascend, would in reality be a descent towards its own place at the centre.

Both these hypotheses he had evidently learned from Aristotle's *De Cælo*,[571] and when thus presented they seem entirely fanciful; so although Dante had reverence for both Pythagoras and Plato it is not surprising that he rejected the systems of both. He adds that it is not his intention to recount here the arguments by which Aristotle confutes them and establishes the truth: it is enough for his readers to know, on so great an authority, that the earth is fixed, and does not revolve, and is at the centre of the World.

Dante has to confess reluctantly that Aristotle made two mistakes in his astronomy: he said that there were only eight spheres, those of the seven planets, and that of the stars, "e che di fuori da esso non fosse altro alcuno;"[572] also he placed the heaven of the sun immediately above that of the moon, that is to say second in order, counting from Earth outward. Aristotle's loyal admirer hastens to add that these serious mistakes "questa sua sentenza così erronea"[573] are explicable and excusable because, as Aristotle himself shows in the *Metaphysics*, when he treated of astrology he was merely following the opinions of others. That is to say, he took his facts from others, and they cannot all be depended upon, but his philosophical deductions are his own, and are never in error. In the above question of the possible movements of Earth, we saw (pp. 101-2) that Aristotle's arguments were almost entirely drawn from metaphysics.

There have been many opinions regarding the number and the positions of the spheres, Dante says, but the truth about them has at length been found. The scheme which he expounds is that of Ptolemy, modified by the Arabs and adapted to the doctrines of the Church by Christian writers. In the *Vita Nuova* he almost seems to imply that the nine moving spheres form an article of faith. "Secondo Tolommeo e secondo la Cristiana verità, nove siano li cieli che si muovono."[574]

Quoting from Albertus Magnus, or Averroës, he apparently credits Ptolemy with the discovery of precession, for he says that it was Ptolemy who added a ninth sphere to the eight of Aristotle, because he perceived that otherwise the outer sphere would have a double movement, and so he felt obliged to assume a sphere beyond the star sphere which should have only the one simple motion of the diurnal east to west revolution. The period of this revolution (the sidereal day) Dante gives as 23 hours and $^{14}/_{15}$ths of an hour, "grossamente assegnando,"[575] *i.e.* 23 hours 56 minutes, which is the correct value ignoring the 4 seconds. We know that it was not Ptolemy in fact who thought it necessary to add a ninth sphere; he was not sufficiently interested in spheres, since to him they were only mathematical abstractions: but the Arabs of Baghdad noted that he had added another *movement*, the discovery of Hipparchus; and they, accepting the spheres as material instruments of celestial motions, felt obliged to assume the existence of the Primum Mobile or 9th sphere.

That Dante should believe the spheres to exist as actual entities was inevitable, for his principal authorities, Greek, Arab, and Christian, all taught this. And there is abundant evidence that he did think of them thus. He calls them bodies:—"omnia corpora,"[576] "questi corpi grandi,"[577] "cerchi corporai,"[578] "il maggior corpo,"[579] etc. They are transparent, one sphere not obstructing the light from another;[580] and composed of æther, "questo etera tondo."[581] They have a certain thickness, and would be visible if near enough, for Dante speaks of the inner margin of the Primum Mobile, and says that it was too far above him for him to have seen it yet, when he was in the Star Sphere.[582]

He gives the order of the eight heavens as in Ptolemy, and adds the ninth, which is only perceived by the diurnal movement of which it is the cause; and it is called by many the Crystalline, that is, the diaphanous or completely transparent Heaven.[583] The order of these spheres has been discovered by observation and reasoning, with the use of the principles of Perspective, Arithmetic, and Geometry. Dante gives as an instance the experience of Aristotle, who with his own eyes saw the moon pass below Mars, and so hide it for some time; and he mentions also how in eclipses it is evident that the moon is below the sun (*i.e.* nearer to us).[584]

To these nine spheres of the astronomers, the Catholic Church, "which cannot lie," adds a tenth, the Empyrean Heaven, which means the heaven of flame or light, and is the abode of the blessed spirits and of God Himself.[585] And here Dante is able to give himself the pleasure of adding that Aristotle seems to have been of this opinion also; "Ed anco Aristotile pare ciò sentire, a chi bene lo intende, nel primo *Di Cielo e Mondo*."[586] He is no doubt alluding to what Aristotle says about the existence beyond the finite universe.[587] The whole passage is thus translated in the "Temple" Dante:[588] "Nor is there any change of any of those things which are ranged above the outmost rotation, but they are unchangeable and passionless, enjoying the superlative existence, and passing in absolute self-sufficiency their eternal life."

The ninth sphere rotates on two poles, which are absolutely fixed, "fermi e fissi e non mutabile, secondo alcuno rispetto,"[589] the eight within it have each two poles which are only relatively fixed (for they move with the movement of the Primum Mobile), and every sphere has an equator, every point on which is equally distant from the two poles of rotation, as anyone may see who will turn round an apple, or any other spherical thing. Any point on the equator moves more quickly than any other point on the sphere, and each part moves more quickly the nearer it is to the equator, more slowly as it is far from it and nearer to the pole, since all the circling is done in the same time, and some parts have a greater, and some a less distance to go, according to their distance from the equator. Dante adds that the nearer any part of a heaven is to its equator, the nobler it is compared with its poles, for it has more movement, and more actuality, more life, and more form, and it touches more of the heaven just above, and in consequence is more noble.[590]

Besides these ten heavens there are others which, strictly speaking, should be included in the number: these are the little spheres which are called "epicycles" by astronomers, and they also have equators and poles of rotation, and are fixed on the large spheres, or heavens. Dante does not enter into any particulars about these, but only mentions (as belonging specially to the subject in hand) that the brilliant star of Venus is carried "on the back" of an epicycle, which is itself fixed on the back of the third Heaven.[591] The apparent movement of Venus is therefore compounded of no less than four movements: that of her epicycle, of her whole heaven (the "deferent" Dante does not name, probably as introducing too technical a word), of the west to east motion of the star sphere (precession), and the diurnal. The epicycle of Venus is also called in this treatise, as we have already seen, "quello suo cerchio che la fa parere serotina e mattutina, secondo i due diversi tempi,"[592] and the "terzo epiciclo"[593] is referred to in *Par*. viii. 3.

A little further on in the same treatise, Dante tells us the periods in which all these nine spheres revolve. It is an interesting passage, and illustrates the poet's vivid way of bringing home to his own mind and those of his readers the effects of the different celestial motions.[594] The ninth sphere, which is the Crystalline or Primum Mobile, causes by its daily revolution the daily revolution of all the other spheres, which are within it; it is, as Beatrice says, "costui, che tutto quanto rape l'altro universo seco."[595] Therefore if we imagine it possible for the motion of this Crystalline Heaven to cease, what will be the effect on the motions of the other heavens? Obviously their diurnal movement from east to west will cease, and we shall only see them revolving, each in its own period, from west to east. There will be no alternation of day and night, or rather, there will be one long day of six months, and one long night of the same length, in any given place all over the earth. In the same way the moon will be invisible for half the month; and all the planets will be invisible to us for half their periods; and as Dante believed only the known hemisphere of Earth to be inhabited, he could say that during one half of each period no one would see the planet at all. He enumerates the half period in each case, as follows:—

"Planets."	Half Period (approximate) given by Dante.	Whole Period given by Alfraganus (El. Ast. ch. xvii.)			
		Years	Months	Days	
Saturn	14½ years	29	5	8	
Jupiter	6"	11	10	14½	about.
Mars	about a year	1	10	22	nearly.
Sun	182 days, 14 hrs.	...		365¼	nearly.
Venus	{about the same { as the sun.	...		,,	
Mercury	,, ,,	...	...	,,	
Moon	14½ days	...	(nearly) 27, 7¾ hrs.		
		(Paris edition, 29 days, 12¾ hrs.)			

If we compare these with the zodiacal periods given by Alfraganus, we see that Dante has taken those of Saturn, Jupiter, and Mars at the nearest whole number of years and halved them, but with the sun he seems to have used that more exact value with which we have already seen that he was acquainted. For to divide the 365¼ days of Alfraganus by 2 would have given him exactly 182 days 15 hours; but dividing 365 days 5 hours, and neglecting minutes gives 182 days 14 hours. The large oscillations of Venus and Mercury would make their periods of visibility very irregular, but the mean period would be, as Dante says, about the same as that of the sun. He has made a slip with the moon, for of course it is her sidereal period of 27 days 7¾ hours which he should have taken, giving a half period of less than 14 days, whereas he has evidently taken the mean synodic period of 29 days 12¾ hours, and ignored the hours. All the printed editions of Alfraganus give the sidereal period in this passage, except the Paris edition of 1546; but that gives the synodic,[596] so it doubtless appeared on some mediæval MSS. Dante must have had the misfortune to possess or consult one of these.

As for the stars, one third of their sphere would not yet have been seen. This is not difficult to understand if we remember what the movement of the star sphere was supposed to be, and how long it was since man had first begun to look upon it, according to Dante's chronology. The movement which makes the equinoctial point shift its place among the stars was thought to be a movement of revolution of the whole star sphere, and its rate to be one degree in a century; the age of the world, we learn from *Par.* xxvi. 118-123, was over sixty centuries when Dante wrote. For Adam said that he had spent 930 years on Earth and 4302 in the Limbo, so that Dante evidently counted 5232 years from the Creation to the Crucifixion, that is (5232-33 =) 5199 to the beginning of our era at the birth of Christ, which is the date given by Orosius; adding 1300 years to this, we find that the age of the world in Dante's day was more than 6000 years. Therefore the star sphere had revolved, since the beginning of the world, through more than 60 degrees.

The first inhabitants of this earth saw half the starry sky or 180°, and since then it had revolved through 60°, which together is equal to 240°; therefore there would still remain 120°, or one third of a revolution, to be made to complete the whole 360°, and bring every part of the star sphere into view. Just before this passage Dante had expressed the same idea in another way. This movement proper to the star sphere, he says, will never have an end, for the end of a revolution is to return to the same point; and this heaven will never do so, for since the beginning of the world it has turned through a little more than the sixth part of its whole revolution (60° = ⅙ of 360°), and we are already in the last age of the world, and indeed are awaiting the consummation of the celestial motion.[597]

If the premises be granted, the conclusions are correct. As regards sun, moon, and planets, they are correct even from the point of view of modern knowledge. For if the diurnal motion (that is Earth's rotation) were to cease, while her other motions, and those of the rest of the solar system remained as at present, her motion round the sun would cause it to rise and set once in a year, and the appearance of moon and planets also would be as described. The vastly distant stars would have no appreciable motion, save that caused by precession, for if the earth revolved round the sun without turning on her axis, the same stars would always remain on the meridian at any given place. Precession, however, would not have the result which Dante supposed. He considered it to be a rotation of the star sphere, but we know that it is merely a nodding. Hence at any given place a star at the pole would describe, in the course of centuries, a small circle in the sky, and those at the equator would move in short lines first north and then south, while stars between would move in ellipses which become narrower as they approach the equator. All round the horizon, therefore, there would be slight very slow changes, bringing a few fresh stars into view, and causing others to disappear. The whole amount of the sky actually seen would vary but little, even if the world should last much longer than Dante expected.

Dante evidently took his value of precession from Alfraganus, for he was one of the few Arab astronomers who accepted Ptolemy's erroneous figure.

In the *Paradiso*, as the poet mounts with Beatrice from sphere to sphere he mentions the number of each one in its order. The moon is "la prima stella,"[598] Mercury "il secondo regno,"[599] Venus "il terzo ciel,"[600] the spirits met in the sun are "la quarta famiglia,"[601] Mars is "questa quinta soglia,"[602] and "più levato"[603] than the last heaven, Jupiter is the "stella sesta,"[604] Saturn "il settimo splendore."[605] The starry heaven is alluded to as "la spera ottava."[606] The Primum Mobile is "il ciel velocissimo,"[607] "il maggior corpo."[608] The Empyrean, "il ciel ch' è pura luce,"[609] is called "l'ultima spera,"[610] but this has no limits and no movement, and therefore no poles:—

" ... In quella sola E ogni parte là dove sempr' era; Perchè non è in loco, e non s'impola."[611]

In *Conv.* II. iv. 35-39 it is thus described:—

"Questo è il sovrano edificio del mondo, nel quale tutto il mondo s'inchiude; e di fuori dal quale nulla è; ed esso non è in luogo, ma formato fu solo nella Prima Mente, la quale li Greci dicono Protonoe."[612]

We must remember, however, that the spheres are sometimes counted in the other direction, from the outermost to the innermost or lowest, and therefore this "ultima spera"[613] is elsewhere spoken of as "il primo cielo,"[614] "il primo giro,"[615] "cœlum primum."[616]

2. THE SPHERES AND THE FOUR ELEMENTS.

The pure region of the spheres, "il paese sincero," is immortal as the spirits themselves;[617] but below the lowest celestial sphere ("la celestial c'ha men salita" *Par.* iv. 39) all is mortal and transitory, as the Greeks and the

Latin poets had said. This is expressed in the Letter to Can Grande, when Dante contrasts the spheres (cœlum) and the elements, and says: "illud incorruptibile, illa vero corruptibilia sunt."[618] It is also implied in Virgil's address to Beatrice:

"O donna di virtù, sola per cui L'umana specie eccede ogni contento Da quel ciel che ha minor' li cerchi sui."[619]

It is only through the spiritual teaching which Beatrice symbolizes that the human race can rise above all that is comprised within the heaven of the smallest sphere. And it is taught by Beatrice herself when, in the heaven of Mercury, she discourses on the mysteries of the creation and redemption. The angels, the souls of men, and the spheres, were created directly by God, she says, and are therefore immortal, but the elements, water, fire, air, earth, and all things that are formed out of them, also the souls or living principles of all animals and plants, were created through intermediate instruments and are mortal. These intermediate instruments are the heavenly bodies. Pure or inchoate matter was created at the beginning by God, and a power infused into the stars to give it form.[620]

Dante follows the classics and agrees with his contemporaries in arranging the four elements in four spheres, which are below the celestial spheres. The sphere of Fire is immediately within that of the moon, below this comes Air, then Water, and lastly the solid sphere of Earth. The contrast between Fire which constantly tends to rise towards the moon, and Earth which sinks towards the centre of the universe, is a favourite thought with him;[621] and lightning striking the earth is described as fleeing its proper place.[622] Aristotle seems to have regarded the four elements as flowing into one another, so that these lowly spheres beneath the moon were not sharply divided, but his mediæval disciples, following the Greek idea to its logical conclusion, conceived them with boundaries as definite as those of the celestial spheres. "Exhalations" could pass from one to another, but each particle of every element tended always to revert to its own place, which was distinct from the others.

And here they encountered a serious difficulty. If the sphere of water is naturally higher than that of earth, how comes it that in one quarter of the globe land rises above the ocean? It was recognised as necessary that this should happen in some part of the world, in order for an opportunity to be given for the elements to combine and to form all those substances and all those living creatures, including man, for which this earth was prepared; it was also agreed that the celestial spheres were the agency by which the Creator had first chosen to "let the dry land appear," and by which it was maintained above the ocean; but the exact nature of this force and the way in which it acted was a question sometimes discussed. Albertus Magnus was of opinion that the rays of the sun and the stars, which had more power in the temperate and tropic zones than in the arctic, had here dried up the ocean, and he quotes Albumassar's statement that this proves the existence of land on the southern side of the equator also. Ristoro, on the other hand, believed that the waters had been drawn up and rolled back from the habitable earth, and he attributes the elevating force to the stars of the north, bringing in his favourite theory that in this hemisphere they are far more numerous, and the constellations, from their upright position, are far more powerful than in the south.

He compares this lifting force of the stars to the attractive power of a magnet over iron, and argues that it must be the waters which are drawn away from the earth, not the earth which is lifted above the water, because the starry influence in its downward course would meet first the sphere of water, and moreover it is lighter and easier to move than Earth. He also thinks that the presence of springs on mountain tops is to be explained by the fact that the ocean is preserved at a higher level, hence water percolating through the porous land, as through a sponge, is forced up to its utmost height.

It does not seem as if Dante had paid much attention to this problem when he was writing the *Convivio*, for although he calls the habitable part of the earth "the uncovered land," "la terra discoperta" (*Conv.* III. v. 73), he speaks of Earth and ocean as if they formed a globe together, "Questa terra ... col mare è centro del cielo,"[623] in the manner of Alfraganus, "Inter sapientes convenit, terram unà cum aqua globosam esse."[624] But if we may believe that Dante is the author of the *Quæstio de Aqua et Terra*—a belief shared by many experts—he became greatly interested in the problem later on. According to the introduction and the final colophon, he attended a discussion on the subject and joined in it himself, when in Mantua, and afterwards in Verona less than two years before his death he wrote and read in public a Latin treatise in which his arguments and conclusions are set forth.

He seems to have had in mind Ristoro's book, and while refuting some of the conclusions agrees with others. As a thirteenth century scholar he admits, as a matter of course, that the centre of the earth is the centre of the Universe and the goal towards which all heavy things tend; that earth as the heaviest element ought to be everywhere nearest this centre, while water ought to occupy the nobler place above it, that is, nearer to the most noble sphere of all which envelops all the others (the Empyrean). He also reasons, in true Dantesque and true mediæval fashion that there must be some place in the universe where all the elements may meet and mingle so that all potential forms of matter may be realized; otherwise the Mover of heaven, in whom all these forms (Plato's Ideas) actually exist, would fail to give complete expression to his goodness, which is impossible.[625]

The author of the *Quæstio* denies, however, that this was accomplished by the removal of part of the sphere of water from the underlying earth. The ocean cannot have been drawn away and heaped up, he says, for water would always flow down again, being fluid and naturally seeking its own level; nor can the sphere of water be eccentric to that of Earth so that the latter emerges from it in one part. The reason for this is interesting since it implies a curious and apparently novel theory of gravity. Aristotle's theory was that all heavy things tend by a natural law towards the centre of the universe, and therefore necessarily range themselves in spheres round it; the writer might therefore have made short work of Water's eccentric sphere by stating briefly that it could not exist; instead of this, however, he suggests a new law of gravity, apparently invented on the spur of the moment, according to which everything which has weight tends towards the centre of its own circumference. If then earth and water had two separate centres, they would fall naturally in different directions, or as he puts it "to different *downs*." Yet we know that the law of gravity is the same for both, and therefore their spheres must be concentric. The centre of the universe must be also the centre of the sea.[626] It had apparently been urged (perhaps in the discussion at Mantua) that since water follows the moon in the tides, its sphere may also imitate hers in being eccentric, but although Dante takes for granted that a connection between moon and tides is proved (and indeed it had been noted by the Greeks),[627] he replies that imitation in one particular need not imply imitation in all. Again, if the sphere of earth emerged from the sphere of water in the way proposed, the outline of the emergent earth must be circular, instead of which it is that of a half-moon, or nearly so, its width from east to west being much greater than its length from north to south.

Dante has been blamed for assigning this shape to the emergent earth, but the comparison of Earth's habitable quadrant with the illuminated quadrant of the moon is perfectly just, as we saw with Ristoro,[628] from whom Dante is perhaps quoting here. The qualifying "vel quasi"[629] is added because the truly habitable regions did not extend over the whole quadrant, but stopped short at the arctic circle.

The existence of springs on mountain tops is explained by the fact that water rises thither not in the form of water but of vapour. Another upholder of Ristoro's theory that Water was rolled away from Earth seems to have brought forward as evidence that sailors can see distant land from the mast which is invisible from the deck, and that therefore it evidently lies below the ocean; but this is confuted by the true and well known explanation that it is the convex shape of the sea which hides the distant land. Our author adds to his other arguments, in the curiously casual way so characteristic of his age, the ordinary observation which would be the first to occur to us, and quite sufficient in itself, viz. that instead of seeing a great wall of ocean rising above our sea-coasts, we invariably see the coast rising above the sea.

His own idea is that though the sphere of Water in general rises above Earth, in one part of the northern hemisphere Earth has been drawn up above the ocean. Earth must be subject to an uplifting influence, as well as to the downward drag of gravity, just as man is susceptible to the influence of reason as well as to the sway of his lower passions. This uplifting influence is not inherent in Earth herself, an upward impulse would be contrary to her own nature; nor can it be in water, or air, or fire, since all these are homogeneous bodies, and their virtue is evenly distributed, so that they could not cause a local and partial effect. It must be therefore in one of the heavens, and it remains for us to enquire which.

It cannot be in the heaven of the moon, for in the moon herself resides whatever influence comes from her sphere;[630] and she goes as far south of the equator as she does north: therefore she would raise land beyond the equator as well as on this side, but this is not so.

Then follows a passage of which commentators can make nothing as it stands:—

> "Nec valet dicere quod illa declinatio non potuit esse propter magis appropinquare terræ per excentricitatem; quia si hæc virtus elevandi fuisset in luna, (quum agentia propinquiora virtuosius operentur) magis elevasset ibi quam hic."[631]

Mr. Wicksteed suggests reading "elevatio" instead of "declinatio," and apparently deletes the "non,"[632] and Dr. Moore seems to read the same way,[633] for both agree in interpreting as follows:—the author is stating an argument that the elevation of land (in only one hemisphere) might be due to the moon's coming nearer to Earth in one part of her orbit, owing to its eccentricity; to which he replies that if that were the cause the elevation would be greater in the south than here in the north, because the moon is nearer to the earth when she goes south.

This is supposed to imply a belief that the moon is always in perigee[634] when south of the equator, whereas the fact that her perigee revolves all round the zodiac was as well known to Ptolemy and Alfraganus as it is to modern astronomers. The interpretation therefore convicts the author of the *Quæstio* of so serious and inexplicable a blunder that these commentators find themselves obliged to consider it as evidence either that it was not written by Dante, or that his knowledge of astronomy was much less extensive than has been supposed. It is suggested that there may have been a popular fallacy in his time that the moon was like the sun in this respect—although to know that the sun has a perigee, and that it is situated in the south, requires some acquaintance with astronomy! However, Dr. Moore has searched Alfraganus and other works in vain to find a suggestion of this kind.

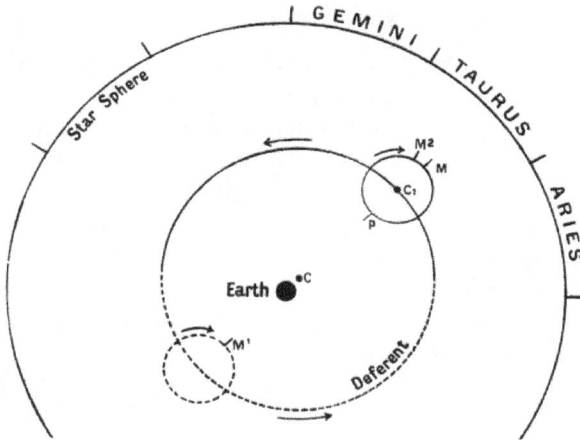

Fig. 52. The moon's epicycle and deferent.

(The dotted line represents the southern half of the deferent).

Yet an explanation and a source of this statement is not far to seek, and instead of proving ignorance it indicates that the author of the *Quæstio* knew his Alfraganus well. The passage is evidently slightly corrupt, but the clue to its meaning lies in the word "eccentricity." In two other places of the *Quæstio* the moon's eccentric "orbis" is alluded to,[635] but if we translate this as "orbit" we shall be introducing modern ideas which meant nothing to Dante or his contemporaries. According to Alfraganus and Ptolemy the moon's orb or sphere[636] was eccentric, but this was not the main cause of her varying distance from Earth. It was the revolution of her epicycle round its centre C, which caused her to move continually from apogee M to perigee P; and as the whole epicycle was meanwhile revolving on the deferent round C, and these two periods were not quite equal, the moon was as often in the perigee of her epicycle when it was south as when it was north. But besides this, the deferent, being eccentric to Earth, had also a perigee and apogee; and Alfraganus says that *the perigee of the eccentric is always in the south.* "Saturnus, Jupiter, atque Mars eccentricorum suorum absidas summas et imas habent declinatas a zodiaco, illas ad boream, hasce ad austrum, secundum eandem semper deflexus mensuram: *quedadmodum res in Luna obtinet.*"[637]

The upper and lower apsides are defined in chapter xii. as the positions of apogee and perigee respectively.[638]

In this quotation the word "declinatas" also supplies us with the meaning of Dante's "declinatio," which need not be changed for "elevatio." He is using it in the astronomical sense of position north or south, like Alfraganus and modern astronomers, and it refers here to the "declinet" of five lines earlier. "Luna ... tantum *declinet* per zodiacum ab æquinoctiali *versus polum antarcticum* quantum versus arcticum."[639]

It is only necessary to delete "quod" and "non potuit," and the whole passage becomes clear and, from the point of view of mediæval astronomy, correct. The author has shown that the moon cannot be supposed to exert any preponderating influence in the northern hemisphere, because her declination south is equal to that in the north. Nor is it of any use, he adds, to say that this declination (south) results from her approaching the earth more nearly on her eccentric orb (see diagram);[640] for (although this does imply an unequal influence) if the elevating influence came from the moon, it would act more powerfully in the south than here in the north, since the nearer an agent is the more powerfully it acts.

He goes on to say that the same reasoning rules out the planetary spheres, and here Alfraganus still supports him, for Saturn, Jupiter, and Mars are said to have the perigees of their eccentrics in the south, just like the moon, while those of Venus and Mercury are continually changing.[641]

As for the Primum Mobile, or ninth sphere, it is uniform throughout, and consequently quite uniform in virtue, therefore there is no reason why it should elevate land here rather than elsewhere. Since, then, there are no more moving bodies except the star sphere, or eighth heaven, it must be from this that the lifting power proceeds. And it is specially adapted to exercise such an influence since it contains a number of stars, all of which have different "virtues." Without repeating Ristoro's naïve assertions about the greater number and the upright positions of the northern constellations, our author bids us remark that star differs from star, constellation from constellation, and that their influences are also diverse, so that those in the north have different virtues from those in the south. The lifting virtue which raises the land must reside in those stars which are between the equator and 67° north, since those are above the habitable earth; but we do not know the nature of the force they exercise, whether it attracts the land as a magnet attracts iron, or forces it up by generating underground vapours, such as have raised some mountains (Aristotle had said in his *Meteorologica* that volcanic hills were formed by pent-up winds beneath them).

After so elaborate a search for a power capable of raising land only in the north and not in the south, we naturally inquire why in the east and not in the west? But to this we only receive the reply that there was not enough material to go round, and that we must not ask presumptuous questions. Let us seek to approach divine and immortal things as nearly as is possible to human nature, but leave those which are too high for us.

This somewhat lengthy description of the contents of the *Quæstio* has been given because, even if it is not Dante's work, it illustrates well the current ideas concerning the Cosmos, its form and forces. So many of the ideas, however, and even the expressions find parallels in the unquestioned writings of Dante, that the internal evidence for its genuineness is strong.[642]

In his poems Dante is faithful to his own descriptions of the universe in prose, and in the visionary journey of the *Divine Comedy* he passes successively through all the spheres which are enumerated in the *Convivio* and the *Quæstio*. The central sphere of earth, the lowest of all the elements, "il suggetto dei vostri elementi,"[643] is completely traversed, from side to side, and on the further side is seen the vast sphere of ocean which envelopes three-quarters of the earth. The Mountain which the poet has placed here (his only innovation) rises out of the water towards the heavens higher than any other, "inverso il ciel più alto si dislaga,"[644] and in climbing it he reaches that region in the sphere of air into which no cold exhalations can rise from earth or sea, neither the "secco vapor"[645] which causes wind and earthquake, nor the "esalazion dell' aqua"[646] from which are formed rain and hail, cloud and dew. Hot exhalations ascend to the sphere of fire, there to become meteors or comets (see p. 103); but the cold vapours rising as high as they can towards the sun which draws them up ("retro il calor"[647]), are unable to ascend beyond the Gate of Purgatory.[648] Therefore Purgatory is exempt from all changes except those caused by that which heaven receives from itself into itself. This was ingeniously conjectured by Venturi to be light, since there is change of day and night in Purgatory; but from the context it seems clear that human souls are meant. Souls come forth from their Creator, whose special abode is the heaven of heavens, and return thither after purification. And it was the rising of the purified soul of Statius, ready to return to heaven, which had been the cause of that trembling of the Mountain of which he is here speaking.

After learning this, Dante is surprised to find, on the summit of the Mountain, that his brows are caressed by a gentle breeze which comes from the east and bends all the trees of the Forest in one direction. Matilda tells him that it is true no atmospheric changes are felt on this Mountain, and it was for this that it was made to rise so high, so that man in the Garden of Eden, on the top of the Mountain, should be untroubled by the disturbances which occur lower down: what he feels now is no fluctuating wind. On this lofty mountain, remote from any other land in the midst of the hemisphere of sea, he feels for the first time the revolution of the atmosphere uninterruptedly following the movement of the celestial spheres.

The idea that Eden was on a lofty mountain, secure from stormy weather, and inaccessible to fallen man, was a familiar one with mediæval writers: it was sometimes located on Adam's Peak in Ceylon, and Bede thought it might rise as high as the moon, but Albertus Magnus says this was only meant figuratively. But the connection with Purgatory, and the thought that here the movement of the heavens might be felt as a soft breeze, is, so far as I know, all Dante's own.

In all the *Divine Comedy* there is no more beautiful allegory than this. Man in the age of innocence, man to-day when purified from sin, is already partly in heaven, though he still walks the earth. And like many another allegory of Dante's, it follows naturally from the science of his day. Aristotle had said that the upper air "flows in a circle, because it is drawn along with the circulation of the Whole," and Aquinas:—

> "Thus the air which exceeds the greatest altitude of the mountains flows round, but the air which is contained within the altitude of the mountains is impeded from this flow by the immoveable parts of Earth."

Purg. xxviii. 103-108 is almost a poetical paraphrase of this:—

" ... In circuito tutto quanto L'aer si volge con la prima volta, Se non gli è rotto il cerchio d'alcun canto: In questa altezza, che tutta è disciolta Nell' aer vivo, tal moto percote E fa sonar la selva, perch 'è folto."[649]

The whole body of the air, everywhere in the world, is turning with the First Revolution (the Primum Mobile), wherever it is not interrupted by any obstruction (of hills etc.); and as at this height the Mountain is altogether free and open to the pure air, this motion strikes the thick forest and makes it murmur.[650]

From this region of pure air, Dante rises with Beatrice through the sphere of Fire, into which already in a dream he had been rapt, on the first night in Purgatory, by the Eagle with golden feathers, when the burning heat of the dream-fire roused him from his sleep.[651] Now he feels no heat, but is amazed at the vast extent of light, greater than any river or lake ever seen on earth, and at the same moment he begins to hear the music of the eternal spheres,[652] for he is approaching the heaven of the moon.

In each one he hears it until the seventh is reached, but here, in pity for his mortal senses, which are not yet strong enough to bear so much divine beauty, not only does Beatrice forbear to smile, but the sweet symphonies of heaven are silent.[653]

Did Dante then, in spite of Aristotle, believe in the music of the spheres? Perhaps the fact that only the *Commedia*, of all his works, makes mention of it indicates that he only thought of it as a poetical fiction, an allegory? Yet it was fully believed in by many in his day, although opinions varied greatly as to what note each planet sang. The favourite idea among Greek and Latin writers seems to have been that the notes made up a musical scale, the lowest sphere sounding forth the lowest note, and the star sphere its octave; but others thought that only the seven planets took part in the harmony, corresponding with the seven strings of the lyre: the moon being at the shortest distance from Earth, was the shortest string and gave the highest note, Saturn was the longest and gave the deepest.

Our own Milton, when he prays, "Ring out, ye crystal spheres," bids the Primum Mobile join in, as well as the star sphere:

"And with your nine-fold harmony Make up full concert to the angelic symphony."[654]

And Shakespear lets every separate star swell the chorus:

"There's not the smallest orb which thou beholdest But in his motion like an angel sings, Still quiring to the young-eyed cherubins."[655]

In the same spirit, but more literally perhaps than these, are we to understand the music of Dante's spheres, which accompanies the singing of the angels:—

"Il cantar di quei che notan sempre Dietro alle note degli eterni giri."[656]

3. DIMENSIONS AND PHYSICAL NATURE OF THE UNIVERSE.

We have seen that Dante believed the circling of the nine moving spheres and their epicycles to be the immediate cause of the movements which he saw in stars and planets. If we ask what were his ideas with regard to their sizes and distances the question is easily answered.

It is a curious fact that although our ideas about the size of anything we see depend upon the distance at which it appears to be, and no one can say how far away the sun looks, most of us nevertheless have a quite definite idea as to how large it looks, about the size of a dinner-plate, a cart-wheel, &c. Cleomedes, in the days of Augustus, quotes Lucretius and the Epicureans as believing that the sun is no larger than it looks, *that is a foot in diameter*; and he rightly remarks, amongst other arguments to the contrary, that if it were only that size it would be invisible unless nearer than the tops of the mountains, yet we see islands and hills projected against it when rising, which shows that its distance is greater. Dante twice quotes the same fallacy, as an instance of the folly of trusting to the impression of our senses, when not corrected by reason, and he tells us what he believes the true size of the sun to be.

In *Conv.* IV.[657] he says:—

> "Thus we know that to most people the sun appears to be a foot wide in diameter; and this is so utterly false that according to the investigation and discovery made by human reason with her attendant arts, the diameter of the sun's body is five times that of the earth's, and a half besides. For whereas the earth has a diameter of six thousand five hundred miles, the diameter of the sun, which when measured by sense-impressions seems to be a foot in extent, is thirty-five thousand, seven hundred and fifty miles." (See also Ep. x. 42-46).

When speaking of the planet Venus he says she is "far off from us, being distant even when she is nearest to us one hundred and sixty-seven times the distance of the distance of the centre of the earth from us, which is three thousand two hundred and fifty miles."[658]

Mercury is spoken of in the *Paradiso* as a small star, "Questa picciola stella,"[659] and in *Conv.* II.[660] Dante says:—"Mercury is the smallest star in the sky; for the length of its diameter is not more than 232 miles, according to Alfraganus, who says that it is one twenty-eighth of the diameter of the earth,[661] which is six thousand five hundred miles."

It is only in the last passage that our poet quotes his authority, but in the others also the diameter or the semi-diameter of the earth is used as the unit of measure, and it is the same as that given by Alfraganus; so are the sizes of Mercury and the Sun, and the least distance of Venus is the value given by Alfraganus for the greatest distance of Mercury, which he asserted to be equal. It seems quite clear, therefore, that the figures given in the tables on pp. 189-90 represent Dante's belief as to the dimensions of the Universe. These passages also are another proof that Dante had not studied Ptolemy's own work, for Ptolemy had said that the distances of Venus and Mercury could not be accurately known.[662]

We do not expect exact measures of distance in Dante's descriptions of his soaring flight through the spheres, but he so constantly speaks of the marvellous speed with which he passes from star to star that we are not required to ignore the stupendous distances between. The one indication he does give us of the space which separates him from earth is full of poetry and symbolic meaning: I refer to Folco's words in the heaven of Venus:—

"Questo cielo, in cui l'ombra s'appunta Che il vostro mondo face" ...[663]

This is the last of the three heavens in which Dante has met spirits whose lives did not reach the perfection of the saints in higher heavens, but were marred by broken vows, earthly ambition, earthly love; and here he is reminded that he has not yet soared high enough to be beyond reach of the cone of shadow thrown by the earth. This is another allegory resting on what was to the poet an astronomical fact. For Earth's shadow was said by Ptolemy and Alfraganus to extend to 268 times the distance of her semi-diameter, and the distance of Venus from Earth varied, according to Alfraganus, between 167 and 1120 times this unit, while the next planet, the sun, at his least distance was 1120 units distant (the same as Venus' greatest). Therefore the spheres of Venus and the two nearer planets Mercury and the Moon, were all within reach of Earth's shadow, but the sun and the rest were far beyond it.[664] Venus and Mercury themselves never could be actually touched by the shadow like the moon, however, because they keep too near the sun for Earth to interpose herself between them and the source of light.

(Though the length of the shadow thrown by Earth is nearly accurate, see pp. 191-92, the planets are a great deal further than Alfraganus supposed; for the least distance of Venus is nearly 6000 times Earth's semi-diameter, and the least distance of Mercury 12,000).

Except for this we have only the passage in the *Paradiso* already referred to, where Dante, standing among the stars of Gemini, found the sphere above him too distant to be seen, but turning his gaze downwards was able to survey all the planets and the earth at the centre of the World. He knew very well that at a distance from Earth 20,000 times her own semi-diameter it would be impossible to see her as a disc,[665] far less to distinguish oceans and continents, hills and river-mouths. It would in fact be like looking at an object one foot in diameter at a distance of two miles. But as we have before observed, his power of vision here was more than human: at every step in this marvellous journey it had grown clearer and stronger, and in the fourth heaven he had been able to look on spirits which shone in the sun more brilliantly than the sun itself, although on earth no eye can look upon the sun.[666]

In the Empyrean, distinctions of space, and impediments to vision vanished altogether.[667]

With regard to the physical nature of the heavenly bodies, Dante and his contemporaries had no means whatever of investigation, and could only profess ignorance or accept one or other of the guesses of their predecessors. Some of the Greek philosophers, as we saw, guessed that the planets were worlds somewhat like the earth, others thought they were composed of fire or of air. But the general belief in the thirteenth century was that planets and spheres alike were composed of a kind of celestial matter, called by Aristotle and the Greeks "æther," by the Arabs "al-acir," an immortal substance which had neither heaviness nor lightness, and was altogether distinct from any of the four elements existing below the sphere of the moon. Beatrice speaks of "this sphered ether," "questo etera tondo." *Par.* xxii. 132.

Dante's description of this celestial substance when he first enters the ethereal world is one of the finest instances of his faithfulness to the teachings of astronomy as he had learned it, combined with poetical imagination, and at the same time his power of using material facts (as he conceived them) to present an allegory of the deepest religious mysteries. If we merely listen to the magic of the words, an impression is conveyed of something mysteriously beautiful and dazzling, unlike anything known on earth; but if we look into the meaning, we find that the globe of the moon has just those strange, contrasting qualities, just the colour and the form which a sphere of ether may be imagined to have. It is white and rounded like a pearl, "polished" as Plato said of the universal orb, it is thick and shining, soft as cloud but hard as diamond; and it offers no more resistance to Dante as he enters into it than does water to a ray of light.[668]

Then follows a discussion concerning the substance not only of the moon but of all the heavenly bodies. On hearing that he has reached the boundaries of the immortal world, and has entered "la prima stella,"[669] Dante eagerly enquires of Beatrice what are those dark markings in the body of this planet which are seen from Earth and have given rise to the fable of Cain. She smiles a little, and bids him tell her first what is his own opinion. He advances the theory which he had taught in the *Convivio*, viz. that the parts of the moon which look dark to us are less dense than the rest, and therefore the rays of the sun when striking them are not stopped and reflected back to us; hence the brightness is less than in other parts of her surface.

"L'ombra ch' è in essa, ... non è altro che
rarità del suo corpo, alla quale non possono
terminare i raggi del sole e ripercuotersi così
come nell' altre parti."[670]

This explanation seems to have been first suggested by Averroës, the
Arab philosopher and astronomer of Cordova, who was known in the
Middle Ages as "The Commentator," from his famous commentary of
Aristotle. The suggestion occurs in his *De Substantia Orbis*, and as it follows
a quotation from Aristotle (to the effect that the moon is more nearly
related to Earth in her nature than to the stars), it was believed by many to
be originally due to Aristotle himself.

The strange circumstance that the moon alone, among all the heavenly
bodies, appeared to have dark shadows on her bright face, had roused
much curiosity already among the Greeks; and Plutarch, in his dialogue
"On the Face in the Moon," mentions the different explanations suggested
in his day. Some thought that the polished surface of the moon reflected,
like a mirror, parts of the earth; the Stoics said that air was enclosed within
her globe of fire and obscured it in certain places, but this view was rejected
with contumely by Plutarch.

"It is a slap in the face to the moon when they fill her with
smuts and blacks, addressing her in one breath as Artemis and
Athena, and in the very same describing a caked compound of
murky air and charcoal fire, with no kindling or light of its own,
a nondescript body smoking and charred like those thunderbolts
which the poets address as lightless and sooty!"[671]

The third hypothesis, which was favoured by most of the speakers in
the dialogue, was that the moon was not a star at all, but a kind of more
beautiful Earth; and it was even suggested that men might dwell on her
who thought their world the only place fit for human beings, looking down
with contemptuous pity on our Earth, as a sort of sediment and slime of
the Universe, appearing through damps and mists and clouds, a place
"unlighted, low, motionless," entirely incapable of supporting moving
breathing warm-blooded animals!

This was not a view which could possibly be held by mediæval
Christians, and in a treatise on Aristotle's *De Cœlo*, attributed to Albert of
Saxony, we find others put forward. The author disproves the old theory of
a mirror-like moon reflecting parts of Earth, and also another that the
moon draws up cold vapours which are seen like clouds on her surface, and
upon which she is nourished—for how should an eternal heavenly body
need nourishment?—and finally he expounds the theory of "The
Commentator," with which he agrees. The markings are rare parts of the

moon's substance, which cannot shine so brightly as the denser parts, and this varied surface he compares with alabaster, in which "the dense and not translucent part is very white, while that which is translucent like glass is obscure and tends to blackness. And if," he adds, "it is asked why the moon is thus dissimilar in her parts, the reply is that this is her nature."[672]

This theory was perhaps the most popular one with mediæval scholars, though there was some difference of opinion as to whether the dense or the rare parts of the moon were those that showed dark. Ristoro perhaps alludes to it, in the passage where he speaks of the moon's markings, but in a very confused way, and the contrast of "polished" and "rugged" surface is what he chiefly lays stress upon.

After upholding this theory in the *Convivio*, Dante rejects it in the *Paradiso*: perhaps because it was more or less bound up with the idea that the moon was partly of an earthy nature, and this was inconsistent with the other theory of Aristotle, so popular among classical writers, that all within and above the sphere of the moon was eternal and heavenly. He is evidently very anxious to convince us of the falsity of his old belief, for he is not content with a simple assertion, but lets Beatrice reason like a learned doctor, using arguments which would appeal to his readers, drawn now from experiments which might be used in a school, and now from the accepted astrological beliefs of the day.

First she reminds him that the numerous stars scattered over the surface of the eighth sphere differ in brightness, like the different parts of the moon's surface, but they differ in the quality (colour?) of their light, as well as the quantity, and we know that there are essential differences between them, because their "virtues" are different. It cannot be, therefore, that they differ simply in density or rarity of their substance.

Then she shows that in no circumstances is the theory of varying density in different parts of the moon capable of explaining the observed variety of brightness. For if the parts which look dark are less dense than others, either they must be so right through the planet, or there must be an alternation of dense and rare matter in those parts, like fat and lean in a material body, one behind the other like pages in a book. (The double metaphor is characteristic of Dante.) Now if the first be true, the sunlight would shine through the moon is those parts, during an eclipse;[673] and if the second, the sun's rays must penetrate to the lower-lying denser part, and be reflected thence just as from the dense parts of the moon's surface. A possible objection that in that case the reflecting surface would be further away, and therefore appear fainter, may be disproved by an experiment. Take two mirrors and place them at an equal distance from you; place a third between them but further away, face the three mirrors, and place a light behind you so that it may be reflected from them all: the distant one will appear smaller, but no less bright than the two near ones. This reasoning is perfectly sound.

Then follows the explanation for which all this is a preparation. It involves an exposition of ultimate causes which we need not enter into here: for the present we need only say that difference of brightness is not caused by differences of much or little of the same substance. There are intrinsic differences between star and star, and between different parts of the moon,[674] but all are manifestations of divine intelligence; just as the different members and faculties of a human body are manifestations of a human soul.

The diversity of the celestial matter is also touched upon in the *Convivio*, where it is said that the epicycle of Venus is "not of the same essence as that which carries it [the deferent or large sphere], although it is more nearly of one nature with it than with the rest."[675] Differences of colour are indicated in the glowing masses of the planets, which are entered by Dante and Beatrice. As with the moon, they do not merely alight upon the surface, but penetrate the planetary ether. "Quel ch' era *dentro* al sol,"[676] "Nel profondo Marte,"[677] "La temprata stella sesta, che *dentro a sè* m'avea ricolto."[678] Mercury is called a pearl—"la presente margarita"[679]—while Mars is "questo fuoco,"[680] and the astronomer poet says he knew the planet by its burning smile, for it seemed to him even more ruddy than its wont.

"Ben m' accors' io ch' io era più levato, Per l'affocato riso della stella, Che mi parea più roggio che l'usato."[681]

And when, without being conscious of any movement, he found that the red light shining on Beatrice had turned to purest white, like a blush fading from a fair face, he knew that the sixth star (Jupiter) had received him within itself.

"E quale è il trasmutare in picciol varco Di tempo, in bianca donna, quando il volto Suo si discarchi di vergogna il carco, Tal fu negli occhi miei, quando fui volto, Per lo candor della temprata stella Sesta, che dentro a sè m'avea ricolto."[682]

This contrast between the colours of Jupiter and Mars is referred to more than once. In *Par.* xxvii., the flaming spirit of St. Peter, flushed with righteous indignation, becomes brilliant as the planet Jupiter, and red as Mars:—

"E tal nella sembianza sua divenne, Qual diverrebbe Giove, s'egli e Marte Fossero augelli, e cambiassersi penne."[683]

And when Dante looks down upon the seven planets he sees

"il temperar di Giove Tra il padre e il figlio."[684]

To understand which we must turn to the oft-quoted fourteenth chapter of the second treatise in the *Convivio*. Here we find that Jupiter "moves between two heavens" which "are antagonistic to its excellent temperateness, that is to say the heavens of Mars and that of Saturn,"[685] and Dante quotes Ptolemy as saying that Jupiter is a star of temperate constitution, between the cold of Saturn and the heat of Mars.

It seems that these ideas of cold and heat, as applied to the planets, were taken literally, not merely as poetical descriptions of colour, nor of astrological significance; for Dante says explicitly in the same chapter of the *Convivio* that Mars is hot like fire, that this is the cause of his colour,[686] and that he dries up and burns things.[687]

Similarly he appears to have believed that Saturn and the moon were literally of a cold nature. "Quel pianeta che conforta il gelo,"[688] in *Canzone* xv. is doubtless Saturn; and the chill of the hour before dawn is described as due not only to loss of heat by the earth but sometimes also to the influence of the cold moon or of Saturn.[689]

Besides her shadowy markings, the moon had another peculiarity which distinguished her from all other heavenly bodies, for she was the only planet then known to vary in apparent size and shape "according to the way in which the sun looks upon her."[690] She was therefore a very dark body, though it was believed that she had some light of her own, and Dante argues in favour of this that she does not become totally invisible during an eclipse.[691] This was a very natural conclusion: it would not readily occur to anyone that the reddish light of the eclipsed moon is sunlight refracted (*i.e.* bent out of its direct course) in passing through the earth's atmosphere.

The sun, although the astrologers made him not very much larger than the brightest fixed stars and the largest planets, was much brighter and hotter than any, and it was universally agreed that all obtained their light from him. Even if, like the Moon, they had some light of their own, and did not only reflect like mirrors, this inherent light had originally come from the sun and was but absorbed sunlight.[692] We meet constantly with this idea in Dante's works. In one of his Odes the sun is said to "donar luce alle stelle;"[693] in the *Convivio* he says, "Il sole ... di sensibile luce sè prima, e poi tutti i corpi celestiali ed elementali allumina."[694] The Morning Star is described as deriving her beauty from the sun;[695] and after the sun has set he shines forth again in the light of all the stars.[696]

So also says Brunetto Latini, "Sans faille li solaus est fondemenz de toutès lumieres et de toute chalor."

4. INFLUENCE OF THE SPHERES ON HUMAN AFFAIRS.

If we are to believe a fifteenth century biographer, we have so far omitted to mention Dante's greatest claim to be considered a proficient in astronomy. "He foretold," says Filelfo, "many of the calamities of Florence, the wars of Italy, and other political changes, all of which would have been impossible without an accurate knowledge of the movements of the celestial bodies."

Filelfo was a poet, and his authority for this surprising statement can be no other than that on which, Dr. Moore says, he mainly depends, and which seldom or never fails him, "his own most airy and lively imagination unembarrassed by any references to documents."[697] The prophecies of the *Divina Commedia* are obviously written after the events, and in no single instance are they supposed to have any connection with the art of the astrologers. On the contrary, it is remarkable how very little we find in Dante's writings about the fate-predicting and character-reading which passed as the chief part of astronomy with the large majority of his contemporaries, and played a considerable part even in the scientific writings of his fellow authors.

Not that he disbelieved in the art: it would have been almost a miracle if he had. The astrologers in the fourth Bolgia were punished with other soothsayers not because they deceived credulous clients, but because they had desired to look too deeply into the future,[698] which was impiety. For this reason the learned Michael Scot, the honoured of Popes, was there, as well as poor Asdente, and Guido Bonatti, and Aruns who from his solitary cave had watched the stars, and prophesied the victories of Cæsar over Pompey.[699] This orthodox sentiment that "judicial astronomy," was apt to be a snare, and lead to practices condemned by the Church, may have been one reason for Dante's avoidance of it; but we cannot read his works without feeling that it was the beauty of the skies, and the keen desire to understand the structure of the universe which inspired him to study astronomy, and that he had no superstitious fear of the heavenly bodies, no greedy wish to make use of his knowledge to pry into the future. There is no evidence that he was acquainted with the details of astrology. The few allusions made in his works to special effects of certain stars or planets are only to those most familiarly known, and parallel instances might be found to many of them in modern poems. With this difference, however, that to-day they are figures of speech: with Dante they represented facts. Venus was to him in serious truth "la stella d'amor,"[700] "lo bel pianeta che ad amar conforta,"[701] and exercised a real power over Folco[702] and Cunizza,[703] and his own heart.[704] He even tells us, in serious prose, that it is the planet and not the whole sphere, which is the source of this influence, and that it is borne to us upon her rays.[705] So also the moon has a special influence, which is communicated by her rays.[706]

To the strong star of Mars the noble warlike nature of Can Grande was partly due;[707] Jupiter is a temperate star, not only in the purity of his colour, but his influence is felt where justice is worked on earth.[708] Saturn's influence is modified when he shines in Leo,[709] the sun has special powers in Aries,[710] and to his presence in Gemini at the time of Dante's birth the poet attributes his literary gift.[711] In *Inf.* xxvi. 23, he again suggests that his mental powers may be due to the influence of a good star, and Brunetto Latini bids him follow his star.[712]

Dante also quotes Albumassar, and Seneca, and the events of his own day, to show that those fiery vapours, meteorites, portend the death of kings and the downfall of states, adding that this is because they are effects of the sovereignty of Mars.[713] This passage explains why the warrior Malaspina is called a vapour drawn by Mars from Valdimagra, which in the tempest of battle at Piceno shall suddenly split the clouds and deal death to the combatants.[714]

The results of the right disposition of sphere with sphere are spoken of in *Canz.* xix. 77-79; all the nine were in perfect accord when Beatrice, herself a nine or miracle of perfection, was to be born;[715] and every star rained its light and its virtue into her beautiful eyes.[716] There is also a charming sonnet, in which the poet tells us how all the seven planets influence his lady.

"Da quella luce che il suo corso gira Sempre al volere dell' empiree sarte, E stando regge tra Saturno e Marte, Secondo che l'astrologo ne spira; Quella che in me col suo piacere aspira, D'essa ritragge signorevol arte: E quei che dal ciel quarto non si parte Le dà l'effetto della mia desira. Ancor quel bel pianeta di Mercuro Di sua virtute sua loquela tinge, E 'l primo ciel di sè già non l'è duro. Colei, che 'l terzo ciel di sè costringe Il cor le fa d'ogni eloquenza puro; Così di tutti e sette si dipinge."[717]

In more serious vein is the discourse in *Par.* ii. on the different virtues of the different stars;[718] and this should be compared with passages in the *Quæstio* and the *Convivio*. "Sciendum quod licet cœlum stellatum habeat unitatem in substantia, habet tamen multiplicitatem in virtute,"[719] sounds like an echo of the words of Beatrice:

"Così l'intelligenza sua bontate Multiplicata per le stelle spiega, Girando sè sopra sua unitate."[720]

And when comparing Nobility with the star-spangled sky, our author says:

"Tante sono le sue stelle che nel cielo si stendono, che certo non è da maravigliare, se molti e diversi frutti fanno nella umana Nobiltà, *tante sono le nature e le potenze di quelle, in una sotto una semplice sustanza comprese e adunate,* nelle quali siccome in diversi rami fruttifica diversamente."[721]

The author goes on to say in the *Quæstio* that the many and diverse stars which we see in the eighth heaven were placed there in order that it might pour down diverse virtues, and that he who perceives this not, let him know that he is outside the pale of philosophy.[722] In another place he tells us that the nearer a star is to the mid-circle of the star sphere, the greater is its virtue:[723] he no doubt means, not the celestial equator, but the zodiac (or more exactly the ecliptic), since this was the mid-circle of the star sphere, parallel to which its slow revolution of 36,000 years was made. This would agree with the view of the astrologers, who paid more attention to the constellations of the zodiac than to any others.

But we may search all Dante's works in vain for any reference to the "houses" or "aspects" of the planets, or for any information about the masculine and feminine, mobile and stable, signs. All such things, which seemed to most writers of his day quite as important as the periods of the planets or the positions of the signs, he entirely ignores. Among the circumlocutions by which he describes the signs, such as "the fair nest of Leda," or "the sign that follows Taurus," we never find "the house of Saturn," or "the sign in which Mars is mildest." Leo is called Mars' own Lion, probably because the "leo audax et ferox" of bestiaries agreed with the fiery colour and nature of Mars:[724] astrologically, his signs were Scorpio and Aries. It is quite astonishing that among all the characteristics of the planets and spheres mentioned in *Convivio* II. chapters xiv. and xv., only two relate to astrology, namely, the temperate disposition of Jupiter between the hot Mars and cold Saturn, and the power of Mars to draw up the hot vapours which become meteors and portend the deaths of kings. All the rest—the markings and phases of the moon, the small size and proximity to the sun of Mercury, the beauty of Venus and her appearance alternately as morning and evening star, the brilliance of the sun and his capability of giving light to all the stars, the position of the sphere of Mars midway between those of the other planets, the purity of Jupiter's colour, the slow movement and distance of Saturn—all these are purely astronomical, to use the modern distinction. So also are the characters of the star sphere—the thousand and twenty-two visible stars, and the indistinguishable stars of the

Galaxy, the pole we see and the other which remains hidden, the conspicuously swift diurnal movement and the almost imperceptible secular revolution of 36,000 years. Of the Primum Mobile it is remarked that it controls the movement of all the rest, and of the Empyrean that it is full of peace.

The symbolism of the spheres assigned in Paradise to the different saints is partly drawn from astrological ideas, for in Venus Dante meets the lovers, in Mars the warriors, in Jupiter the just rulers. But those who were not eternally faithful to their vows are in the moon because she has the lowest sphere,[725] the contemplative saints are in the highest planet, the theologians who enlightened the faithful are in the sun, and the ambitious are in the little planet which ever keeps closest to the sun. The heaven of "many and diverse stars" shows the poet a host of saints, and the Primum Mobile, which controls the movement of the other spheres, reveals to him the Angels who direct them.

It seems, then, that Dante was not interested in technical astrology, and had never made a study of it; and I think we may adduce this additional evidence that he had not learned astronomy in a university, or under Cecco d'Ascoli. It is only necessary to look at the list of books prescribed at Bologna[726] to feel sure that Dante had not taken a regular course there. The list, it is true, belongs to a somewhat later time, but there is no reason to suppose that astrology became of more importance in the late fourteenth century than at its beginning.

Nevertheless, however little Dante may have cared for the details of astrology, we shall never understand all that the skies meant to him unless we realize that a general belief in their influence on man was deeply rooted in his mind. For him it was a tremendous fact, and one which pervades his writings, that stars and spheres are the instruments of God's providence, and are ordained by the First Mover to mould the destinies of Earth.[727] It is their movements which manifest His Will;[728] they are the hammers,[729] earth the metal, they are the seals,[730] and earth the wax.

Three times the stars are invoked to bring justice and righteousness on earth.[731] Through the spheres the elements were evolved out of the first chaotic matter, and everything on earth took form; all life and motion is generated by them.[732] Man's soul is a direct creation by God, but the moving heavens play an important part in the formation of his material body, his mental faculties, and his disposition.[733] Were it not for the action of the spheres, children would be precisely like their parents, their whole nature being governed by the law of heredity alone: it is the different influences of the skies that give them different natures.[734] This, says Beatrice, is perhaps what Plato meant when he said that souls came from different stars and returned thither: if his meaning was that the influences of the stars is so great that to them is due the praise or blame which we distribute to men, there is some truth in the saying.[735]

Some truth only—"alcun vero"; for the stars are not independent powers, for good or evil, which men are unable to resist. This view is expressly combated in *Purg.* xvi. 58-84. Dante has asked the spirit of the courtly Venetian Marco Lombardo why the world is so full of wickedness now, for some say it is the effect of the heavens, and others seek an earthly cause. Marco groans at the blindness of men, who accuse heaven for the results of their own misdeeds: it is true that the heavens influence our lower natures, our instincts, and desires; but the soul is free, and can control them. This is in accordance with the teaching of St. Augustine and Aquinas. In one passage of the *De Monarchia*, Dante goes even further than this, and asserts that the influences of the spheres are only good, and that evil results from imperfections in the material on which these perfect instruments work.[736]

Conceptions such as these give us a new idea of mediæval astrology. They may be mistaken, but at least they are grand, and not unworthy of a philosopher and a Christian.

5. THE MOTIVE POWER.

Dante's ideas regarding the laws and origin of motion in the universe were derived from Aristotelian metaphysics and Church doctrines. He believed that there were three kinds of motion proper to three kinds of bodies: *i.e.* heavy bodies tended towards the centre of the universe, which is also the centre of the earth; light bodies tended upwards to the region of fire just below the moon; and bodies of ethereal substance—the stars, planets, and spheres—moved eternally in circles round the centre. Thus Beatrice, when speaking of the instinct implanted in all created things, which bears them through the great ocean of being each to his destined haven, says:—

"Questi ne porta il foco inver la luna, * * * * * Questi la terra in sè stringe ed aduna."[737]

And Virgil:—

"Il foco movesi in altura, Per la sua forma, ch' è nata a salire Là dove più in sua materia dura."[738]

That is to say the form of a flame fits it to rise to that place where, being in its own element, it may last longer than on Earth. Conversely, the centre of the earth is described as the bottom of the whole universe, and the point towards which all weights are attracted.[739] These opposite forces Dante calls gravity and levity.[740]

The most interesting illustration of this force of gravity is the description, in the last canto of the Inferno, of passing the centre of the earth, "il punto al qual si traggon d'ogni parte i pesi."[741] Virgil turned here "con fatica e con angoscia,"[742] and bid Dante hold him fast, "ansando com' uom lasso."[743] It was natural to suppose that the force of gravity would increase as one neared the centre, though the fact is that it decreases. In deep mines it is less than at the surface of earth; and if we could go to much greater depths we should find that we could stand in any position, and walk in any direction with almost equal ease, till at the exact centre the sense of weight and of "up" and "down" would vanish altogether. This is because it is the whole mass of the earth which attracts, and when we penetrate beneath the surface the part above is no longer dragging us down, and its effect is annulled. But at the surface gravity acts as if the force were all concentrated at the centre, and falling bodies increase their speed as they approach the surface: before the days of Newton, therefore, it was thought that the whole of the attracting force was actually situated at the centre of the earth. Brunetto Latini speaks of gravity increasing as this point is approached, and Dante, like Aristotle, calls it the goal of gravity.[744]

Between the physical and the spiritual universe as conceived by Dante, there is a very close correspondence. Souls burdened with the weight of sin fall towards the centre of the universe, sinking deeper the heavier is their guilt, and at the bottom of the whole universe is the personification of all evil. Repentant souls, as they are freed from their sins, rise higher and higher, and the ascent becomes always easier, until at last they soar heavenward as naturally and spontaneously as flames of fire. In Paradise the motion of the pure spirits is inconceivably rapid, and their mystic dances expressive of joy are always in circles.

Motion in circles was the only appropriate kind for the heavenly bodies,[745] but the actual force which originated and preserved their motions was intelligence and will. When Dante sang, "Voi che intendendo il terzo ciel movete,"[746] he was not addressing beings created merely by his poet's imagination. His own commentary on the words shows us that they were as real to him as the planets themselves. He can know nothing of them through the senses, yet is as certain of their existence as one with closed eyes is of light.[747] These beings move the spheres not through any material contact, but solely by the power of thought: they are immaterial immortal intelligences, commonly called Angels, and Dante identifies them with the Forms of Aristotle, the Ideas of Plato, and the gods and goddesses of the Pagans, "avvegnachè non così filosoficamente intendessero quelle come Plato."[748] As the hammer beats out forms in metal, and thus expresses what is in the mind of the smith, and as he works from the artist's design, so the movements and "virtues" of the sacred spheres mould earthly events, but these movements and virtues are the expression of controlling angels, and they are carrying out the Will of the First Mover. And as the joy of a happy mortal shines out in the eyes, so the joy of the angels shines out in stars and planets.[749]

There were nine Orders in the angelic hierarchies, according to the teaching of the Church, the chief authority on this subject being a book *De Celestia Hierarchia*, supposed to have been written by Dionysius the Areopagite, a disciple of St. Paul, who had learned from the apostle what he had seen when rapt to the third heaven. Aquinas assigns to one of these nine orders the guidance of stars and planets, and to others the rule over earthly affairs. But since the latter follow the celestial motions, there was no need to separate the two functions, and Dante laid down in the *Convivio* that a few Angels from each Order preside over the movements of each sphere. With Aristotle, he agreed that there is one Intelligence for each movement, so that for Venus (for instance) there are three movers to direct respectively the circling of the Epicycle, the Deferent, and the movement of Precession which every planet shares with the stars; but whether there is a fourth mover to guide the diurnal motion of Venus or if all the inner spheres are swept along by the motion of the Primum Mobile, he would not presume to decide.[750]

When he wrote the *Paradiso*, he no longer thought that a few Angels were set apart for this ministry: he implies that all the angels in each order united in the act of moving a whole heaven and perhaps he means to include the beatified human spirits as well.[751] He also slightly changes the correspondence between the angelic orders and their respective spheres.[752]

Swiftness of motion in a sphere corresponds with fervour of adoration in the Movers: hence the highest and swiftest of the moving heavens, which is the Primum Mobile, is moved by the highest Order, the Seraphim: and the lowest and slowest, which is that of the moon, by the lowest order, the Angels.[753]

All these movements are a proof of the existence of a First Mover, Himself unmoved, as Aquinas had said, following Aristotle, and the first words of Dante's confession of faith in Paradise are an echo of this:—

"Io credo in uno Iddio, Solo ed eterno, che tutto il ciel muove, Non moto, con amore e con disio."[754]

Here, as often elsewhere, Dante identifies the heavens with their Movers, and speaks as if the spheres themselves were moved by love and longing. In the tenth heaven, or Empyrean, this longing was completely fulfilled, hence this heaven was for ever at rest.[755]

VIII.
MEDIÆVAL AND MODERN VIEWS
OF THE UNIVERSE.

Let us now gather into one picture the details of mediæval astronomy which Dante has taught us, and try to see the universe as it presented itself to men's minds in his days.

Motionless, at the centre of all things, is the earth, a massive globe 20,400 miles in circumference.[756] Only the northern part of one hemisphere is inhabited; the rest is covered by ocean. It is surrounded by air, into which rise the exhalations from land and sea which cause winds, rain, etc. Above the height to which these rise, and above the tops of the highest mountains, the air sweeps uninterruptedly round the earth from east to west, sharing the movement of the skies. Above this again is the region of pure fire: all fire on Earth's surface tends to rise into it, and hot dry vapours ascending here burst into flame, and are seen from Earth as shooting stars and comets.

Bounding the sphere of fire is the lowest of the celestial spheres, that of the moon; and here we pass abruptly from the world of the four elements, transitory, changing, to the celestial world, eternal, changeless. Sphere rises above sphere, carrying moon and sun and planets, until we reach the vast orb in which are set all the stars.

All these bodies are composed of a mysterious imponderable substance called æther; all are perfectly spherical, and all their movements are circular. The diurnal rotation which we see is caused by an outermost sphere which contains no star: it communicates its own motion to all the spheres within it. The zodiacal motions of sun, moon, and planets, and the precession of the equinoxes, are caused by the proper motion of each sphere, rotating from west to east. The oscillating movements of the five planets and the varying speed of the moon are caused by small rotating spheres called epicycles, on which planets and moon are fixed, the epicycles in their turn being attached to the large spheres.

The distance from the earth to the nearest heavenly body, which is the moon, is 64⅛ times the radius of the earth; the distance of the fixed stars, more than 20,000 times the radius of Earth.

Mercury, the smallest of the heavenly bodies, is only 232 miles in diameter,[757] the moon and Venus are both rather more than a quarter the size of the earth, Mars is a little larger than Earth; next in order of size come the stars, from the faintest visible increasing up to those of the second magnitude; Saturn and Jupiter are much larger than Earth, and the fifteen brightest (first-magnitude) stars larger still; the sun is the largest body in the whole universe, having a diameter 5½ times Earth's diameter. It is immensely bright and hot, is the source of light to the whole universe, and on Earth is the chief giver of light, heat, and life, and the chief measure of time.

The moon and stars are also useful in this way to Earth, and upon her are poured all the influences of the celestial spheres, the sun, moon, and planets: these are the instruments by which every event on Earth is brought to pass.

It is their nature to move in circles, as it is Earth's nature to remain motionless at the centre of the universe; but all their motions are controlled by the wills of immortal angels, which are subject to the Will of the First Mover.

The sensible universe is eternal, but it is finite and measurable, and of definite form,[758] being bounded by a sphere. Beyond this in every direction is infinity.

Such was the mediæval conception of the Universe, as we gather from learned mediæval books, and as we find it vividly pictured in Dante's glowing verse. How do we conceive it now?

The heavens are the same to our eyes as to Dante's. The blue dome of the sky still arches over our heads, and in it we may see the stars shine out after the sun has set; we may see Venus making all the orient smile at early dawn, and the light of the moon not yet risen paling the stars of Scorpio. We may also see with our bodily eyes what was only visible to the keen eyes of Dante's mind, the rising sun travelling to our left, and the southern stars circling round the southern pole, while the stars of the north sink below the ocean floor.[759]

But to one who knows anything of modern astronomy the ideas suggested are widely different. Most striking is the absence of that sharp contrast between Earth and heaven which our forefathers felt. Earth is not to us small, dark, and inert, made of the basest material and sunk, like the dregs of the Universe, to its lowest depths. Earth is a heavenly body, as the Pythagoreans rightly guessed, a beautiful big planet, shining brightly by reflected sunlight and moving swiftly, like her sister-planets. To Venus she is the brightest star in the sky; to the moon she shows a far larger disc than the moon shows to us, and markings more beautiful and wonderful, brilliant caps of snow at either pole, oceans coloured blue-green, and continents varying in tint, whenever the dazzling white clouds part to reveal her surface.

Earth is the largest planet among her neighbours, Mercury, Venus, and Mars; just as Jupiter is largest among the giant planets—Saturn and the more recently discovered Uranus and Neptune. In the gap between these two groups modern astronomers have discovered hundreds of tiny planets, to which Earth is a monster, for many are only a few miles in diameter. The meteorites which shoot across our view, "startling quiet eyes," we still regard as a kind of conflagration in our upper atmosphere, but instead of vapours rising from Earth's surface we recognise them as visitants from inter-planetary space, caught by Earth's great mass as they pass too near, and flaming up with the heat caused by their sudden rush through the air. They exist in shoals of thousands and millions, some small as pebbles, some like great rocks, but all belonging to our system and pursuing definite paths like planets. The same must be said of comets, which have just as little to do with a hypothetical fire-sphere round Earth, but travel in regular periods, some long, some short; several small ones are seen every year, and occasionally a great splendour like Halley's draws near to us in its orbit, yet even these seem to consist of very small quantities of matter. Other members of our celestial family are the moons of Mars, Jupiter, and the rest, of which more are still being discovered. The Earth-Moon system is, however, unique in that our satellite is much nearer our own size than the proportionately tiny companions of other planets and the pair must look like a beautiful "double" moving through the stars and continually revolving round one another.

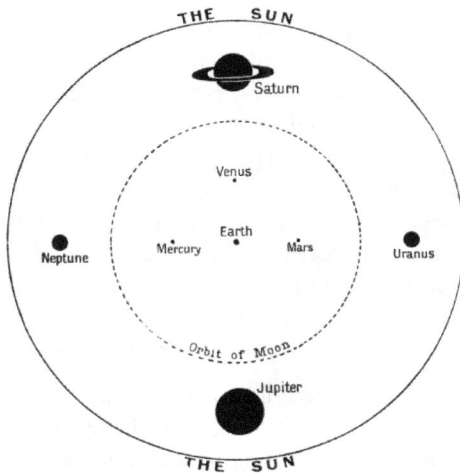

Fig. 53. Comparative sizes of the Sun and his satellites. *p. 490*

All these bodies are made, not of a mysterious ether, but of something just as wonderful, beautiful, and incomprehensible—the same stuff as

Earth. Most probably all developed in long-past ages out of a nebula, or mass of gas which for some unknown reason was intensely hot. Small bodies, like the Moon and Mars, cooled off the most quickly, and have lost or are losing their air and their water; larger bodies, like Jupiter, have still some heat, and are probably in a more or less fluid condition. On the sun an inconceivable heat still rages; everything is in a gaseous state, and this globe which men used to think eternally changeless is the scene of the fiercest turmoil; currents are always ascending and descending, clouds of fire are flung up thousands of miles high, and patches which look black compared with the surrounding surface and indicate a mysterious commotion are constantly forming, developing, and disappearing. The sun is enormously larger than Dante thought, larger than all the planets put together, large enough for our moon's whole orbit to fit comfortably inside. It is hot enough and brilliant enough to light all the planets: this, which was only a guess of mediæval astronomers, we know as a fact to-day, and the planets which are suitably placed show it to us, when seen in the telescope; Venus, for instance, changing her phase "according as the sun looks upon her," exactly like the moon. And Earth must show similar phases as seen from Mars.

For Aristarchus was right: the sun is the centre of the planetary movements, including those of the planet Earth.

We saw how Ptolemy's system hinted at this, in the significant connection between each planet's movement and the movement of the sun; but a strong argument against it was that if Earth were really moving round the sun we ought to see a yearly displacement in the positions of all the stars. Modern instruments have at last made it possible to perceive this displacement. About a hundred stars, our nearest neighbours, describe a very minute orbit in the sky, which in every case is an exact reflection of the sun's apparent orbit round the earth. Unless, therefore, we are going to provide all these stars, as well as the planets, with epicycles, which must all revolve in the same period, in the same direction, and in parallel planes, we must conclude that the movement is really Earth's yearly movement round the sun. These miniature star-orbits also show us the stupendous distances which lie between Earth and the nearest stars, while by far the greater number are too remote to show any displacement at all.

Aristarchus and those other ancient astronomers were also right who thought that Earth was turning on her axis, and thus producing by her own motion the alternation of day and night, and the apparent circling of the stars. When we look now at stars hastening across the sky, they no longer present themselves to our thought as fixed on a sphere with a radius of 20,000 Earth-radii, which is revolving round us at a tremendous speed; we feel that we are looking into space at glowing globes which are at various and almost inconceivable distances from us; also that all those we see, all those that Hipparchus numbered, are but an insignificant fraction of all those now known to exist. All have their own real motions, as well as the apparent motions produced by Earth, but these are in different directions, towards what goal we cannot at present tell. Compared with them, Earth is indeed insignificant, in size and also in brightness, for while she shines with reflected light they glow as the sun does and are his peers.

Nevertheless, with them too we have a feeling of kinship, for analysis of their light shows that they too are made of earth-stuffs: they are not eternal nor changeless, for very many vary constantly in their light, and they are at different stages of development, some appearing very young, while others are beginning to burn dim and low, and some can only be inferred to exist, for they seem to be wholly dark.

More wonderful perhaps than anything else, we find that the force which binds the whole of our sun's family together is nothing strange and unknown, but the same familiar force of gravity, which the Greeks knew so well but thought only applicable to Earth. Its rule includes the seemingly unruly comets and meteors, and it extends as far as our search can reach, to the uttermost star.

This sense of unity throughout the world, which is always growing stronger, is comforting. And we allow ourselves to wonder whether intelligent life has not developed elsewhere, as well as on this globe that we inhabit. Perhaps each planet that attends our sun is fitted at some period to be the abode of life: surely among the millions of stars many must have attendant planets, and on some of them even now are living, thinking, beings.

Dante's world was easier to think about than ours. Encompassed by an infinity which gave the imagination free scope, the material universe within was neatly rounded off, as it were, complete, finished. Only in the central spot was any development taking place, and thither were directed all effects of the circling spheres. Our universe is vague, vast, mysterious, without known limits or centre, offering problem after problem to the thinker. Dante's marvellous moon-substance is simple compared with the baffling nature of our æther; Dante's Angel-Movers are intelligible compared with the amazing mystery of universal gravitation.

Yet when we lift our eyes from earth to heaven, we share the feeling of our mediæval forefathers, of the ancient Greeks, of the earliest men whenever and wherever they became men. The unerring courses of the stars speak to us, "the unperturbed to the perturbed," of perfect harmony, untouched by chance or arbitary conflicting wills. Now, however, we do not believe this is because heaven is essentially different from earth: it is because we see there only the grand outlines, only the working of great laws, of a system which includes our own earth. Could we but lose sight of the details here, and view the history of man as we view the stars, the harmony would be as grand.

HYMN TO ZEUS.

Most glorious of the Immortals, many-named, Almighty for ever, Zeus, Ruler of Nature, that governest all things with law, Hail! for lawful it is that all mortals should address thee. For we are thy offspring, taking the image only of thy voice, as many mortal things as live and move upon the earth. Therefore will I hymn thee, and sing thy might for ever. For thee doth all this universe that circles round the earth obey, moving whithersoever thou leadest, and is gladly swayed by thee. Such a minister hast thou in thine invincible hands—the two-edged, blazing, imperishable thunderbolt. For under its stroke all nature shuddereth, and by it thou guidest aright the universal Reason, that roams through all things, mingling itself with the greater and lesser lights, till it has grown so great, and become supreme king of all. Nor is aught done on the earth without thee, O God, nor in the divine sphere of the heavens, nor in the sea, Save the works that evil men do in their folly. Yea, but thou knowest even to find a place for superfluous things, and to order that which is disorderly, and things not dear to men are dear to thee. Thou dost harmonize into one all good and evil things, that there should be one everlasting Reason of them all. And this the evil among mortal men avoid and heed not wretched, ever desiring to possess the good; yet they nor see nor hear the Universal Law of God, which, obeying with all their heart, their life would be well. But they rush graceless each to his own aim, Some cherishing lust for fame, the nurse of evil strife, Some bent on monstrous gain, Some turned to folly and the sweet works of the flesh, Hastening indeed to bring the very contrary of these things to pass. But thou, O Zeus, the all-giver, dweller in the darkness of cloud, lord of thunder, save thou men from their unhappy folly, Which do thou, O Father, scatter from their souls, and give them to discover the wisdom, in whose assurance thou governest all things with justice; So that, being honoured, they may pay thee honour, Hymning thy works continually, as it beseems a mortal man, Since there can be no greater glory for men or gods than this, Duly to praise for ever the Universal Law.

CLEANTHES (3rd century B.C.).
Translation by T. W. Rolleston.

[1] "The fair things that Heaven holds." Inferno XXXIV. 137, 138.

[2] Greek *gnomon*, an interpreter. A pole set up in order to show the length of shadow thrown by the sun.

[3] Greek *planetes*, a wanderer. This name was originally given to Mercury, Venus, Mars, Jupiter, Saturn, and also to sun and moon, for it indicated all the known heavenly bodies which changed their places among the stars. In modern usage it is not applied to the sun, but only to his satellites, of which many more are now known.

[4] Fractions omitted.

[5] Conv. IV., xxiii. 56, 57.

[6] Dr. C. Hose, in "Travel and Exploration," for Feb. 1910, quoted in "Nature," Feb. 17, 1910.

[7] Journal of the British Astronomical Association, June 24, 1909, report of a lecture on Chinese astronomy by E. B. Knobel, F.R.A.S.

[8] Acts, xvii. 28.

[9] *The Phainomena of Aratos, done into English verse by Robert Brown*, lines 1-13.

[10] *Ibid.*, 373-382.

[11] Ptolemy says he made a few changes, as his predecessors had done. (Delambre, *Histoire de l'Astronomie Ancienne*, ii. 261).

[12] *Origine de tous les cultes, ou Religion universelle*, by C. F. Dupuis.

[13] *Journal of the British Astronomical Association*, "The Oldest Astronomy," July 1898, June 1899, April 1904 May 1909; *The Observatory*, December 1898; *Knowledge*, October 1904; and elsewhere.

[14] *The Phainomena of Aratos*, by Robert Brown. See also his *Eridanus, River and Constellation; Primitive Constellations*, and other works.

[15] We are not sure what limits Aratus intended to set in the south to Centaur and Argo, and notably to the River Eridanus, which used to flow beneath the Sea-Monster (Cetus), joining the Water poured out by Aquarius. It changed its bed (like the Euphrates, of which it is perhaps the heavenly counterpart), and now has left the Sea-Monster high and dry, while on its ancient banks a chemist's furnace and a sculptor's workshop have been set up.

The Celestial Equator of Aratus fails to agree with the Equator of B.C. 2084, not only in passing over the head of Orion, instead of through his belt, as Brown himself points out, but also in running through the eye of the Bull, instead of his "crouching legs alone," so this part is altogether too far north. The Equator some 1200 years later agreed better here, and equally well elsewhere, except in the opposite part of the sky where it was then too northerly for Aratus, leaving Corvus to the south of the line. Much the same may be said about the tropical circles. Either Aratus was careless, or the globe from which he took his descriptions was incorrect: in any case, there results an uncertainty of many centuries and many degrees in date and latitude.

[16] Epping and Strassmeier, *Astronomisches aus Babylon*, Kugler's *Babylonische Mondrechnung*, and *Babylonische Sternkunde*. Schiaparelli's two monographs on Babylonian Astronomy, from which much of the information here given is derived, are chiefly based on these works.

[17] King, History of Sumer and Akkad, p. 246.

[18] Schiaparelli, *L' Astronomia nell' Antico Testamento*, chap. vii.; Wellhausen, *History of Israel*, chap. iii.

[19] Sayce and Bosanquet identify Dilgan with Capella, not with part of Aries, and consider that a date of about B.C. 2000 is indicated— (*Monthly Notices* xxxix, 454). But in any case the method of calendar formation is the same.

[20] Sayce and Bosanquet understand Capella here also.

[21] If Taurus was originally considered the first constellation of the zodiac, instead of Aries, of which there are some indications, the change may well be explained by this change of method. It does not necessarily imply that the equinox was in Taurus when our zodiac was invented. It was near ω Arietis in B.C. 1000.

[22] A Babylonian treatise on astronomy recently published by the trustees of the British Museum supports Kugler's view that truly scientific methods were not adopted before the sixth century B.C. This treatise formed the subject of a lecture given by Mr L. W. King before the Society of Biblical Archæology on Feb. 19, 1913.

[23] Gen. i. 6, 7; vii. 11.

[24] Ezek. xxxii. 18, 24.

[25] *De Cœlo* I. 3, and II. 1.

[26] But see note, pp.75, 76.

[27] His life was saved by his illustrious pupil, Pericles, of whom the story is told that on one occasion, just as his army was embarking for an expedition, the sun was eclipsed, and his pilot was terrified. Pericles snatched off his cloak, and held it so as to hide himself from the man's eyes. "Is that terrible? is that an evil omen?" he cried. "Then do not fear the disappearance of the sun, for it is just the same, only the thing that hides it is larger than my cloak."

[28] Some late followers said that Pythagoras, alone amongst men, could hear the music of the spheres.

[29] See Heath's *Aristarchus of Samos*, pp. 187-189 and 251, 252. I very much regret that as Mr. Heath's book was only published this year, I have been unable to make use of it while writing of early Greek astronomy. I can now only advise any readers who may be interested in my brief sketch of this period to read Mr. Heath's history, where they will find the opinions of modern writers summarized and discussed, and also the full text (in English) of the most ancient and reliable sources of information. It is a great encouragement to find that my statements are in agreement with his in nearly all essential points, but readers will mark the following important differences:—

1. Anaximander's heavens are said to have been spherical, not hemispherical, and this seems to be clearly proved by the evidence quoted from ancient writers.

2. Anaxagoras, not Thales, is said to have been the first to explain correctly the cause of solar eclipses and of the moon's phases, viz. that the moon is an opaque body, shining only by reflected sunlight, and periodically hiding the sun from us when she passes in front of it. Mr. Heath regards the authorship of Anaxagoras as conclusively proved: readers will be able to judge of this from his quotations. Personally they seem to me to prove no more than that Anaxagoras agreed with others on this point, and was the first to express it clearly in writing. It is difficult to see why Mr. Heath denies that Parmenides held the same views before Anaxagoras: Parmenides' own words seem to prove it, and his theory that the moon was composed of air and fire mingled is rather in favour of it than otherwise. He surely meant that the moon was not wholly bright, like the sun; yet that she had some light of her own must have seemed evident from the faint illumination we see during total lunar eclipses and on the part of her surface not lighted by the sun. (See Dante's views, p. 402 of this book.)

The connection between her phases and her distance from the sun in the sky is so extremely obvious that I can hardly think the Greeks drew no inference from it until the fifth century B.C., and I cannot see why we should refuse to credit Thales with the discovery attributed to him that her light came in some way from the sun. Gruppe acutely observes that the reason why Thales' pupil Anaximander did not accept the true explanation of lunar phases and solar eclipses may have been because he felt it necessary to have a theory which would apply equally well to eclipses of the moon; and as he believed in a flat earth he could not advocate the true explanation here. This was why he invented a new theory (viz. that both sun and moon were fire shining through holes in hollow rings, and that the occasional stopping up of these holes caused both lunar and solar eclipses, and also the lunar phases).

But Parmenides had learned the Pythagorean doctrine of Earth's spherical form, hence he was able to accept the older theory that the moon obtains her light from the sun, and sometimes eclipses the sun by her opaque spherical body, for he could have added that the moon is eclipsed in like manner by the opaque spherical body of the earth.

[30] I follow the translation of Jowett.

[31] Compare *Par.* vii. 64-66.

[32] See his famous description of the eight spheres, on each of which stands a siren, singing, while the whole system turns upon a diamond spindle, the end of which rests upon the knees of Necessity. This book was not known in the Middle Ages.

[33] ειλλομενη.

[34] The period in which a planet is seen to revolve round the zodiac, and return to the same star, varies greatly, because complicated by its retrograde movements; but if the average of a sufficient number of periods be taken, it coincides for Mercury and Venus with the sidereal year; for Mars, Jupiter, and Saturn, with the period in which each is actually revolving round the sun (its "sidereal period").

[35] Geminus.

[36] This varying velocity is due to the fact that all celestial orbits are not true circles, but ellipses, which was first discovered by Kepler (1609 A.D.).

[37] *De Cælo* II. 10.

[38] Thus, (he adds) time also is threefold, for we have Beginning, Middle, and End. Therefore we apply three to Divine things, and also in common speech we call two "both," and only say "all" when we

reach three, following Nature's law. The Pythagoreans say "The all and all things are bounded by the number three."—*De Cælo* I. 1.

[39] *Par.* iv. 1-3.

[40] See p. 75, note.

[41] *Conv.* II. iii. 59-65.

[42] It is also mentioned in a compilation of philosophers' opinions, probably made in the fifth century A.D. by Stobæus, who is very likely quoting Plutarch.

[43] See page 101.

[44] Indian astronomers also refused to accept the doctrine of Aryabhata. Varâha Mihira (sixth century A.D.) says:—"Others maintain that the earth revolves and not the sphere: if that were the case, falcons and other birds could not return from the ether to their nests."

[45] Spheres all centring in one point.

[46] Mahaffy, *The Progress of Hellenism in Alexander's Empire*, p. 119.

[47] See p. 26.

[48] *Almagest,* Bk. viii.

[49] Greek *peri* near, *apo* away from, *ge* Earth.

[50] *Syntaxis,* book VII.; Delambre, *Histoire de l'Astronomie Ancienne* Vol. II. page 247 (1817 edition).

[51] Venus had passed her "inferior conjunction with the sun" on Sept. 15.

[52] It amounts to 5 degrees of longitude in 300 years.

[53] This is evident from the way he treats them, picking up an epicycle or a deferent just as best suits the purpose in hand and explaining sometimes that either would answer equally well.

[54] Taken from the *Almagest* catalogue, as given in Delambre's History.

[55] Fifteen stars.

[56] Lockyer, *Dawn of Astronomy*, p. 196.

[57] The changing latitudes of the planets, for instance, which gave Ptolemy much trouble, are much more easily explained when it is

granted that they partly depend upon Earth's motion in an orbit whose plane is slightly inclined to the planes of their orbits.

[58] It therefore really approached more nearly the sidereal year, although the cycle was based on the tropical year.

[59] *De Mon.* II. vii., "Temple Classics" edition.

[60] *De Senectute.*

[61] The *average* year was the same, 365¼ days, in the old 8-year cycle of the Greeks, and also in the Calippic cycle, which did not come into practical use. The average year of the Metonic cycle was longer, and therefore departed further from the true tropical year.

[62] See p. 45.

[63] *Ezekiel*, v. 5.

[64] Arabic *gib* = Latin *sinus*, a fold; *i.e.* the chord folded in two.

[65] The Arabian mile was equal to 4000 "black cubits," and if this is the Egyptian and Babylonian cubit, the values are rather too large, being in round numbers 26,500 and 8,500 English miles, instead of 25,000 and 8000.

[66] The Catalogue of Hipparchus is said to have contained 1080 stars, but Ptolemy's has only 1022.

[67] Earth's diameter, and even Earth's distance from the sun, is too small a unit. Light, travelling 186,000 miles a second, takes 4¼ years to reach us from the nearest star.

[68] It is impossible to measure the diameter of any star, even with the help of the most powerful telescopes, but in the case of a double star at a known distance the movement of the components as they travel round their common centre of gravity enables us to determine the gravitational force they exercise on each other, and thus their combined mass; and their spectra give some idea of their density. For instance, the mass of the double star Alpha Centauri is nearly twice that of our sun; and as the components appear to be about equal to each other, and both show a spectrum resembling that of the sun, we may conclude that Alpha Centauri consists of two stars, each of which has about the same diameter as our sun. Arcturus has a diameter far greater, some say ten times, some not less than twenty-five times as great as the sun!

[69] The mean length of Earth's shadow (which varies a little with her distance from the sun,) is 857,000 miles, or 216 times her semi-diameter.

[70] It was also translated from Arabic into Latin at Toledo, in 1175.

[71] *Paradise Lost*, II. 418

[72] *Paradise Lost*, VIII. 34-38.

[73] *Paradise Lost*, VIII. 77-84.

[74] All quotations from Salimbene's Chronicle are taken from Coulton's *From Saint Francis to Dante*.

[75] *Croniche Fiorentine*, Bk. VII. par. 80.

[76] *Inf.* xx. 118.

[77] *V. E.*, I. xii. 20-35.

[78] Villani, *Croniche Fiorentine*, VI. 1 and 24.

[79]

"That other, round the loins So slender of his shape, was Michael Scot, Practised in every sleight of magic wile." *Carey*.

[80] Boccaccio's *Vita di Dante*; Translation from Toynbee's *Dante Alighieri* (Oxford Biographies), pp. 92, 93.

[81] *Inf.* xv. 23 *et seq.*

[82] *Inf.* xv. 119.

[83] "If thou follow thy star," *Inf.* xv. 55.

[84] "First refuge," *Par.* xvii. 70.

[85] *Par.* xvii. 71, 72.

[86] Imbriani, *Dante a Padova*.

[87] "*Torno a Ravenna e de lì non mi parto* (I am going back to Ravenna, and shall not leave it again), is a line in the *Acerba* which Cecco d' Ascoli puts into the mouth of Dante, as though from a letter written to himself from the divine poet at the time" (about the year 1319). *Dante and Giovanni del Virgilio*, by Wicksteed and Gardner, p. 84.

[88] *Inf.* xxxi. 136-141.

[89] *Conv.* II. xiii. 22-26.

[90] *Conv.* III. ix. 146-157.

[91] *Conv.* II. xv. 73-77.

[92] *Conv.* II. iii. 36-52, *V. N.* xxx.

[93] Averroës, in his commentary on Aristotle's *De Cœlo*, says that the ancients believed the eighth, or starry, heaven, to be the outermost, but that Ptolemy assumed a ninth, "because he said that he had discovered a slow motion along the signs of the zodiac in the fixed stars." Albertus Magnus, in his *De Cœlo et Mundo*, Book II., says also that the ancients, including Aristotle, believed that there were only eight heavens, but that Ptolemy, so far as he can understand, believed in ten, on philosophical not mathematical grounds (compare *Conv.* II. iii. 40, 41). Albertus accepted the theory of "trepidation," and thought this was the only movement which ought to be assigned to the star sphere; there remained, therefore, two motions, which affect all the planetary spheres and the star sphere, for which two more spheres must be assumed, a ninth sphere for precession, and a tenth, the

Empyrean. Dante never mentions trepidation, and evidently did not believe in it: he needed only nine moving spheres, therefore, but counts the Empyrean as a tenth heaven.

[94] *Conv.* II. xiv. 198-202. *Ibid.* 249-253.

[95] "Ptolemy says in the book above cited."

[96] *Par.* xiii. 1-13.

[97] See Schiaparelli's letter in Lubin's *Dante e gli Astronomi Italiani.* The name is, however, also used as a sub-title in the printed edition of Christmann, Frankfort, 1590, which was based not on the translation of Gerard but of Johannes Hispalensis of Seville; and Toynbee thinks that this Frankfort edition represents most nearly the version of Alfraganus used by Dante. It is the only one of the five printed editions which gives the same figure for the diameter of Mercury as

Alfraganus" in *Romania* xxiv. 95, and Moore, *Studies in Dante* iii. p. 3, *note.*

[98] "That glorious philosopher to whom Nature most fully revealed her secrets," *Conv.* III. v. 54-56; "almost divine," *Conv.* IV. vi. 133; "supreme and highest authority," *Ibid.* 52.

[99] Moore, *Studies in Dante* I. (Scripture and Classical Authors in Dante), from which much of the information in this chapter has been taken.

[100] *Conv.* II. iii. 19-21.

[101] *Conv.* III. v. 62-65.

[102] "My master."

[103] *Conv.* III. v. 32.

[104] *Inf.* ii. 76-78; *Par.* xxii. 134-138.

[105] *Conv.* ii. xiv. 174-176.

[106] Phaëthon, *Conv.* II. xv. 53-55, *Purg.* xxix. 118-120; Latona, *Purg.* xx. 130-132; the Horses of the Sun, *Conv.* IV. xxiii. 134-139, etc., etc.

[107] *Conv.* III. v. 115-117.

[108] "A man of supreme excellence." *Conv.* II. v. 21, 22.

[109] *Conv.* III. xiv. 76-79.

[110] *Par.* iv. 22-24, 49-60.

[111] *Conv.* III. xi. 39; *Inf.* iv. 137.

[112] *Conv.* III. xi. 22-33; II. xiv. 144-147; III. v. 29-44; III. xi. 41-47.

[113] *Conv.* II. xiv. 34, 35.

[114] *Conv.* II. xv. 56.

[115] *Qu.* xviii. 38, 39.

[116] *Conv.* III. ii. 37.

[117] *Conv.* II. xv. 77, II. xiv. 32.

[118] *Conv.* II. xiv. 170-174.

[119] *Inf.* iv. 80, 81, 90, 131-144.

[120] "The advocate of the Christian centuries." (*Par.* x. 199). Orosius is also mentioned by name in *Conv.* III. xi. 27; *V. E.* II. vi. 84; and *De Mon.* II. ix. 26.

[121]

"He who is nearest to me on the right My brother and master was, and he Albertus Is of Cologne, I Thomas of Aquinum." (*Par.* x. 97-99).

[122] Toynbee, *Dante Dictionary*; the source also of many other details given in this chapter.

[123] Albert in *Conv.* III. v. 113-115, vii. 26-28; IV. xxiii. 125-6; Aquinas in *Conv.* II. xv. 125-6; IV. viii. 3-6, xv. 125-130, xxx. 26-30; *De Mon.* II. iv. 5-8; and see *Purg.* xx. 69.

[124] *V. E.*, I. xiii. 1-11.

[125] *Inf.*, xxxii. 81, and x. 85, 86.

[126] *Purg.*, iii. 112-129.

[127] "Replied after this fashion."

[128] *Li Louvres dou Trésor*, Chabaille, Paris 1863.

[129] *Ibid.*

[130] "Here beginneth the book of the *Composition of the World together with its Causes*: written by Ristoro of Arezzo in that most noble city."

[131] "Here endeth this book, in the year of our Lord one thousand two hundred and eighty-two. Rudolph Emperor at this date. Martin IV. resident Pope." Amen.

[132] "Alfraganus said in the 8th chapter"; "Alfraganus bears witness in the 22nd chapter of his book."

[133] "The famous Ptolemy."

[134] "Who was a very great teacher of astrology."

[135] "An Arabian philosopher of Baghdad, 1058-1111."

[136] "The uncovered earth," *i.e.* not hidden under the ocean.

[137] Bk. VI. cap. xi.

[138] *Conv.* I. i. 125, 126. Compare *Conv.* IV. xxiv. 1-13.

[139] *V. N.* ii. 9-12, xxx. 13-24.

[140] *V. N.* ii. 1-12.

[141] *V. N.* xlii. 47.

[142] *V. N.* xxx. 1-6.

[143] *V. N.* xlii. 30.

[144] *Conv.* II. xiii. 22-26.

[145] *Son.* xxxvi. 2; *Canz.* xx. 89.

[146] *Canz.* xix. 117; *Canz.* ix. 16, 17.

[147] *Canz.* xx. 89; *Son.* xxviii. 11.

[148] *Son.* xxviii.; *Canz.* xv. 4, 7.

[149] *Canz.* xv. 3, 29, 41.

[150] *Son.* xxviii. 2.

[151] *Son.* xxviii. *Canz.* xix. 77, *Ball.* vi. 11, 12, *Canz.* xv. 41., *Son.* xxvi. 14.

[152] *Conv.* I. i. 111-113 and 125-127.

[153] *Conv.* I. i. 67-86.

[154] "As the Philosopher says at the beginning of the *First Philosophy*, 'All men naturally desire to have knowledge.' The reason of this may be that everything, being impelled by foresight belonging to its own nature, tends to seek its own perfection. Wherefore inasmuch as knowledge is the final perfection of our soul in which our final happiness consists, all men are naturally subject to the desire for it." *Conv.* I. i. 1-11.

[155]

"Oh ye whose intellectual ministry Moves the third heaven."—*Carey.*

[156] "The sun sees not, though circling all the world."

[157] The spherical form of Earth, and the action of gravity at the earth's surface, were commonplaces with the Greeks, as we have seen in Part I. of this book. Posidonius, Strabo, and other classical writers speak of the tides as following the revolution of the heavens, and having periods similar to those of the moon; Albertus Magnus and Aquinas ascribe them to the influence of the moon, and so does Dante himself in *Par.* xvi. 83.

[158] See Moore, *Studies in Dante*, II. "The Genuineness of the *Quæstio de Aqua et Terra*," for a complete discussion of the question.

[159] *V. N.* xliii. 3-7.

[160] *Conv.* II. xiv. 244-217. "It is noble and lofty because of its noble and lofty subject, which is the movement of the heavens; it is lofty and noble because of its certainty, which is without flaw."

[161] See p. 156.

[162] "The great wheels," "eternal wheels," "starry wheels."

[163] "Swift, almost as the heaven ye behold." *Par.* ii. 21.

[164] "Against the course of the sky." *Par.* vi. 2.

[165]

... "That sphere, Which aye in fashion of a child is playing." *Purg*. xv. 2, 3. (*Longfellow*).

[166] "Under a poor sky." *Purg*. xvi. 2.

[167]

"And as advances, bright exceedingly, The handmaid of the sun, the heaven is closed, Light after light, to the most beautiful." *Par*. xxx. 7-9. (*Longfellow*).

[168]

"As at evening hour Of twilight, new appearances through heaven Peer with faint glimmer, doubtfully descried." *Par*. xiv. 70-72. (*Carey*).

[169] In Kenneth Grahame's delightful book, full of sympathy with Nature, *The Wind in the Willows*. The moon rose when it was "past ten o'clock," and "sank earthwards reluctantly and left them" before dawn.

[170] H. G. Wells, *The Time Machine*.

[171] *Inf*. xv. 18, 19; *Purg*. xviii. 76-81; *Purg*. x. 14, 15, *Cf*. ix. 44.

[172] *Purg*. xxix. 53, 54.

[173] *Qu*. xx. 61-63.

[174] "Now she shines on one side, and now on the other, according to the way the sun looks upon her." *Conv*. II. xiv. 77-79.

[175]

"At what times both the children of Latona, Surmounted by the Ram and by the Scales, Together make a zone of the horizon, As long as from the time the zenith holds them In equipoise, till from that girdle both Changing their hemisphere disturb the balance, So long, her face depicted with a smile, Did Beatrice keep silence." *Par*. xxix. 1-8.

[176] "Many moons." *Canz*. xx. 89.

[177] "And the first heaven is not grudging to her."—*Sonetto* xxviii.

[178] *Purg*. xix. 1, 2.

[179] *Inf.* xx. 127-129.

[180] *Purg.* ix. 1-9.

[181] *Conv.* IV. xvi. 89-93.

[182] *Inf.* x. 80.

[183] *Inf.* xx. 126.

[184] *Purg.* xxix. 78; *Ep.* vi. 54.

[185] *Par.* xxiii. 26.

[186] *Par.* x. 67; xxii. 139; xxix. 1.

[187] *Purg.* xxiii. 120.

[188] *Purg.* xx. 132.

[189] *Par.* ii. 25-36.

[190] *De Mon.* I. xi. 35-37.

[191] *Par.* x. 67-69; and *Purg.* xxix. 78.

[192] *Purg.* xxix. 53, 54; and *Par.* xxiii. 26.

[193] "And this moon, because of her inferiority, is rightly called feminine."

[194] *Ecl.* ii. 1-4.

[195] *Purg.* iv. 62, 63; *Purg.* iv. 59; and xxix. 117, 118; *Par.* i. 38; *Canz.* xix. 114; *Par.* xxii. 116.

[196] "The perfection and beauty of his shape." *Canz.* xix. 76.

[197] *V. N.* xlii. 29; *Canz.* ix. 2; *Conv.* II. xiv. 126, 127; *Purg.* xvii. 52, 53; *Par.* i. 54; and x. 48, etc.

[198] *Inf.* i. 17, 18.

[199]

"O pleasant light, my confidence and hope! Conduct us thou," he cried, "on this new way." *Purg.* xiii. 16, 17. (*Carey*).

[200] *Par.* xxii. 55, 56; *Inf.* vii. 122; *De Mon.* II. i. 37-39; and *Par.* ii. 106-108; *Canz.* xi. 37; *Conv.* III. xii. 59, 60, etc. etc.

[201] *Canz.* ix. 5.

[202] *Inf.* i. 41-43.

[203] *Purg.* xix. 10, 11.

[204] *Par.* xxiii. 1-9.

[205] *Inf.* ii. 127-129.

[206] *De Mon.* II. i. 36-41.; *Canz.* ix.

[207] *Canz.* xix. 96-114.

[208] *Ep.* v. 10; and vii. 19, 20, 25.

[209] "A sun rose upon the world." *Par.* xi. 50.

[210] *Par.* xi. 52-54.

[211] "O sun that healest all imperfect vision," *Inf.* xi. 91.

[212] "The sun of my eyes." *Par.* xxx. 75 (See also *Par.* iii. 1-3).

[213] *Conv.* III. xii. 52-63.

[214] "The Sun of the angels." *Par.* x. 53.

[215] "That Sun which enlightens all our company." *Par.* xxv. 54.

[216] "The Sun which satisfies it." *Par.* ix. 8.

[217] "I have lost the sight of that high Sun whom thou desirest." *Purg.* vii. 25, 26. Compare *Par.* xxx. 126; xv. 76; xviii. 105.

[218] "The path of the sun." *Purg.* xii. 74.

[219] "Shining more brightly and with slower steps, the sun had gained the circle of midday." *Purg.* xxxiii. 103, 104.

[220] "Nine times already since my birth had the heaven of light returned to the selfsame point almost, as concerns its own revolution."
V. N. ii. 1-4. (*Rossetti*).

[221] "I have dwelt with Love since my ninth revolution of the sun." *Son.* xxxvi. 1, 2.

[222] See diagram on p. 276.

[223]

"O glorious stars ... With you was born, and hid himself with you, He who is father of all mortal life, When first I tasted of the Tuscan air." *Par.* xxii. 112-117. (*Longfellow*).

At this date the sun entered the *constellation* of Gemini on June 1 (Old Style), but was in the *sign* from May 11 to June 11, and it is always to the signs that Dante refers in the *Divine Comedy*. The anonymous fourteenth century commentator known as "l'Ottimo" interprets this passage as indicating the time "between the middle of May and the middle of June."

[224]

"Ere January be unwintered wholly By the centesimal on Earth neglected." *Par.* xxvii. 142-143. (*Longfellow*).

[225]

"In that part of the youthful year wherein The sun his locks beneath Aquarius tempers, And now the nights draw near to half the day." *Inf.* xxiv. 1-3. (*Longfellow*).

[226] "Night that opposite to him revolves." (*Longfellow*).

[227]

"The Scales, that from her hands are dropped When she reigns highest." *Purg.* ii. 5, 6. (*Carey*).

[228]

"And he: Now go, for the sun shall not lie Seven times upon the pillow which the Ram With all his four feet covers and bestrides. Before that such a courteous opinion ..." *Purg.* viii. 133-136. (*Longfellow*).

[229] "I have come to that part of the wheel." *Canz.* xv.

[230] Like all mediæval writers, Dante includes the sun and moon among the seven planets. The others do not cast perceptible shadows, except Venus and Jupiter at their brightest.

[231]

"I to that point in the great wheel have come, Wherein the horizon, when the sun doth set, Brings forth the twin-starred heaven to our sight; And Love's fair star away from us doth roam, Through the bright rays obliquely on it met In such wise that they veil its tender light; That planet which makes keen the cold of night Shows himself to us in the circle great, Where each star of the seven casts little shade." (*Plumptre*).

[232] "The Wheel which, when the sun sets, brings forth for us on the horizon the jewelled sky."

[233] *Sulla Data del Viaggio Dantesco* p. 90, note.

[234] Comparing *Conv.* II. ii. 12, xiii. 49-52, and IV. i. 60-62, we learn that in August 1293 (*vide infra*, p. 314), Dante first became acquainted with the Lady Philosophy; that in the early part of 1296 he was completely under her spell; and that some time afterwards she for a while estranged herself from him.

[235]

"Scattered and faded now is all the foliage Which had burst forth, beneath the power of Aries, To beautify the world, the grass is withered." *Canz.* xv. 40-42.

[236]

"This everlasting spring Nocturnal Aries never can despoil." *Par.* xxviii. 116-117.

[237]

"Thereafterward a light among them brightened So, that if Cancer one such crystal had Winter would have a month of one sole day." *Par.* xxv. 100-102. (*Longfellow*).

[238] *Ep.* ix. 46-49.

[239] *Inf.* iii. 23.

[240] *Inf.* xvi. 82, 83.

[241] "Resounded through the air without a star." *Inf.* iii. 23.

[242] "The fair things that heaven holds." *Inf.* xxxiv. 137, 138.

[243] *Purg.* viii. 85.

[244] *Purg.* xxvii. 89, 90.

[245] "Beautiful stars," *Inf.* xvi. 83.

[246] "Thence issuing we beheld again the stars."

[247] "Pure and disposed to mount unto the stars."

[248] "The Love that moves the sun and the other stars."

[249] *Purg.* ix. 4; *Purg.* viii. 89; *Purg.* i. 25; *Par.* xxiii. 26.

[250] *Par.* ii. 130, 142-144.

[251] *Par.* x. 76.

[252] *Par.* xxi. 28-33.

[253] *Inf.* ii. 55.

[254] *Par.* xxv. 70. See also *Conv.* II. xvi. 4-12, where the writings of Boëthius and Cicero, and all instructive books, are called stars full of light.

[255] *Par.* xxiv. 147.

[256]

"Even as remaineth splendid and serene The hemisphere of air, when Boreas Is blowing from that cheek where he is mildest, Because is purified and resolved the wrack That erst disturbed it, till the welkin laughs With all the beauties of its pageantry: Thus did I likewise, after that my Lady Had me provided with her clear response, And like a star in heaven the truth was seen." *Par.* xxviii. 79-87. (*Longfellow*).

[257] "The shining star." *Par.* xxiii. 92.

[258]

"O Trinal Light, that in a single star Sparkling upon their sight so satisfies them, Look down upon our tempest here below!" *Par.* xxxi. 28-30. (*Longfellow*).

[259] *Purg.* xxix. 91.

[260]

"Of those long hours wherein the stars above Wake and keep watch, the third was almost nought." *V. N.* iii. 81, 82. (*Rossetti*).

[261] *Inf.* vii. 98, 99.

[262]

"Like unto stars neighbouring the stedfast poles." *Par.* x. 78.

[263] *Purg.* viii, 86, 87.

[264] "These stars all revolve round the same point, and the nearer a star is to this point, the smaller is the circle that it makes, and the slower its motion appears." *El. Ast.* cap. ii.

[265] *Purg.* i. 22-27.

[266]

"My insatiate eyes Meanwhile to heaven had travelled, even there Where the bright stars are slowest, as a wheel Nearest the axle." *Purg.* viii. 85-87. (*Carey*).

[267]

"And he to me: The four resplendent stars Thou sawest this morning are down yonder low, And these have mounted up to where those were." *Purg.* viii. 91-93. (*Longfellow*).

[268] *Conv.* II. xv. 10-14, and 96-104.

[269] "The glorious Lady."

[270] *V. N.* ii. 9-15.

[271]

"I say that the starry heaven displays a multitude of stars to us, for as the Sages of Egypt have perceived, including the last star which appears to them in the south, they reckon one thousand and twenty-two starry bodies, of which I am now speaking."

Conv. II. xv. 18-22. (*Jackson*).

[272]

"You are to know that the Sages measured the places of all the fixed stars as accurately as possible with their instruments, as far south as they could see in the third climate.... The number of all the stars which he was able to measure is one thousand and twenty-two."

[273]

"We see in it (the starry heaven) a difference in the magnitude of the stars and in their light."

Qu. xxi. 19-21.

[274]

"Lights many the eighth sphere displays to you Which in their quality and quantity May noted be of aspects different." *Par.* ii. 65-66. (*Longfellow*).

[275] "The Ram." *Purg.* viii. 134; *Par.* xxix. 2.

[276] "The sign which follows Taurus," *Par.* xxii. 110, 111; "The eternal Twins," xxii. 152; "The fair nest of Leda," xxvii. 98.

[277] "The Balance." *Purg.* ii. 5.

[278] "The cold creature." *Purg.* ix. 5.

[279] "The Goat of the sky." *Par.* xxvii. 69.

[280] "The celestial Carp." *Purg.* xxxii. 54.

[281] *Purg.* iv. 61.

[282] "The burning Lion's breast." *Par.* xxi. 14.

[283] "Greater Fortune."

[284]

"Gems ... set in the shape of that cold animal Which with its tail doth smite amain the nations." (*Longfellow*).

[285] *Par.* xiii. 11, 12.

[286]

"A voice, That made me seem like needle to the star, In turning to its whereabout." *Par.* xii. 29, 30. (*Carey*).

[287] "The needle which guides mariners, for by the virtue of the heavens it is attracted and turned towards that star which is called the North Star." *Composizione del Mondo*, Bk. VII. part iv. ch. 2.

[288] *Inf.* xxvi. 127-129.

[289] *Purg.* i. 30.

[290] *Purg.* i. 26.

[291] *Purg.* viii. 89.

[292] "Four bright stars, four sacred lights."

[293]

"We are nymphs here, and in heaven we are stars." *Purg.* xxxi. 104-106.

[294] *Purg.* xxxi. 111.

[295] Antonelli thinks the four stars were α and β Crucis, α and β Centauri, all of which had been mentioned by Ptolemy, and all lie near the circle which marks the limit of circumpolar stars in the supposed latitude of Purgatory (32° south). The three stars he says were ζ Navis, Canopus, and Achernar:—Antonelli, *Accenni alle Dottrine Astronomiche nella Divina Commedia.*

[296] *Inf.* xi. 113, 114; *Purg.* i. 30.

[297]

"The Wain, that in the bosom of our sky Spins ever on its axle, night and day." *Par.* xiii. 7-9. (*Carey*).

[298]

"Fled is every bird that seeks the warmth, From European lands which never lose The seven cold stars." *Canz.* xv. 27-29.

[299]

"Seven cold oxen." *De Mon.* II. ix. 96.

[300]

"To duty there Each one convoying, as that lower doth The steersman to his port." *Purg.* xxx. 4-6. (*Carey*).

[301] *Purg.* xxx. 1-3.

[302]

"If the barbarians coming from some region That every day by Helice is covered, Revolving with her son whom she delights in, Beholding Rome and all her noble works Were wonder-struck...." *Par.* xxxi. 31-35. (*Longfellow*).

[303] "Those under the sway of the seven cold oxen."

[304] I do not know whether this comparison originated with Dante, but it was well known to Spanish sailors two centuries later. In the

Arte of Navigation which was "Englished out of the Spanyshe," by Richard Eden in 1561, Beta and Gamma of Ursa Minor are referred to as "two starres called the Guardians, or the mouth of the horne."

[305] *Par.* xiii. 1-28.

[306] *Par.* viii. 52, 53.

[307] *Par.* v. 136, 137.

[308] *Par.* viii. 16.

[309] *Par.* x. 76, 40-42.

[310] *Par.* xiv. 97-101.

[311] *Par.* xv. 13, 14.

[312] *Par.* xxi. 32, 33; xxiii. 26, 27.

[313] *Par.* xxii 23; xxiv. 11, 12.

[314] *Par.* viii. 20, 21; and xxviii. 100-102.

[315]

"Saw I many little flames From step to step descending and revolving, And every revolution made them fairer." *Par.* xxi. 136-138. (*Longfellow*).

Compare 80, 81 and 39; and *Par.* xxiv. 10, 11.

[316]

"As soon as singing thus those burning Suns Had round about us whirled themselves three times, Like unto stars neighbouring the steadfast poles." *Par.* x. 76-78. (*Longfellow*).

[317] *Par.* xxi. 80, 81; xii. 3; xviii. 41, 42.

[318] *Par.* xxiv. 22-24, x. 73, and many others.

[319]

"What time abandoned Phaëton the reins, Whereby the heavens, as still appears, were scorched." *Inf.* xvii. 107-108. (*Longfellow*).

[320]

"Even as, distinct with less and greater lights, Glimmers between the two poles of the world The Galaxy that maketh wise men doubt, Thus constellated in the depths of Mars Those rays described the venerable sign That quadrants joining in a circle make." *Par.* xiv. 97-102. (*Longfellow*).

[321] "The Galaxy, that is, the white circle commonly called St. James's Way."

[322] "And in the Galaxy this heaven has a close resemblance to Metaphysics. Wherefore it must be known that the Philosophers have had different opinions about this Galaxy. For the Pythagoreans affirmed that the sun at one time wandered in its course, and in passing through other regions not suited to sustain its heat, set on fire the place through which it passed; and so these traces of the conflagration remain there. And I believe that they were influenced by the fable of Phaëton, which Ovid tells at the beginning of the second book of the *Metamorphoses*. Others (as for instance Anaxagoras and Democritus) said that the Galaxy was the light of the sun reflected in that region. And these opinions they confirmed by demonstrative reasons. What Aristotle may have said about it cannot be accurately known, because the two translations give different accounts of his opinion. And I think that any mistake may have been due to the translators, for in the *New Translation* he is made to say that the Galaxy is a congregation, under the stars of this part of the heaven, of the vapours which are always being attracted by them; and this opinion does not appear to be right. In the *Old Translation* he says that the Galaxy is nothing but a multitude of fixed stars in that region, stars so small that they are not separately visible from our earth, but the appearance of whiteness which we call the Galaxy is due to them. [And it may be that the heaven in that part is more dense, and therefore retains and reproduces that light] and this opinion Avicenna and Ptolemy appear to share with Aristotle. Therefore, since the Galaxy is an effect of those stars which cannot be perceived except so far as we apprehend these things by their effect, and since Metaphysics treats of primal substances which in the same way we cannot apprehend except by their effects, it is plain that there is a close resemblance between the starry heaven and Metaphysics."

Conv. II. xv. 44-86. (*Jackson*).

[323] "That most brilliant star, Venus." *Conv.* II. iv. 88.

[324] "The brightness of her appearance, which is more lovely to behold than that of any other star." *Conv.* II. xiv. 112, 113.

"Sweet colour of oriental sapphire, Which was gathering in the serene aspect Of the sky, pure even to the first circle, To my eyes restored delight, So soon as I had come forth from that dead air, Which had troubled eyes and breast. The fair planet that inspires love Was making all the orient smile, Veiling the Fishes which were in her train."

Alternative rendering of the first three lines:—

"Sweet colour of oriental sapphire, Which was diffused over the tranquil scene, From mid-heaven even to the first circle." *Purg.* i. 13-21.

[326] First, or prime, circle.

[327] *Conv.*, II. iv. 1-3.

[328] Literally, "was assembling," or "was being collected."

[329] "From the middle."

[330] "Of the air."

[331] "From the east there shone upon the Mountain Cytherea, who in the flame of love seems to be always burning." *Purg.* xxvii. 94-96.

[332] "Veiling the Fishes" (the zodiacal constellation).

[333] "Her appearance, now in the morning, and now in the evening." *Conv.* II. xiv. 114, 115.

[334]

"The star That woos the sun, now following, now in front." *Par.* viii. 11-12. (*Longfellow*).

[335]

"I saw how move themselves, Around and near him, Maia and Dione." *Par.* xxii. 143-144.

[336]

"That fair planet, Mercury." *Son.* xxviii. 9.

[337] "Mercury ... as it moves is more veiled by the rays of the sun than any other star." *Conv.* II. xiv. 99-100.

[338]

... "The sphere That veils itself from men in alien rays." *Par.* v. 128, 129.

[339] "This Fire." *Par.* xvi. 38.

[340] "The burning smile of the star." *Par.* xiv. 86.

[341] "Mars shows red." *Purg.* ii. 14.

[342]

"This Mars ... his heat is like the heat of a fire ... his colour is as if he were on fire." *Conv.* II. xiv. 162-165.

[343] "Sweet star." *Par.* xviii. 115.

[344] "The torch of Jove." *Par.* xviii. 70.

[345] "Amongst all the stars it shows white, as if silveredover." *Conv.* II. xiv. 202-204.

[346]

"Jupiter Seemed to be silver there with gold inlaid." *Par.* xviii. 95-96. (*Longfellow*).

[347] "One is the slowness of its movement through the twelve signs; for twenty-nine years and more, according to the writings of the astrologers, are required for its revolution: the other is that it is high above all other planets." *Conv.* II. xiv. 226-231.

[348] *Par.* xxi. 18.

[349] *Par.* xxi. 25.

[350] *Par.* xxi. 13.

[351] "Circling the world." *Par.* xxi. 26.

[352] "Beneath the burning Lion's breast." *Par.* xxi. 14.

[353] *Purg.* ii. 13-15.

[354]

"Towards us came the being beautiful, Vested in white, and in his countenance Such as appears the tremulous morning star." *Purg.* xii. 88-90. (*Longfellow*).

[355] "He who drew beauty from Mary, as the Morning Star does from the Sun."
Par. xxxii. 107, 108.

[356] "All the seven." *Son.* xxviii. 14, and *Par.* xxii. 148.

[357] "The oblique circle which carries the planets."

[358] *Par.* xvi. 34-39.

[359] "About a year."

[360] "Three," for "thirty."

[361] *Conv.* II. vii. 88, 89.

[362] "The star of Venus had twice revolved in that circle of hers which makes her appear as evening and morning star, according to her two seasons, since the translation of that holy Beatrice who lives in heaven with the angels and on earth in my soul, when that Gentle Lady, of whom I made mention at the end of the 'New Life,' appeared first before my eyes, escorted by Love, and took some place in my mind." *Conv.* II. ii. 1-12.

[363] "Venus [ambitum epicycli peragit] anno Persico 1, mensibus 7, et diebus prope 9," that is, the period of Venus on her epicycle is $365 + 210 + 9 = 584$ days nearly, according to Alfraganus. The modern mean value is also 584 days.

[364] See Lubin's *Dante e gli Astronomi Italiani*. The period of 225 days may be easily deduced from Ptolemy's system, for it is the time in which the epicycle of Venus would make an absolute revolution round its centre, the diameter becoming parallel to its former position. But the Greeks invariably reckoned the period as the time in which it revolved relatively to Earth, that is 584 days.

[365] *Ep.* viii. 158, 159.

[366]

"The while, little by little, as I thought, The sun ceased, and the stars began to gather." *V. N.* xxiii. 176, 177. (*Rossetti*).

See also the prose description just before, lines 35-37.

[367] *Par.* xxix. 97-102.

[368] *Par.* xxvii. 35, 36.

[369] *Par.* xxv. 118-121.

[370] *Qu.* xx. 3-5, 26-29.

[371] "Blazing brilliantly like comets." *Par.* xxiv. 12.

[372] *Inf.* xxviii. 16-17; *Purg.* iii. 112-132.

[373] *V. E.* II. vi. 48.

[374] "I, who saw it clearly."

[375] *Naturales Quæstiones*, Bk. I.

[376] *Conv.* II. xiv. 168-171, and *Purg.* v. 37.

[377] *Par.* xv. 16-18.

[378] "Some ignorant people think that they are stars which fall from heaven and vanish." *Comp. del Mondo*, VII. v.

[379] "Early in the night."

[380] "Midnight." *Purg.* v. 38.

[381]

"Vapours enkindled saw I ne'er so swiftly At early nightfall cleave the air serene, Nor, at the set of sun, the clouds of August, But upward they returned in briefer time, And on arriving with the others, wheeled Towards us." *Purg.* v. 37-41. (*Longfellow*).

(By "vapours that cleave the clouds of August," flashes of lightning without thunder are meant. Aristotle believed both these and meteorites to be ignited vapours).

[382]

"As through the pure and tranquil evening air There shoots from time to time a sudden fire, Moving the eyes that steadfast were before, And seems to be a star that changeth place, Except that in the place where it is kindled Nothing is missed, and this endureth little." *Par.* xv. 13-18. (*Longfellow*).

[383] "The star."

[384] As this is the value given by Alfraganus, we must here understand Arabian miles. The distances from Rome to the north and south pole are therefore probably equal to 3500 to 9750 English miles, and both are a little too large, because the half circumference of Earth is too large. The proportion is about right, however, for Rome is nearly three times as far from the south pole as the north, her latitude being 42° N. Alfraganus placed her in the fifth climate, at the northern boundary of which he said the pole was elevated 43½°.

[385] *Qu.* xix. 36.

[386] Adopting the reading of Dr. Moore: "nella mezza terra, alla mezza terza," that is, "at the equator at middle-tierce." See *Studies in Dante* III. 107, 108.

[387] "For now, after what has already been said, the rest may be understood by whomsoever has a noble mind, to which it is well to leave a little labour." (*Cf. Par.* x. 22-25).

[388] *Purg.* xiv. 148-151.

[389] "O unspeakable Wisdom who hast thus ordained, how poor is our intellect to understand Thee! And you, for whose benefit and pleasure I am writing, in what blindness you live, not lifting up your eyes to these things, but keeping them fixed on the slough of your folly."

[390] "And the sky revolves like a mill-stone." *El. Ast.* ch. vii.

[391] "There the sky will revolve, with all its stars,
mill-stone fashion." *Comp. del Mondo.* I. xxiii.

[392] "It follows that *Maria* must see this sun 'circling the world' like a mill." *Conv.* III. v. 142-147.

[393] "A winding path, which the learned call a spiral."

[394] "Lucan, who was well known to Dante, had observed that shadows cast by the sun in the southern hemisphere travel to the right instead of to the left, and fall southwards when with us they fall to the north." Moore, *Studies in Dante* i. p. 239.

[395]

"If their pathway were not thus inflected, Much virtue in the heavens would be in vain, And almost every power below here dead." *Par.* x. 16-18. (*Longfellow*).

[396]

"And now remain, reader, on your bench, thinking over this!" *Par.* x. 22, 23.

[397] Thus Petrarch: "Le stelle vaghe e lor viaggio torto." ("The wandering stars and their winding way.") *Sonetto de Morte di Madonna Laura.*

[398] *Cf. Conv.* III. v. 76, 191.

[399]

"Which always remains between the sun and the winter." *Purg.* iv. 81.

[400] *Studies in Dante* iii. p. 166.

[401] *Purg.* xxi. 46-48; *Purg.* xix. 38.

[402] *Purg.* xxviii. 143.

[403] "The land where shadows are lost." *Purg.* xxx. 89.

[404] "The great dry land." *Inf.* xxxiv. 113.

[405] "The unpeopled world." *Inf.* xxvi. 117.

[406] "Behind the sun." *Ibid.*

[407] *Conv.* III. v. 117, 118.

[408] Refraction makes the pole visible a little before one reaches the equator, but such refinements need not be considered in dealing with a popular work like the *Convivio*.

[409]

"There riseth up from Ethiopia's sands A wind from far-off clime which rends the air, Through the sun's orb that heats it with its ray. The sea it crosses; thence, o'er all the lands Such clouds it brings that but for wind more fair O'er all our hemisphere 'twould hold its sway; And then it breaks and falls in whitest spray Of frozen snow and pestilential showers." *Canz.* xv. 14-21. (*Plumptre*).

[410] "Now heats it."

[411] "This hemisphere."

[412] Just as in *Par.* xxviii. 80.

[413] *Qu.* xix. 53-57, and repeated in xxi. 36-40.

[414] *Luke* xxiii. 44.

[415] *Par.* xxix. 97-102.

[416] Dante believed the death of Christ to have taken place when it was noon in Jerusalem. *Conv.* IV. xxiii. 105-106.

[417] *Purg.* iv. 137-139.

[418] *Purg.* xxvii. 1-5.

[419] Orosius says: "Europæ in Hispania occidentalis oceanus termino est, maxime ubi apud Gades insulas Herculis columnæ visuntur.... Asia ad mediam frontem orientis habet in oceano Eoo ostia fluminis Gangis."

[420] *Qu.* xix. 38-52.

[421] Moore, *Studies in Dante* iii. p. 124.

[422]

"The strait pass, where Hercules ordained The boundaries not to be o'erstepped by man." *Inf.* xxvi. 107-109. (*Carey*).

[423] *Inf.* xxvi. 106-142.

[424] *Ptolemy's Geography*, Bk. I.

[425] See p. 176.

[426] *Esdras* II. vi. 42.

[427] Beazley, *Dawn of Geography* iii. 28, 29.

[428] Corvino and Marco Polo made the voyage in the same year, 1292, but in reverse directions.

[429] *Purg.* iii. 25.

[430] *Purg.* xv. 1-6.

[431] *Purg.* iv. 68-71; xxvii. 1-5.

[432] *Purg.* xxviii. 142. Opinions differed as to its exact site, and some placed it in the ocean beyond the eastern limit of the habitable earth. In *V. E.* I. viii. 6-10, Dante says that the root of the human race was planted in eastern lands, but this refers to Adam's home after the expulsion from Paradise.

[433] *Par.* xxx. 1-3.

[434] *Par.* ix. 82-87.

[435] *Conv.* III. vi. 7-32.

[436] *Conv.* IV. xxiii. 50 to end.

[437] *Comp. del Mondo*, I. xxii.

[438] "The sixth hour, that is, the middle of the day, is the most noble of all the day, and the most virtuous."

[439] *Luke* xxiii. 44-46. Dante understands this to mean that death took place at about the sixth hour, not the ninth.

[440]

"From the first hour to that which cometh next (As the sun changes quarter) to the sixth." *Par.* xxvi. 141-142. (*Carey*).

[441] It is easily seen that this is correct, if we divide the 360° of the moon's path through the zodiac by the 27 days and eight hours in which she traverses it, and returns to the same star. Brunetto Latini says "La lune s'esloigne *dou Soleil* chascun jor xiii degrez po s'en faut," but he is wrong, having apparently forgotten that the sun is also moving in the same direction, at the average rate of nearly 1° daily.

[442] *Inf.* xxix. 11.

[443] *Purg.* xxiii. 5, 6.

[444] *Par.* xxxii. 139.

[445] "Midway in the journey of our life." *Inf.* i. 1.

[446] "The sweet season." *Inf.* i. 43.

[447]

"Aloft the sun ascended with those stars That with him rose when Love Divine first moved Those its fair works." *Inf.* i. 38-40. (*Carey*).

[448] "Those stars."

[449] *Purg.* xi. 108; *Conv.* II. xv. 12-14.

[450] *Conv.* III. vi. 28-30.

[451] On the day of the spring equinox the sun rises and sets a few minutes after six o'clock, because we set our clocks not by the real sun but the more convenient "mean sun."

[452] The clock-hours in this column are not to be regarded as if taken from a railway time-table! Sunrise is at 6 a.m. within a few minutes only, for the reasons above stated, viz.: our clocks are

regulated by the "mean" sun, and it is not necessarily the exact day of the equinox when the vision begins.

[453] Or "almost at late midnight."

[454] Or "nigh me:" "presso" may mean either.

[455] "The Lady who rules here." *Inf.* x. 80.

[456] "Lower Hell." *Inf.* viii. 75.

[457] *Cf. Par.* ii. 48-51.

[458] The retardation is not likely to be more than the average of 50 minutes, and may be less, because the moon is in Libra, and therefore going south. This tends to diminish the interval between one moonset and the next in the northern hemisphere, just as the days get shorter when the sun goes south in autumn.

[459] Some think the interval between this reference and the last almost too short, but the words do not indicate that Malacoda spoke on the stroke of seven! and the moon may have set at about 6.30.

[460] "Darkness of Hell." *Purg.* xvi. 1.

[461] "The fair planet which kindles love."

[462] "Departed, as he came, swiftly."

[463] If the meaning is that the sun is now 50° *above the horizon*, this would indicate a later hour, nearing midday, for the sun does not rise vertically in this latitude, and reaches only 58° at noon at the time of the equinox. But the first explanation is the more probable.

[464] "The shore."

[465] See table on p. 361.

[466] "We must not assume, as some commentators have done, that the signs rise at equal intervals of two hours, although each circles the star sphere in 24 hours. See fig. 46, where the dotted line shows how *e.g.* Cancer, circling parallel with the celestial equator, will rise at a point considerably north of east, and having had to traverse more than 90 degrees since Aries rose due east, will not rise until about 7 hours later, *i.e.* 1 p.m. Conversely Scorpio, following Libra, will rise south of east, and a little *less* than 2 hours later. The moon's retardation, therefore, is less in this part of the zodiac than it would otherwise be, and the hour is probably nearer eight than nine."

[467] "Superlatively obscure."

[468] "Out of the arms of her lover."

[469] "Cold creature."

[470] "The climax of the day."

[471] See Moore, *Studies in Dante* iii. pp. 75-84, for a detailed discussion of this passage. Several commentators have held that lines 1 to 6 describe the dawn of day *elsewhere*; and it is true that it would be nearly 6 a.m. and Pisces would be on the horizon in Italy when the hour was nearly 9 p.m. in Purgatory.

[472] "Vespers there."

[473] "Here."

[474] See for instance the "Carte Pisane," and the Central Mediterranean map of Vesconte, dating from about 1300 and 1311 respectively, in Beazley's *Dawn of Geography*, iii. Latitudes and longitudes are not given, but from certain centres lines radiate to all points of the compass, like great spiders' webs.

[475] *De Mon.* II. iii. 87-90. etc.

[476] "It is bounded *on the east and north* by the Tyrrhenian sea, which lies towards the port of Rome," Moore, *Studies in Dante* iii. p. 72.

[477] *Purg.* xxxii. 56, 57.

[478]

"My more than father said unto me, Son, Come now, because the time that is ordained us More usefully should be apportioned out." *Purg.* xxiii. 4-6. (*Longfellow*).

[479] *Purg.* xxi. 20-27.

[480] *Purg.* xiii. 22-23.

[481] On an Astronomical Point in Dante's *Purgatorio*, by P. H. Cowell, F.R.A.S., *The Observatory*, December 1906.

[482] "That circles opposite to him." *Purg.* ii. 4.

[483] The signs follow one another *on the meridian* at intervals of exactly 2 hrs.

[484] "At the hour."

[485] "I turned to the east."

[486] "Pure and ready to rise to the stars." *Purg.* last line.

[487] "The climax of the day."

[488] It has been suggested to read the line *Par.* i. 44. "Tal foce; e quasi tutto là era bianco," transferring the "quasi" (almost) so that the meaning should be "almost was wholly white that hemisphere," and to interpret that it is now morning of the day following the events in the last Canto of the *Purgatorio*. But if so, Dante would have spent a whole night in the Earthly Paradise without mentioning it, or explaining this long delay after he had become "pure and ready to rise to the stars." (For the meaning of "foce," the "passage," see later, p. 400).

[489] *Par.* i. 46, 47.

[490]

"Love that rules the heavens, *with thy light* Thou didst raise me." *Par.* i. 74, 75.

[491] "Turned her eyes again towards heaven."

[492] "Beatrice gazed upwards, and I at her."

[493]

"Turned again with yearning to that part where the world is most living." *Par.* v. 86, 87.

[494] *Inf.* i. 38-40.

[495] *Conv.* II. iv. 52-62.

[496] "That part."

[497] "He who is father of all mortal life." *Par.* xxii. 116; see also *De Mon.* I. ix. 6, 7.

[498] See *Par.* x. 7-21.

[499] Longfellow says:—"Looking down from the terrace of Monte Cassino upon the circular threshing-floor of stone in a farm lying below, I first felt the aptness of Dante's phrase. This very scene may have suggested it to him."

[500]

"So my lady stood, erect and intent, turned towards that place under which the sun shows least haste." *Par.* xxiii. 10-12.

[501] See *Par.* iv. 34-39.

[502] *Purg.* xxix. 12 and 34.

[503] "Region."

[504] *Par.* xxiii. 29, 30.

[505] *Par.* xxvii. 64-66.

[506] Compare *Purg.* xv. 1-5, where the course the sun still has to run between vespers and sunset is described as equal to the space between the third hour and sunrise.

[507]

"The threshing-floor that maketh us so proud, To me revolving with the eternal Twins, Was all apparent made from hill to harbour." *Par.* xxii. 151-3. (*Longfellow*).

[508] Della Valle boldly assumes that they *were* over the same meridian, by a poetical licence, although at the same time the sun was in a different sign. Dante only mentions the latter fact, he thinks, in order to show that he was a few degrees north of the sun (Gemini being more northerly than Aries); therefore he could see over the edge, as it were, of the sun-lighted hemisphere of Earth. This is desperately subtle.

It is, however, the only way in which the passages can be reconciled with his further assumption, shared by some other commentators, that Dante, in his flight through the spheres, simply ascended without any movement in longitude except that he was carried round by the daily revolution of the spheres. All the planets, therefore, were ranged one above the other, in the sign of Gemini, and it was always noon on the earth below his feet, since that was the hour at which he ascended from the Earthly Paradise, and his movement was the same as the sun's. (Here Della Valle is inconsistent, however, for he maintains that the ascent was made in the early morning.) But this is a very artificial conceit, and not indicated by Dante. He implies that Beatrice, in leading him to each sphere, chose the part of it which he was to enter (see *Par.* xxvii. 102), and he tells us that Saturn was in Leo, Venus in Pisces, and the sun in Aries.

[509] *Il Paradiso di Dante dichiarato ai giovani*, by Angelo de Gubernatis.

[510] "From the hills to the river-mouths."

[511] "Love which moves the sun and the other stars." *Par.* last line.

[512] Alfraganus, as we saw, gives the daily movement of the sun eastward in the zodiac as about 59 minutes of arc. (A degree contains 60 minutes).

[513] *Purg.* xxv. 2, 3.

[514] *Inf.* xi. 113; *Purg.* i. 21, xix. 4-6.

[515] *Inf.* i. 38-40.

[516] *Par.* i. 38, 39.

[517] "That part."

[518] "Those stars."

[519] *Par.* x. 7-33.

[520] *Par.* xxvii. 86, 87.

[521] *Purg.* i. 21.

[522] *Purg.* xxvii. 95.

[523] "The Lion's breast." *Par.* xxi. 14.

[524] "The Lion's Heart."

[525] "Beneath the breast."

[526] *Cf. Purg.* xxxii. 56, 57, where the sun is spoken of as under another star.

[527] "To his Lion has come." *Par.* xvi. 37, 38.

[528] "Sometimes by Saturn."

[529] "With a better course and better star." *Par.* i. 40.

[530] "Rises upon mortals." *Par.* i. 37.

[531] *Inf.* x. 100-105.

[532] *Conv.* IV. xxiii. 96-98.

[533] Moore, *Studies in Dante* iii. p. 146.

[534] See p. 177.

[535] *V. N.* xliii. 1-11.

[536] Exactly at midday, moreover, according to one system then in vogue, though it seems to have been more usual to begin the day at sunrise.

[537] "A thousand two hundred *and one*, added to sixty-six."

[538] "Midway in the journey of our life."

[539] "The highest point," the "summit."

[540] "Until the summit of my life." *Conv.* I. iii. 24, 25.

[541] "Youth."

[542] "Is in truth the summit of our life."

[543] Farinata alludes to the two Guelf banishments of 1248 and 1260. *Inf.* x. 42-50.

[544] *Conv.* I. iii. 22-24; *Inf.* xxiii. 94, 95, etc.

[545] *V. N.* ii. 1-15.

[546] *V. N.* xxx. 4-13.

[547] *Purg.* xxx. 124, 125; *Cf. Conv.* IV. xxiv. 11-13.

[548] Imbriani: *Quando nacque Dante?* and *Che Dante probabilissimamente nacque nei MCCCLXVII.*

[549] *Purg.* ii. 91-103.

[550] For this point and a full discussion of the date, especially from the astronomical data, see Angelitti's *Sulla Data del Viaggio Dantesco*, and *Sull' anno della Visione Dantesca*, and Moore's "The Date assumed by Dante for the Vision of the 'Divina Commedia,'" *Studies in Dante* III.

[551] "This hundredth year." *Par.* ix. 40.

[552] *Purg.* viii. 74.

[553] *Par.* xvii. 80, 81.

[554]

"Up starting suddenly he cried out: How Saidst thou? '*he had?*' Is he not still alive? Does not the sweet light strike upon his eyes?" *Inf.* x. 67-69.

[555] Toynbee's *Dante Dictionary*, under "Cavalcanti."

[556] Moore, *Studies in Dante* III. p. 146.

[557] "The Astronomy of Dante," in *Studies* III., and *Time References in the Divina Commedia.*

[558] Note on *Purg.* i. 19-21 in the "Temple" edition of Dante.

[559] *Par. Lost* vii. 375-378.

[560] See p. 407.

[561] Preface to *Almanach Dantis Aligherii*, Boffito and Melzi d' Eril.

[562] It is important to observe that this suits the description of the "Lion's breast" better than the 20th and 21st degrees (Almanach positions for 1301), if Dante, as usual in the *Divina Commedia*, identifies the constellation with the sign.

[563] The middle of March.

[564] This was written in 1910.

[565]

"Oh ye whose intellectual ministry Moves the third heaven." (*Carey*).

[566] *Conv.* II. iii. 8-14.

[567] *Par.* x. 1-12.

[568] *Inf.* iv. 131.

[569] *Conv.* III. v. 54-56.

[570] *Conv.* III. v. 29-52. See Part I. of this book *Pythagoras and his Followers*, and *Plato*, for a description of their astronomical systems.

[571] See Part I. "Aristotle."

[572] "And that outside the latter there was no other." *Conv.* II. iii. 24, 25.

[573] "This entirely mistaken opinion of his." *Ibid.* 28.

[574] "According to Ptolemy and according to Christian Truth, nine is the number of the moving heavens." *V. N.* xxx. 16-18.

[575] "Roughly speaking." *Conv.* II. iii. 36-48.

[576] "All bodies." *Ep.* x. 445.

[577] "These vast bodies." *Par.* viii. 99.

[578] "Material circles" (here compared with the Angels, who are purely spiritual). *Par.* xxviii. 64.

[579] "The greatest body." *Par.* xxx. 39.

[580] *Purg.* iii. 29, 30.

[581] "This rounded ether." *Par.* xxii. 132.

[582] *Par.* xxiii. 112-117.

[583] *Conv.* II. iv. 1-13.

[584] *Conv.* II. iii. 52-65.

[585] *Conv.* II. iv. 13-16, 30-32. See also *Ep.* x. 448-452.

[586] "And Aristotle also, to one who rightly understands him, seems to mean the same thing, in his first book of *The Heaven and the Earth.*" *Conv.* II. iv. 32-34.

[587] See p. 97.

[588] *The Convivio of Dante*, "Temple" Edition, note on II. iv.

[589] "Firm and fixed and not moveable, from any point of view." *Conv.* II. iv. 50, 51.

[590] *Conv.* II. iv. 68-75.

[591] *Conv.* II. iv. 78-104.

[592] "That circle of hers which makes her appear as evening and morning star, according to the two different periods." *Conv.* II. ii. 3-5.

[593] "Third epicycle."

[594] *Conv.* II. xv. 132-155.

[595] "That one which sweeps the whole of the rest of the universe along with it." *Par.* xxviii. 70, 71.

[596] Toynbee, "Dante's Obligations to Alfraganus in *Vita Nuova* and *Convivio*," in *Romania* xxiv. 95.

[597] *Conv.* II. xv. 108-118.

[598] "The first star." *Par.* ii. 30.

[599] "The second realm." *Par.* v. 93.

[600] "The third heaven." *Par.* viii. 37.

[601] "The fourth family." *Par.* x. 49.

[602] "The fifth threshold." *Par.* xviii. 28.

[603] "More lofty." *Par.* xiv. 85.

[604] "Sixth star." *Par.* xviii. 68, 69.

[605] "The seventh splendour." *Par.* xxi. 13.

[606] "The eighth sphere." *Par.* ii. 64.

[607] "The swiftest heaven." *Par.* xxvii. 99.

[608] "The greatest body." *Par.* xxx. 39. Compare *V. N.* xlii. 47. "Oltre la spera che più larga gira," ("beyond the sphere which has the largest circle)."

[609] "The heaven of pure light." *Par.* xxx. 39.

[610] "The last sphere." *Par.* xxii. 62.

[611]

" ... Within that one alone Is every part where it has always been; For it is not in space, nor turns on poles." *Par.* xxii. 65-67. (*Longfellow.*)

[612] "This is the sovereign edifice of the world, in which all the world is enclosed, and beyond which is naught; and it exists not in space, but received form only in the Primal Mind, which the Greeks call Protonoe." See also *Ep.* x. 442-447.

[613] "Last sphere."

[614] "The first heaven." *Purg.* xxx. 1.

[615] "The first circle." *Par.* iv. 34.

[616] "First heaven." *Qu.* iv. 7.

[617] *Par.* vii. 130-132 compared with 67, 68. See also *Par.* i. 64 and 76.

[618] *Ep.* x. 435-437.

[619]

"O Lady of Virtue, thou alone through whom The human race exceedeth all contained Within the heaven that has the lesser circles." *Inf.* ii. 76-78. (*Longfellow*).

[620] *Par.* vii. 124-143. Compare with Plato's account of the creation of living beings (Part I., *Plato*).

[621] *Par.* i. 115, 117; *De Mon.* I. xv. 38-41; and III. vii. 30, 31, etc.

[622] *Par.* i. 92, 130-134.

[623] "This earth, together with the sea, is the centre of heaven." (*Conv.* III. v. 63-65.)

[624] "It is agreed among learned men that earth and water together form a globe." (*El. Ast.* iii.).

[625] *Qu.* xviii. 28-54. This may be compared with *Par.* xxix. 13-18.

[626] *Qu.* xv. 10, 11.

[627] See p. 134.

[628] See p. 247.

[629] "Or nearly." *Qu.* xix. 61-63.

[630] Compare with *Conv.* II. vii. 88-100, where the poet says that the influence from each sphere comes on the rays of the planet in that sphere, also *De Mon.* III. iv. 139, 140.

[631] "Nor does it avail anything to say that that declination could not take place because of approaching nearer the earth through eccentricity; for if this elevating virtue were in the moon, since agents act the more powerfully the nearer they are, she would elevate there rather than here." *Qu.* xx. 65-71.

[632] Note on this passage in "Temple" Dante.

[633] *Studies in Dante,* ii. p. 339.

[634] Nearest Earth.

[635] *Qu.* vii. 4, 6, and xxiii. 52.

[636] "Orbis" and "sphæra" are used indifferently by Alfraganus to indicate the celestial spheres: thus, in ch. xii. he says "*Orbes,* qui stellarum omnes motus complectantur, numero esse octo," and on the next page, "unamquamque harum *sphærarum* octo."

[637] "The upper and lower apsides of the eccentrics of Saturn, Jupiter, and Mars, have a certain declination from the zodiac, the former to the north, the latter to the south, the amount always remaining constant: *and it is just the same with the moon.*" *El. Ast.* xviii. par. 6.

[638] In the edition of Golius ch. vii. par. 5, the position of greatest distance is called the perigee, of least the apogee, but this mistake is corrected in the next paragraph.

[639] "The moon's declination in the zodiac from the equator towards the south pole is as great as towards the north." *Qu.* xx. 61-63.

[640] The deferent was the equator of her orb or sphere.

[641] Ristoro also defines apogee and perigee ("auge" and "opposito d'auge") as effects of movements on an epicycle, and says that the deferent is "declined" from the ecliptic ("declinato della via del sole,")

part being north and part south, but he does not say whether its perigee is north or south. *Comp. del Mondo.* I. xii. xiv., and III. vii.

[642] For a thorough discussion of this question see Moore's *Studies in Dante* ii. pp. 303-374.

[643] "The foundation of your elements." *Par.* xxix. 51.

[644] "Rises highest towards heaven out of the sea." *Purg.* iii. 15.

[645] "Dry vapour."

[646] "Exhalations from the water."

[647] "Behind [*i.e.* following] the heat."

[648] *Purg.* xxviii. 97-99 and xxi. 52-53. The gate is meant by "i tre gradi" ("the three steps"), and "ove si serra" ("where it is locked"). See *Purg.* ix. 76, 106-108.

[649]

" ... The universal atmosphere Turns in a circuit with the primal motion, Unless the circle is broken on some side: Upon this height, that all is disengaged In living ether, doth this motion strike, And make the forest sound, for it is dense." (*Longfellow*).

[650] Scartazzini takes the passage as referring to Eden only, and interprets that all the air *here* is moving evenly and noiselessly unless it meets with any obstacle *such as this forest*, when a murmuring sound results.

[651] *Purg.* ix. 30-33.

[652] *Par.* i. 76-81.

[653] *Par.* xxi. 58-63.

[654] *Hymn on the Nativity.*

[655] *Merchant of Venice.*

[656]

"The song of those who sing for ever After the music of the eternal spheres." *Purg.* xxx. 92, 93. (*Longfellow*).

[657] *Conv.* IV. viii. 51-64.

[658] *Conv.* II. vii. 104-108.

[659] "This small star." *Par.* vi. 112.

[660] *Conv.* II. xiv. 92-98.

[661] The Frankfort printed edition and some Latin MSS of Alfraganus agree with this: the edition of Golius gives $^1/_{18}$.

[662] *Almagest*, Bk. IX.

[663]

"This heaven, where ends the shadowy cone Cast by your world." *Par.* ix. 118, 119. (*Longfellow*).

[664] Rabbi Abraham ben Chija, in his *Sphæra Mundi*, written about 1100 A.D., says that Earth's shadow extends as far as the distance of Mercury, not of Venus (Delambre, *Histoire de l'astronomie ancienne*, and Dreyer, *Planetary Systems*); but this does not agree with the figures given by Alfraganus.

[665] Her diameter would look no larger than the diameter of Saturn, as we see it when nearest to him; but like Saturn she would be visible as a bright point, because shining by reflected sunlight. Ancient and mediæval astronomers never realized this, however, but thought of Earth as a dark body, receiving light but giving none.

[666] *Conv.* II. xiv. 126, 127; *Par.* x. 40-42, 48.

[667] *Par.* xxii. 67; xxxi. 19-24; xxx. 118-123.

[668] *Par.* ii. 25-45.

[669] "The first star."

[670] "The shadow which is in her is nothing but rare parts of her substance, which cannot stop the rays of the sun and reflect them back like the other parts." *Conv.* II. xiv. 72-76.

[671] Plutarch, *On the Face in the Moon*, translation by Prickard.

[672] See Paget Toynbee, *Le Teorie Dantesche sulle Macchie della Luna*, in *Giornale Storico della Letteratura Italiana*, vol. xxvi. pp. 156-161.

[673] Albert of Saxony notes this objection, but says the rarity is not sufficiently great to make these parts transparent.

[674] This is in agreement with modern belief. Some parts of the moon's surface reflect sunlight less well than others, and therefore look dark, just as a peaty soil looks darker than limestone.

[675] *Conv.* II. iv. 94-96.

[676] "That which was *within* the sun [*sc.* the spirits]." *Par.* x. 41.

[677] "In the depths of Mars." *Par.* xiv. 100, 101.

[678] "The temperate sixth star, which *within itself* had received me." *Par.* xviii. 68, 69.

[679] "The present pearl." *Par.* vi. 127.

[680] "This fire." *Par.* xvi. 38.

[681]

"Well was I 'ware that I was more uplifted By the enkindled smiling of the star That seemed to me more ruddy than its wont." *Par.* xiv. 85-87. (*Longfellow*).

[682]

"And such as is the change, in little lapse Of time, in a pale woman, when her face Is from the load of bashfulness unladen, Such was it in mine eyes, when I had turned, Caused by the whiteness of the temperate star, The sixth, which to itself had gathered me." *Par.* xviii. 64-69. (*Longfellow*).

[683]

"He in semblance such became As Jove might be, if he and Mars were birds, And interchanged their plumes." *Par.* xxvii. 13-15. (*Carey*).

[684] "The temperate light of Jove, between his father and his son." *Par.* xxii. 145-146.

[685] *Conv.* II. xiv. 196-198.

[686] *Cf.* Ristoro, *Comp. del Mondo*, III. iii.

[687] *Conv.* II. xiv. 161-165.

[688] "The planet that strengthens the cold." *Canz.* xv. 7.

[689] *Purg.* xix. 1-3.

[690] *Conv.* II. xiv. 78, 79.

[691] *De Mon.* III. iv. 140-142.

[692] See quotations from Albertus Magnus and Roger Bacon in Moore's *Studies*, iii. p. 45, *note*.

[693] "Bestow light on the stars." *Canz.* xix. 117.

[694] "The sun illumines first himself and then all celestial and elemental bodies with visible light." *Conv.* III. xii. 54-56.

[695] *Par.* xxxii. 107, 108.

[696] *Par.* xx. 1-6; compare xxiii. 28-30.

[697] Moore, *Dante and his early Biographers*, p. 95.

[698] *Inf.* xx. 38.

[699] *Inf.* xx. 115-118, 46-51.

[700] "The star of love." *Canz.* xv. 4.

[701] "The fair planet which incites to love." *Purg.* i. 19.

[702] *Par.* ix. 94-96.

[703] *Par.* ix. 32, 33.

[704] *Conv.* II. *Canz.* 1-6.

[705] *Conv.* II. vii. 88-100.

[706] *De Mon.* III. iv. 139, 140.

[707] *Par.* xvii. 76-78.

[708] *Par.* xviii. 115-117.

[709] *Par.* xxi. 13-15.

[710] *Par.* i. 40-42; *Canz.* xv. 41.

[711] *Par.* xxii. 112-117. The Ottimo, commenting this passage, says that Gemini "is the house of Mercury, who signifies, according to the astrologers, literature, science, and learning. And in this direction it disposes those who are born when it is in the ascendant, and more powerfully if the sun is in it."

[712] *Inf.* xv. 55.

[713] *Conv.* II. xiv. 170-180.

[714] *Inf.* xxiv. 145-150.

[715] *V. N.* xxx. 22-24.

[716] *Ball.* vi. 11, 12.

[717]

"From that bright star which moveth on its way For ever at the Empyrean's will, And between Mars and Saturn ruleth still, E'en as the expert astrologer doth say, She who inspires me with her beauty's ray Doth subtle art of sovereignty distil; And he whose glory doth the fourth heaven fill Gives her the power my longing soul to sway. And that fair planet known as Mercury Colours her speech with all its virtue rare, And the first heaven its boon does not deny. She who the third heaven ruleth as her share Makes her heart full of utterance pure and free; So all the seven to perfect her agree." *Sonnet* xxviii. (*Plumptre*).

[718] *Par.* ii. 64-72, 130 to end.

[719] "It should be borne in mind that although the starry heaven has unity in substance it nevertheless has multiplicity in virtue." *Qu.* xxi. 12-14.

[720]

" ... E'en so The intellectual efficacy unfolds Its goodness multiplied throughout the stars, On its own unity revolving still." *Par.* ii. 136-138. (*Carey*).

[721] "So many are the stars which spread themselves over her sky, that surely we cannot wonder if they make many and diverse fruits grow on human nobility, *so many are their natures and potencies, concentrated and united in one simple substance*; and on them as on diverse branches she bears fruit in diverse ways."
Conv. IV. xix. 45-52. (*Jackson*).
See also *Conv.* IV. xxi. 62-66.

[722] *Qu.* xxi. 17-19.

[723] *Conv.* II. iv. 75-77.

[724] But see Tozer's *English Commentary on Dante's Divina Commedia*, note on *Par.* xvi. 37. "In mediæval astrology Mars was one of the Lords of the Lion."

[725] *Par.* iii. 55-57.

[726] See p. 202.

[727] *De Mon.* II. ii. 15-18, 25, 26, 36-38; III. ii. 30-32; xvi. 91-101; *Ep.* v. 133-135; *Par.* ii. 121, viii. 97-99.

[728] *Ep.* v. 124, 125.

[729] *Par.* ii. 127, 128.

[730] *Par.* ii. 130-132; xiii. 73-75; viii. 127, 128; i. 41, 42. Ristoro of Arezzo uses the same expression, *Comp. del Mondo*, Bk. VII. part I. chapter 2.

[731] *Purg.* vi. 100; xx. 13, 14; xxxiii. 40-45.

[732] *Par.* vii. 133-141; xiii. 65, 66;
Conv. III. xv. 159-161; II. xv. 152-154; IV. xxiii. 50-52.

[733] *Conv.* II. xiv. 28-30; IV. ii. 58-61; *Par.* xxvi. 128, 129.

[734] *Par.* viii. 127-135.

[735] *Par.* iv. 49-60.
See also *Ecl.* ii. 16, 17; and
Conv. IV. xxi. 17-19, 25-27.

[736] *De Mon.* II. ii. 21-23; *Cf. Par.* xiii. 64-78.

[737] "This carries fire towards the moon, this compresses and binds Earth together." *Par.* i. 115, 117. See also *De Mon.* III. vii. 30, 31, and I. xv. 38-41; *Conv.* III. iii. 10-13; *Qu.* xviii. 11, 12, and many others.

[738]

"The fire doth upward move By its own form, which to ascend is born, Where longest in its matter it endures." *Purg.* xviii. 28-30. (*Longfellow*).

[739] *Inf.* xxxii. 8, and 74; xxxiv. 110, 111. See also *Par.* xxxiii. 22, 23; *Inf.* ix. 28, 29; *Qu.* iii. 6-9, etc.

[740] *De Mon.* I. xv. 46-48; *Qu.* xii. 39-42.

[741] "The point towards which weights are drawn from every direction." *Inf.* xxxiv. 110-111.

[742] "Laboriously and painfully." *Inf.* xxxiv. 78.

[743] "Panting like a tired man." *Inf.* xxxiv. 83.

[744] *Qu.* xvi. 54, 55.

[745] This is nowhere stated, but everywhere taken for granted, and in *Conv.* II. xiv. 211 the circle is said to be the most perfect of figures.

[746] "Oh ye whose intellectual ministry moves the third heaven." *Conv.* II. canzone; and *Par.* viii. 37.

[747] *Conv.* II. v. 119-126.

[748] "Although they did not think of them as philosophically asPlato." *Conv.* II. v. 5-37, 94, 95; vi. 154-159.

[749] *Par.* ii. 127-129, 142-144.

[750] *Conv.* II. v. and vi.

[751] *Par.* viii. 34-37.

[752] *Cf. Par.* xxviii. 98 to end, with
Conv. II. vi. 43-55, 106-109.

[753] *Par.* xxviii. 34-39, 64-75.

[754]

"In one God I believe, Sole and eterne, who moveth all the heavens, With love and with desire, Himself unmoved." *Par.* xxiv. 130-132. (*Longfellow*).

[755] *Conv.* II. iv. 13-30;
Ep. x. 442-452, and 472-488;
Par. i. 121-123.

[756] Probably equal to 26,500 English statute miles.

[757] 300 English statute miles.

[758] *Conv.* IV. ix. 23-25.

[759] Compare with Dante's accurate descriptions a poem by Kipling in "A School History of England":—

"South and far south below the line Our Admiral leads us on. Above, *undreamed-of planets* shine." (!)

CPSIA information can be obtained
at www.ICGtesting.com
Printed in the USA
BVHW071552210521
607866BV00003B/552